KV-633-009

The Ciba Foundation for the promotion of international cooperation in medical and chemical research is a scientific and educational charity established by CIBA Limited—now CIBA-GEIGY Limited of Basle. The Foundation operates independently in London under English trust law.

Ciba Foundation Symposia are published in collaboration with Excerpta Medica in Amsterdam

Excerpta Medica, P.O. Box 211, Amsterdam

Oxygen Free Radicals
and Tissue Damage

Oxygen Free Radicals and Tissue Damage

Ciba Foundation Symposium 65 (new series)

1979

Excerpta Medica

Amsterdam · Oxford · New York

ISBN Excerpta Medica 90 219 4070 1
ISBN Elsevier/North-Holland 0444 90063 2
Published in May 1979 by Excerpta Medica, P.O. Box 211; Amsterdam
and Elsevier/North-Holland, Inc., 52 Vanderbilt Avenue, New York, N.Y. 10017

Suggested series entry for library catalogues: Ciba Foundation Symposia
Suggested publisher's entry for library catalogues: Excerpta Medica

Ciba Foundation Symposium 65 (new series)

389 pages, 45 figures, 85 tables

Library of Congress Cataloging in Publication Data
Main entry under title:

Oxygen free radicals and tissue damage.

(Ciba Foundation symposium; new ser., 65)
Bibliography: p.
Includes indexes.
1. Active oxygen in the body—Congresses. 2. Active oxygen—Toxicology—Congresses. 3. Pathology, Cellular—Congresses. 4. Radicals (Chemistry)—Physiological effect—Congresses. I. Series: Ciba Foundation. Symposium; 65.
QP535.01093 574.1'921 79-11805
ISBN 0-444-90063-2

Printed in The Netherlands by Casparie, Heerhugowaard

Contents

Participants

Symposium on Oxygen Free Radicals and Tissue Damage held at the Ciba Foundation, London, 6–8th June, 1978

I. FRIDOVICH (*Chairman*) Department of Biochemistry, Duke University Medical Center, Durham, North Carolina 27710, USA

A. C. ALLISON Division of Cell Pathology, MRC Clinical Research Centre, Watford Road, Harrow, Middlesex HA1 3UJ, UK

B. H. J. BIELSKI Department of Chemistry, Brookhaven National Laboratory, Upton, New York 11973, USA

M. CHVAPIL Department of Surgery, College of Medicine, The University of Arizona Health Sciences Center, Tucson, Arizona 85724, USA

G. COHEN Department of Neurology, The Mount Sinai Medical Center, One Gustave L Levy Place, New York, New York 10029, USA

R. R. CRICHTON Unité de Biochimie, Institut Lavoisier, Université Catholique de Louvain, Place Louis Pasteur 1, 1348 Louvain-la-Neuve, Belgium

T. L. DORMANDY Department of Biochemistry, Whittington Hospital, London N19 5NF, UK

L. FLOHÉ Chemie Grünenthal GMBH, Steinfeldstrasse 2, 5190 Stolberg, Federal Republic of Germany

R. J. FLOWER The Wellcome Trust Research Laboratories, Langley Court, Beckenham, Kent BR3 3BS, UK

B. D. GOLDSTEIN Departments of Environmental Medicine and Medicine, New York Medical Center, 550 First Avenue, New York, New York 10016, USA

O. HAYAISHI Department of Medical Chemistry, Kyoto University Faculty of Medicine, Sakyo-ku, Kyoto 606, Japan

H. A. O. HILL Department of Inorganic Chemistry, University of Oxford, South Parks Road, Oxford OX1 3QR, UK

H. KEBERLE Pharma Research, CIBA-GEIGY, CH-4002 Basel, Switzerland

S. J. KLEBANOFF Department of Medicine RM-16, University of Washington School of Medicine, Seattle, Washington 98915, USA

J. McCORD Department of Biochemistry, University of South Alabama, Mobile, Alabama 36688, USA

A. M. MICHELSON Institut de Biologie Physico-Chimique, Fondation Edmond de Rothschild, 13 rue Pierre et Marie Curie, 75005 Paris, France

B. REITER National Institute for Research in Dairying, Shenfield, Reading RG2 9AT, UK

D. ROOS Department of Blood Cell Chemistry, Central Laboratory of the Netherlands Red Cross Blood Transfusion Service, Plesmanlaan 125, Amsterdam-W, The Netherlands

A. W. SEGAL Clinical Research Centre, Division of Cell Pathology, Watford Road, Harrow, Middlesex HA1 3UJ, UK

T. F. SLATER Department of Biochemistry, School of Biological Sciences, Brunel University, Kingston Lane, Uxbridge, Middlesex UB8 3PH, UK

L. L. SMITH Central Toxicological Laboratory, Imperial Chemical Industries Ltd., Alderley Park, Nr Macclesfield, Cheshire SK10 4TJ, UK

A. STERN Department of Pharmacology, New York University School of Medicine, New York, New York 10016, USA

R. J. P. WILLIAMS Department of Inorganic Chemistry, University of Oxford, South Parks Road, Oxford OX1 3QR, UK

R. L. WILLSON Department of Biochemistry, School of Biological Sciences, Brunel University, Kingston Lane, Uxbridge, Middlesex UB8 3PH, UK

K. H. WINTERHALTER Laboratorium für Biochemie I, Eidgenössische Technische Hochschule, Universitätstrasse 16, CH-8092 Zürich, Switzerland

Editor and Symposium Organizer: DAVID W. FITZSIMONS* The Ciba Foundation

---

**Present address:* Bureau of Hygiene and Tropical Diseases, Keppel Street, London WC1E 7HT

Chairman's introduction

IRWIN FRIDOVICH

Department of Biochemistry, Duke University Medical Center, Durham, North Carolina

Like Janus, oxygen has two faces, one benign and the other malignant. Molecular oxygen is toxic to virtually all life forms and this toxicity becomes obvious on exposure to concentrations significantly greater than the ambient fifth of an atmosphere. Our margin of safety is evidently a narrow one. We possess defences against oxygen toxicity which are sufficient to meet the ordinary demands, but which can easily be overwhelmed. In the current view, some biological reduction of oxygen occurs by the monovalent pathway and necessarily produces, first, superoxide radical (O_2^-) and hydrogen peroxide (H_2O_2) and then, if these are not efficiently scavenged, hydroxyl radical ($OH^•$) and possibly singlet oxygen ($^1\Delta_g$) as well. These products are all reactive substances and would not be well tolerated by living cells. The hydroxyl radical, in particular, is incredibly reactive and its production must be minimized. Since it is the third intermediate in the monovalent pathway of oxygen reduction, its production can be avoided by efficiently removing the first two, namely O_2^- and H_2O_2.

O_2^- is scavenged by superoxide dismutases; H_2O_2 is similarly eliminated by catalases and by peroxidases. Catalases and peroxidases have been known for a long time and have been well studied from a mechanistic point of view (Saunders *et al.* 1964; Sies 1974). Historically this is the case because they are haemoproteins which give rise to spectroscopically-distinct intermediates during their catalytic cycle. The first of these intermediates, called compound I, was initially taken to be a Michaelis complex between H_2O_2 and the enzyme. Considerable effort was devoted to the demonstration that compound I is really a bivalently oxidized form of the enzyme; probably an Fe(IV)-porphyrin π-cation radical (Schonbaum & Lo 1972). Catalases and peroxidases are related enzymes. Catalase can, in fact, act as a peroxidase towards a few electron donors, among which are ethanol, formate and nitrite. Catalase, acting in the catalatic mode, is most effective against relatively high concentrations of H_2O_2, whereas peroxidases are more effective against low levels of H_2O_2. Catalase has the advantage of not consuming any reductant other than H_2O_2. It could be called a hydrogen peroxide dismutase. Peroxidases, in contrast, use up some reductant. A particularly important and peculiar peroxidase is the glutathione peroxidase (Arias & Jakoby 1976). It uses glutathione as the reductant, but can use hydrogen peroxide or alkyl hydroperoxides as the oxidant. It can thus scavenge lipid hydroperoxides. Its peculiarity lies in its being a seleno-enzyme. It is appropriate that glutathione peroxidase will be discussed at this symposium, since it is definitely one of the important defences against oxygen toxicity.

Imposition of a stringent selection on a varied biota seems likely to result in independent adaptations. The oxygenation of the biosphere, by the forebears of present blue-green algae, was such a selection pressure and it is not surprising, from this point of view, that there is a variety of enzymes capable of dealing with hydrogen peroxide, that is, catalases with haem and pseudocatalases thought to contain flavin in place of haem and some peroxidases based on haem and others on selenium. A similar situation applies to the superoxide dismutases. There are superoxide dismutases based on copper and zinc (McCord & Fridovich 1969), on manganese (Keele *et al.* 1970) and on iron (Yost & Fridovich 1973). The distribution of these enzymes is a fascinating study in its own right. Thus, prokaryotes have been found to contain the manganese enzyme or the iron enzyme or both of these at once. *Escherichia coli*, for example, contains only the iron enzyme, when grown anaerobically, but makes the manganese enzyme, as well, when exposed to oxygen. The copper-zinc enzyme is characteristically found in the cytosol of eukaryotic cells. It is easily distinguished from the other superoxide dismutases, even in crude extracts, because it is inhibited by cyanide. Numerous bacteria have been surveyed and only one was found to contain the copper-zinc enzyme (Puget & Michelson 1974) and that one has been living as a symbiont of leiognathid fish for millions of years. One is sorely tempted to suggest that this symbiotic bacterium obtained the gene for the copper-zinc enzyme from the host fish. Representative eukaryotes have also been found to contain the manganese-enzyme and, at that, most often within the matrix of their mitochondria. Once again the temptation to cast a retrospective eye upon evolution is irresistible. In this case the relationship between the mitochondrial and the bacterial superoxide dismutases supports the symbiotic origin of these organelles. Striking homologies of amino acid sequence between the bacterial and the mitochondrial enzymes have been noted (Steinman & Hill 1973).

The defences against oxygen toxicity can be overwhelmed in several ways. One of these is to apply hyperbaric oxygen and another is to use compounds which increase the extent of the monovalent pathway of oxygen reduction. Compounds such as methyl viologen (paraquat), or the quinone antibiotics, can be reduced by enzymes within the cell and can then rapidly and spontaneously oxidize. They effectively increase the monovalent reduction of oxygen, at the expense of the tetravalent reduction by cytochrome oxidase. Thus, paraquat added to a suspension of *E. coli* in a TSY medium increases the cyanide-resistant respiration without changing the net respiration. It thus appears that paraquat and hyperbaric oxygenation have an effect in common and that is to increase the rate of production of O_2^- within cells. Appropriately *E. coli* responds to these two stresses in a like manner — it increases its production of the manganese superoxide dismutase (Hassan & Fridovich 1977). When rapid synthesis of this protective enzyme is possible, the bacteria survive these stresses but, when enzyme synthesis is prevented, the cells succumb. The importance of superoxide dismutases is thus demonstrated. Paraquat is a widely used herbicide and its mechanism of killing action is of more than academic interest, as will be discussed by Dr Smith (pp. 321–326).

It is an ill wind that blows no good. This old folk saying applies to the monovalent pathway of oxygen reduction. Thus, although primarily an unavoidable threat to cellular integrity, which must be minimized and against which defences are essential, there are circumstances in which the monovalent reduction of oxygen serves a useful purpose. A case in point is the polymorphonuclear leucocyte, which, on activation, markedly increases its consumption of oxygen. This respiratory burst, whose peculiarity was first recognized on the basis of its resistance to cyanide, is essential for the microbicidal action of these phagocytes. It now seems clear that both O_2^- and H_2O_2 are produced during the respiratory burst and are important parts of the antimicrobial armamentarium of these specialized cells (Babior 1978). It is appropriate that Drs Roos, Segal and Klebanoff will discuss the oxygen metabolism of these phagocytes. O_2^- also seems to be an intermediate in the action of certain oxidases which have been referred to as superoxidases. These enzymes share the property of being inhibited by superoxide dismutase and of being activated by O_2^-. Indoleamine dioxygenase (Hirata & Hayaishi 1975) and 2-nitropropane dioxygenase (Kido *et al.* 1976) appear to belong in this category.

It is wise, in designing defences, to provide a back-up system. The trial-and-error method of evolution must, by its nature, result in mechanisms which seem wise in retrospect, because only the successes are retained. In the case of defences against oxygen toxicity wisdom is evident. Inadequacy in the front-line defences would result in attack on the polyunsaturated fatty acids of the membranes. The damage would be minimized by antioxidants which can break free-radical reaction chains. α-Tocopherol is such an antioxidant and its insolubility in water guarantees that it will partition into the membranes, where it would be most effective. Dr Slater will discuss the importance of antioxidants.

References

ARIAS, I. M. & JAKOBY, W. B. (eds.) (1976) *Glutathione Peroxidase: Metabolism and Function*, Raven Press, New York

BABIOR, B. M. (1978) Oxygen-dependent microbial killing by phagocytes. *N. Engl. J. Med. 298*, 659–668, 721–725

HASSAN, H. M. & FRIDOVICH, I. (1977) Regulation of the synthesis of superoxide dismutase in *Escherichia coli:* induction by methyl viologen. *J. Biol. Chem. 252*, 7667–7672

HIRATA, F. & HAYAISHI, O. (1975) Studies on indoleamine 2,3-dioxygenase. I. Superoxide anion as substrate. *J. Biol. Chem. 250*, 5960–5966

KEELE, B. B. JR., MCCORD, J. M. & FRIDOVICH, I. (1970) Superoxide dismutase from *Escherichia coli* B. *J. Biol. Chem. 245*, 6176–6181

KIDO, T., SODA, K., SUZUKI, T. & ASADA, K. (1976) A new oxygenase, 2-nitropropane dioxygenase of *Hansenula mrakii. J. Biol. Chem. 251*, 6994–7000

KLEBANOFF, S. J. & ROSEN, H. (1979) The role of myeloperoxidase in the microbicidal activity of polymorphonuclear leucocytes, in *This Volume*, pp. 263–284

MCCORD, J. M. & FRIDOVICH, I. (1969) Superoxide dismutase: an enzymic function for erythrocuprein (hemocuprein). *J. Biol. Chem. 244*, 6049–6055

PUGET, K. & MICHELSON, A. M. (1974) Isolation of a new copper-containing superoxide dismutase, bacteriocuprein. *Biochem. Biophys. Res. Commun. 58*, 830–838

ROOS, D. & WEENING, R. S. (1979) Defects in the oxidative killing by microorganisms by phagocytic leucocytes, in *This Volume*, pp. 225–262

SAUNDERS, B. C., HOLMES-SIEDLE, A. G. & STARK, P. B. (1964) *Peroxidase*, Butterworth, Washington

SCHONBAUM, G. R. & LO, S. (1972) Interaction of peroxidases with aromatic peracids and alkyl peroxides. *J. Biol. Chem. 247*, 3353–3360

SEGAL, A. W. & ALLISON, A. C. (1979) Oxygen metabolism by stimulated human neutrophils, in *This Volume*, pp. 205–224

SIES, H. (1974) Biochemistry of the peroxisome in the liver cell. *Angew. Chem. Int. Ed. Engl. 13*, 706–718

SLATER, T. F. (1979) Mechanisms of protection against the damage produced in biological systems by oxygen-derived radicals, in *This Volume*, pp. 143–157

SMITH, L. L., ROSE, M. S. & WYATT, I. (1979) The pathology and biochemistry of paraquat, in *This Volume*, pp. 321–341

STEINMAN, H. M. & HILL, R. L. (1973) Sequence homologies among bacterial and mitochondrial superoxide dismutases. *Proc. Natl. Acad. Sci. U.S.A. 70*, 3725–3729

YOST, F. J. JR. & FRIDOVICH, I. (1973) An iron-containing superoxide dismutase from *Escherichia coli. J. Biol. Chem. 248*, 4905–4908

The chemistry of dioxygen and its reduction products

H. ALLEN O. HILL

Inorganic Chemistry Laboratory, South Parks Road, Oxford

Abstract The electronic properties of dioxygen are discussed with reference to its reactions in biological systems. In particular, the importance of its reduction to the superoxide ion is stressed. The physical and chemical properties of the superoxide ion are described and their relevance to its function in biochemical processes is emphasized. The relationship to other intermediates such as the hydroxyl radical is discussed.

It is a truism that molecular oxygen (or dioxygen as it is properly called) is an oxidant. Most elements and the majority of compounds, including those in living organisms, are thermodynamically unstable with respect to oxidation by dioxygen. Fortunately there exists a significant kinetic constraint on dioxygen in its role as an oxidant. The electronic structure of dioxygen (Fee & Valentine 1977; Hill 1978) is conveniently described by the one-electron molecular orbital diagram (Fig. 1) which indicates that the ground state, $^3\Sigma_g{}^-$, has two unpaired electrons, a reflection of the Pauli Principle. In contrast the first two excited states, $^1\Delta_g$ and $^1\Sigma_g$, have all electrons paired, as indeed do most other compounds. This crucial difference between dioxygen and most other molecules provides a kinetic barrier to reaction since a concomitant change in spin state is required and this change is slow relative to the lifetime of the collision complex. These constraints do not apply to reactions with single electrons, hydrogen atoms or other atoms or molecules containing unpaired electrons. The chemistry of dioxygen is, therefore, mostly a chemistry of one-electron transfer reactions, and hence our interest in oxygen radicals.

The successive addition of electrons to dioxygen generates the intermediates shown in Fig. 2. The relationship between these molecules is succinctly encapsulated in the oxidation-state diagram (Fig. 3), which provides a convenient method of displaying the oxidation-reduction potentials, measured relative to the element in its standard state. The more positive the gradient of

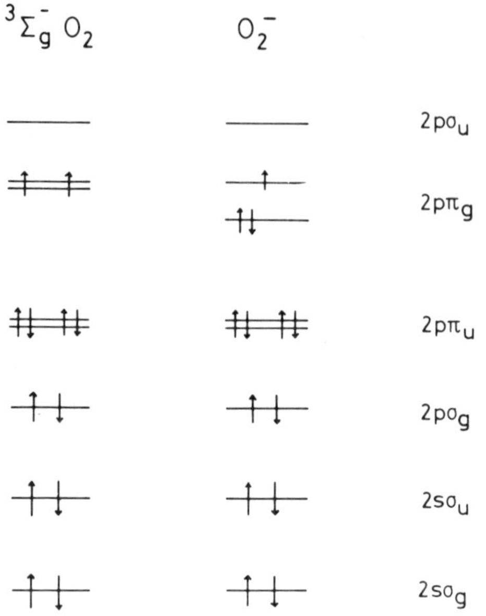

$^3\Sigma_g^-$ O_2 O_2^-

$2p\sigma_u$

$2p\pi_g$

$2p\pi_u$

$2p\sigma_g$

$2s\sigma_u$

$2s\sigma_g$

FIG. 1. A schematic one-electron molecular orbital diagram for dioxygen and the superoxide ion.

the line joining the points representing two species, the more powerful the couple is as an oxidant. Thus, O_2 ($^3\Sigma_g$), O_2 ($^1\Delta_g$), O_2 ($^1\Sigma_g$), O_2^-, HO_2, H_2O_2 and HO_2^- are all oxidants relative to water. Most powerful of all is the hydroxyl radical; so we are rightly concerned with its generation in biological systems.

Both H_2O_2 and O_2^- are unstable with respect to disproportionation, the former to H_2O and O_2, the latter to H_2O_2 and O_2. Since anions are stabilized by protons or cations, equilibria involving them, including disproportionation, will be sensitive to pH, metal ions and solvent. Therefore, the biological environment in which, for example, O_2^- and H_2O_2 are generated, will have an important influence on their stabilities and reactions.

The implication of the superoxide ion in many biological processes involving dioxygen (Michelson *et al.* 1977) has stimulated much research into its properties and reactions. The additional electron in the antibonding orbital results in a lengthening of the O–O bond. The sole unpaired electron provides an important method of investigation in electron paramagnetic resonance spectroscopy. The g-values of the *free* ion are $g_{\parallel} = 4$; $g_{\perp} = 0$ (Känzig & Cohen 1959) but the magnetic properties of the ion are markedly perturbed by, and thus sensitive to, the chemical environment. Of the two g-values, $g_{\parallel}$ is the more responsive, varying from 2.436 for O_2^- in KCl to 2.04 in N,N'-dimethylformamide; $g_{\perp}$

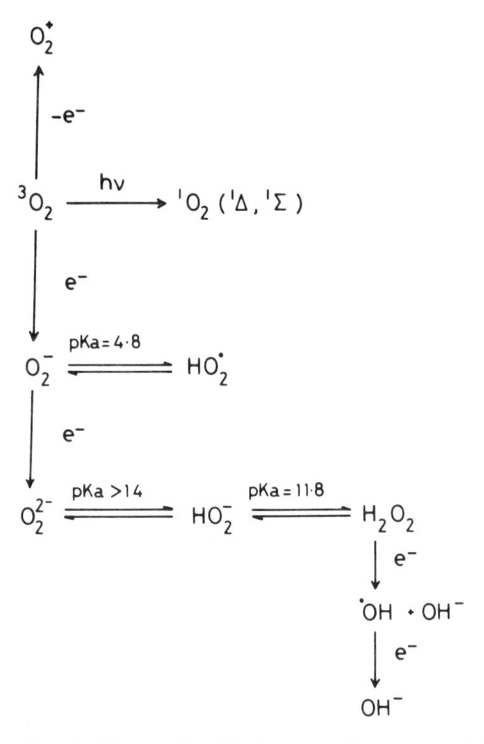

FIG. 2. The relationship between dioxygen and its oxidation and reduction products.

is usually about 2.00. The superoxide ion can be prepared by several methods (Lee-Ruff 1977), of which the most common are the direct reaction of dioxygen with electropositive metals, the electrolysis of dioxygen and pulse radiolysis. Many methods involving biological materials are now known of which the most convenient are the oxidation of xanthine by xanthine oxidase (Knowles *et al.* 1969) and the oxidation of reduced flavin generated by photolysis (Massey *et al.* 1969). It can also be formed by the oxidation of hydrogen peroxide by cerium(IV) salts or by periodate.

Until the renaissance of interest in the chemistry of the superoxide ion its only notable reaction was that of self-destruction. The disproportionation has been thoroughly investigated and the following data are available (Bielski & Allen 1977):

$$HO_2 + HO_2 \rightarrow H_2O_2 + O_2 \qquad k_1 = 7.6 \times 10^5 \text{ dm}^3 \text{ mol}^{-1} \text{ s}^{-1} \qquad (1)$$
$$HO_2 + O_2^- \rightarrow HO_2^- + O_2 \qquad k_2 = 8.9 \times 10^7 \text{ dm}^3 \text{ mol}^{-1} \text{ s}^{-1} \qquad (2)$$
$$O_2^- + O_2^- \rightarrow O_2^{2-} + O_2 \qquad k_3 < 0.3 \text{ dm}^3 \text{ mol}^{-1} \text{ s}^{-1} \qquad (3)$$

The overall second-order rate constant is greater at the pH corresponding to

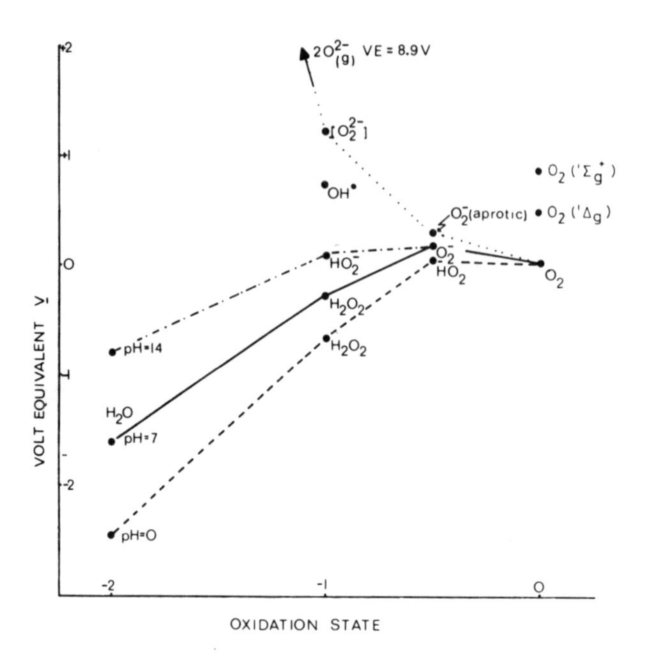

FIG. 3. The oxidation-state diagram for dioxygen: *V*, the volt equivalent (VE), is the standard electrode potential at 25.0°C (*versus* the normal hydrogen electrode) multiplied by the oxidation state (the *formal* charge per atom of element in a *compound* or *ion*, assuming that each oxygen atom has a charge of −2 and each hydrogen atom a charge of +1; the oxidation state of the elemental form is zero). The potentials shown are, therefore, those of the cell:

Pt, H_2 | a_{H^+} = 1 ‖ Standard couple pO_2 = 1 atm, pH solvent as indicated | Pt

The data are from Fee & Valentine (1977).

the pK_a of HO_2; this is consistent with reaction (2) having the fastest rate. The identification (McCord & Fridovich 1969) of the cupreins as extremely effective catalysts for the disproportionation has revolutionized the biology of dioxygen (see Michelson *et al.* 1977). All aerobic, many aerotolerant, and even a few anaerobic organisms contain superoxide dismutases; this fact suggests that the consequences of the formation of the superoxide ion in living organisms may be deleterious to their well-being. It is not immediately obvious from the known reactions (Table 1) of the superoxide ion with organic compounds (Lee-Ruff 1977) why this should be so though the formation of epoxides (Dietz *et al.* 1970) from some alkenes may arouse suspicion. It can react as a reductant, for example, of cytochrome *c* and semiquinones; as an oxidant, for example, of adrenalin, quinones and ascorbic acid; or as a nucleophile, for example, with activated alkyl or aryl halides and even metal complexes. Just as the physical properties of O_2^- are sensitive to the environment, so the chemical reactions depend on, for instance, solvent (most importantly whether it is protic or aprotic)

TABLE 1

Some reactions of superoxide in aqueous and aprotic media

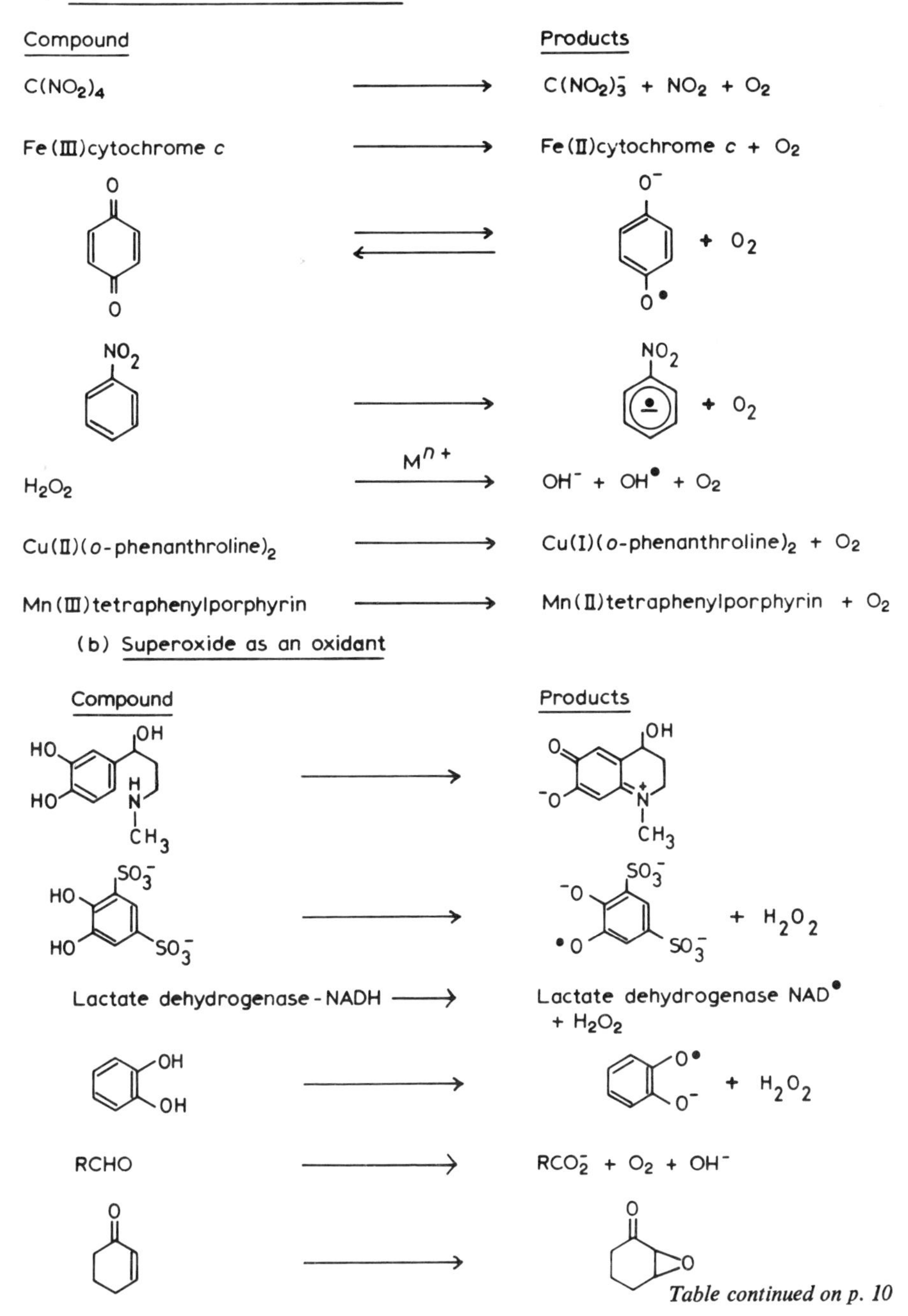

Table continued on p. 10

TABLE 1 (*Cont.*)

(c) Superoxide as a nucleophile

Compound		Products
Fe(III)protoporphyrin dimethyl ester perchlorate		O_2Fe(II)protoporphyrin dimethyl ester + ClO_4^-
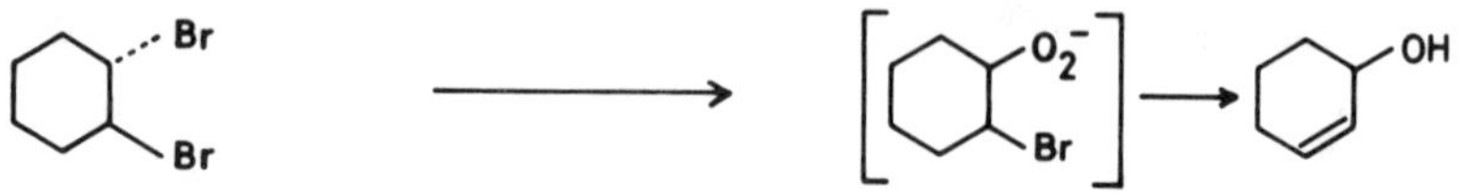 Br, Br	→	O_2^-, Br → OH
2RX	→	$RO_2R + O_2 + 2X^-$
RX	→	ROH
Aquo Co(III)cobalamin	→	Superoxo Co(III)cobalamin
 I, I	→	O, O
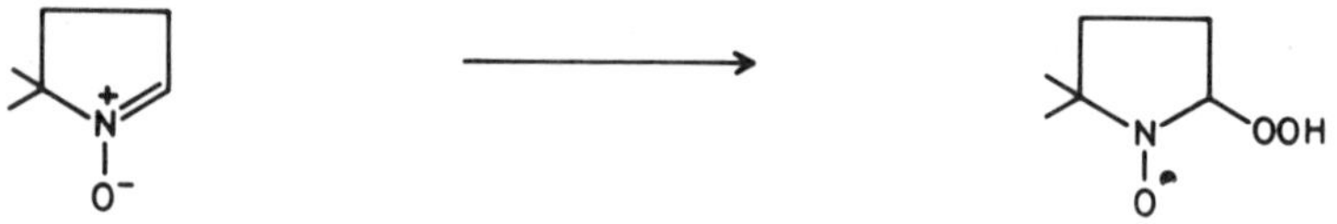 $\overset{+}{N}$, O^-	→	N, OOH, $O^{\bullet}$

and metal ion concentration. Most suspicion must alight on the incestuous reactions between O_2^-, other reduction products of dioxygen and redox-active metal ions. For example, it is possible that the disproportionation of the superoxide ion or even hydrogen peroxide yields dioxygen in one of its excited states. If this happened in biological systems, the effects might be horrendous.

$$O_2^- + H_2O_2 \rightarrow O_2 + OH^- + OH^{\bullet} \quad (4)$$

$$Fe^{III} + O_2^- \rightarrow Fe^{II} + O_2 \quad (5)$$

$$Fe^{II} + H_2O_2 \rightarrow Fe^{III} + OH^- + OH^{\bullet} \quad (6)$$

The same is true, only more so, of the so-called Haber–Weiss reaction (4). This may require catalysis by redox-active metal ions (Cohen 1977), as in reactions (5) and (6), for instance. The hydroxyl radical can wreak havoc; its reactions are discussed elsewhere in this book. So, too, are products formed by the reaction of O_2^- with metal ions. The reaction of O_2^- with, for instance, the iron(II) ion (reaction 7) may yield a powerful oxidant product.

$$Fe^{2+} + O_2^- \rightarrow [FeO_2]^+ \quad (7)$$

It is an attractive postulate (Fridovich 1974) that, to avoid the generation of species as destructive as for example $OH^{\cdot}$, three efficient enzymes — peroxidases, catalases and superoxide dismutases — evolved to prevent the formation of the precursors, H_2O_2 and O_2^-. These same species are just those which result from the irradiation of oxygenated aqueous solutions and it is intriguing to consider that, as well as having to contend with the inadvertent chemical generation of such species as a consequence of making use of dioxygen as the terminal electron acceptor, early primitive organisms, and not so primitive later ones, may have had to evolve a protection against the ravages of radiation. Without defence mechanisms against the toxic effects of that otherwise obligatory element, oxygen, the consequences would have been, and presumably are still, calamitous.

ACKNOWLEDGEMENTS

I am grateful to my colleagues, Professor W. H. Bannister, Drs J. V. Bannister and M. P. Esnouf, and Messrs A. E. G. Cass, M. R. Green and A. McEuan for help. The support of the Medical Research Council is gratefully acknowledged. This is a contribution from the Oxford Enzyme Group.

References

Bielski, B. H. J. & Allen, A. O. (1977) Mechanism of the disproportionation of superoxide radicals. *J. Phys. Chem. 81*, 1048–1050

Cohen, G. (1977) In defense of Haber–Weiss, in *Superoxide and Superoxide Dismutases* (Michelson, A. M., McCord, J. M. & Fridovich, I., eds.), pp. 317–321, Academic Press, London & New York

Dietz, R., Forno, A. E. J., Larcombe, B. E. & Peover, M. E. (1970) Nucleophilic reactions of electrogenerated superoxide ion. *J. Chem. Soc. B*, 816–820

Fee, J. A. & Valentine, J. S. (1977) Chemical and physical properties of superoxide, in *Superoxide and Superoxide Dismutases* (Michelson, A. M., McCord, J. M. & Fridovich, I., eds.), pp. 19–60, Academic Press, London & New York

Fridovich, I. (1974) Superoxide and evolution. *Horiz. Biochem. Biophys. 1*, 1–37

Hill, H. A. O. (1978) The superoxide ion and the toxicity of molecular oxygen, in *New Trends in Bioinorganic Chemistry* (Williams, R. J. P. & da Silva, J. R. F., eds.), pp. 173–208, Academic Press, London & New York

Känzig, W. & Cohen, M. H. (1959) Paramagnetic resonance of oxygen in alkali halides. *Phys. Rev. Lett. 3*, 509–510

Knowles, P. F., Gibson, J. F., Pick, F. M. & Bray, R. C. (1969) Electron spin resonance evidence for enzymic reduction of oxygen. *Biochem. J. 111*, 53–58

Lee-Ruff, E. (1977) The organic chemistry of superoxide. *Chem. Soc. Rev. 6*, 195–214

McCord, J. M. & Fridovich, I. (1969) Superoxide dismutase. An enzymic function for erythrocuprein (hemocuprein). *J. Biol. Chem. 244*, 6049–6055

Massey, V., Strickland, S., Mayhew, S. G., Howell, L. G., Engle, P. C., Matthews, R. G. & Sullivan, P. A. (1969) The production of superoxide anion radicals in the reaction of reduced flavins and flavoproteins with molecular oxygen. *Biochem. Biophys. Res. Commun. 36*, 891–897

Michelson, A. M., McCord, J. M. & Fridovich, I. (eds.) (1977) *Superoxide and Superoxide Dismutases*, Academic Press, London & New York

Discussion

Crichton: Weber *et al.* (1978) have confirmed that the oxygen in oxyhaemoglobin (or, to be more exact, oxyerythrocruorin) from *Chironomus* is bonded end-on, with the dimensions one would expect. They maintained a reducing environment of hydrogen sulphide while continuing to use oxygen.

Cohen: Walling (1975) has stated that, kinetically, the ferryl radical $[FeO]^{2+}$ is equivalent to the hydroxyl radical. I am not certain what is meant by that; can you clarify?

Hill: If hydroxide ion is eliminated from the $[FeO_2]^+$ species, the product can be described as a ferryl system (i.e. reaction 1). When written as $[Fe^{IV}O]^-$, it may be considered as equivalent to the hydroxyl radical.

$$[Fe^{II}O_2{}^- \leftrightarrow Fe^{III}O_2{}^{2-}] + H^+ \rightarrow [Fe^{IV}O^- \leftrightarrow Fe^{V}O^{2-}] + OH^- \quad (1)$$

Cohen: Do you suspect that these species are formed in biological systems?

Hill: There is considerable argument about this. I have betrayed several of my prejudices already and I suspect that the metal peroxide system can be reactive without the necessary intervention, as an intermediate, of the ferryl complexes. I guess that iron(III) peroxide, or iron(II) superoxide (depending on how one writes it), can react in a concerted way, so that the lifetime of any ferryl intermediate is vanishingly short.

Stern: How does the diffusibility in water compare with the reactivity of these various super-reactive species?

Hill: The diffusion coefficients will differ. I cannot see how diffusibility can be significant for reactivity.

Stern: Presumably, these reactive species are generated in biological systems and somehow the cell can compensate for their generation. In certain conditions, some of these reactive species may be produced in greater concentrations than usual; they may not overwhelm the enzymic systems necessarily because they are more reactive, but because they are present in greater concentrations and have greater chance of getting to and damaging the biological structures.

Hill: The only relevant factor is concentration of the reactive species. If, for example, the protective devices were reduced in concentration or inaccessible to the reactants, some of your comments might be pertinent.

Stern: What are the chances of the reduced oxygen products reacting with themselves to form a relatively inactive product rather than diffusing a certain distance away?

Hill: That depends on the concentration; with high enough concentration, the 'self-reaction' (disproportionation) will be significant. As Professor Fridovich has pointed out, that is one of the key features of having a dismutase: superoxide may react in many ways before it dismutates in the second-order reaction. The enzyme removes even very low concentrations of superoxide.

Klebanoff: How rapid is the interaction between hydroxyl radicals to form hydrogen peroxide compared to the reaction with other species with which they react readily?

Willson: The rate constant for the recombination of hydroxyl radicals is about 5×10^9 dm^3 mol^{-1} s^{-1}; it is almost diffusion controlled.

Hill: But that is true of the reaction of hydroxyl radicals with practically anything. The activation energy for reaction seems to be minimal.

Willson: Hydroxyl radicals react rapidly with most organic components ($k = 10^8$–10^{10} dm^3 mol^{-1} s^{-1}). There are some exceptions, however: the glycine zwitterion ($k = 10^7$ dm^3 mol^{-1} s^{-1}), oxalic acid ($k < 10^7$ dm^3 mol^{-1} s^{-1}) and urea ($k < 10^6$ dm^3 mol^{-1} s^{-1}). The recombination of $OH^\bullet$ to yield H_2O_2 is unlikely in biological systems except when they are exposed to ionizing radiation.

Hill: As with superoxide, it depends on the concentration.

Fridovich: In an organic system such as that inside cells the number of competing reactions is overwhelming.

Bielski: Complexes such as the one you mentioned, Dr Hill, between iron (II) and O_2 could be powerful oxidizing agents. For example, although O_2^- reacts slowly with NADH ($k \leqslant 27$ dm^3 mol^{-1} s^{-1}; Land & Swallow 1971), when complexed with Mn^{2+} ($Mn^{2+} + O_2^- \rightarrow [MnO_2]^+$; Bielski & Chan 1978), it reacts rapidly with NADH and sets up a chain reaction in which up to 80 NADH molecules are oxidized per one of O_2^- (Curnutte *et al.* 1976; B. Bielski & P. Chan, unpublished results).

Hill: The best biological example is probably cytochrome P450 in which the metal-oxygen complex is $[Fe^{II}O_2]^-$.

In the oxidation-state diagram (Fig. 3) I tried to emphasize that consideration of the chemistry of simple systems in water could be very misleading; the reactivity of complexes of dioxygen and its reduction products will depend on solvent, the metal, proton concentration etc.

Michelson: I am glad that you did not suggest that dismutation of superoxide gives rise to singlet oxygen: it has been amply proved that singlet oxygen is not formed. For example, D_2O should increase light emission (due to the excited singlet oxygen); it does not. The emission wavelength (λ_{max} 435 nm) does not correspond to any species due to singlet oxygen (Henry & Michelson 1977). Professor Fridovich's group has shown definitely that many of these effects are due to carbonate radicals (Kellogg & Fridovich 1975).

Hill: Is there any example of singlet oxygen being involved in biological systems, other than those which are photochemical?

Michelson: As far as I know all the instances are photochemical. I cannot cite any biochemical reaction that certainly gives rise to singlet oxygen. Foote *et al.* (1978) have shown conclusively that singlet oxygen is *not* produced by human leucocytes or rat lung macrophages using ^{14}C-labelled cholesterol.

Fridovich: There have been some non-photochemical investigations in which, for example, polyunsaturated fatty acids or lipid membranes rich in polyunsaturated fatty acids were attacked by systems generating both O_2^- and H_2O_2 (e.g. the xanthine oxidase system). Both catalase and superoxide dismutase protected, but hydroxyl radical scavengers did not effectively protect; however, known singlet scavengers were very effective. It was also shown in the same system that 2,5-dimethylfuran could be converted into the expected diacetylethylene. So there are reasons for *suspecting* that in these cases singlet oxygen was produced.

Michelson: Once lipid peroxides are formed many other things can happen which could give rise to similar results to those expected from singlet oxygen. I have investigated superoxide dismutation with scavengers, a host of molecules capable of indicating singlet transfer, triplet transfer, and so on, and frankly singlet oxygen is not produced.

A second point is that, despite the doubts of organic chemists (Lee-Ruff 1977), the best proof of involvement of O_2^- in biological processes is the inhibition by superoxide dismutase.

McCord: In as much as reports of inhibition by superoxide dismutase have led to conclusions that O_2^- is the reactive species when catalase will also inhibit, Dr Hill is right in suggesting that superoxide may not be the actual active species.

Michelson: I fully agree; especially as in many systems carbonate radicals are the effective species.

The Haber–Weiss reaction is thermodynamically possible but slow compared with the competing reactions. Can you list the relative kinetics of the disappearance of hydroxyl radicals for example by reaction with O_2^- or with H_2O_2?

Hill: The value of the rate constant for the reaction (2) is given (Schwarz 1962) as 4.5×10^7 $dm^3\, mol^{-1}\, s^{-1}$. Although the reaction (3) is often quoted, the rate does not appear to have been measured.

$$OH^{\bullet} + H_2O_2 \rightarrow O_2^- + H_2O + H^+ \quad (2)$$

$$OH^{\bullet} + O_2^- \rightarrow O_2 + OH^- \quad (3)$$

Goldstein: Is there any overall rule about what will happen in a non-polar medium such as the membrane in terms of these various incestuous reactions you mentioned or will it be necessary to dissect out each one to ascertain the polarity effects?

Hill: Those reduced oxygen species that are anions will be destabilized by aprotic media. Anions in aprotic media are very reactive towards electrophiles. However, there are several species of the superoxide anion: the naked ion, O_2^- (probably not unlike that generated in the gas phase); the solvated superoxide ion, $O_2(H_2O)_n^-$; the conjugate acid, HO_2. In addition 'ion-pairs', $[M^{n+} O_2^-]^{(n-1)+}$, are possible. (We call them ion-pairs if the counter-ion is, say, calcium or barium but we call them complexes if the counter-ion is, say, iron(III).) In Fig. 3 I indicated that the oxide ion, O^{2-}, has a very high oxidation potential. However, the chemistry of the oxide ion is always that of its complexes with metal ions or protons. If these anions are not stabilized by positive charges, then thermodynamically they are much more reactive and, probably, for many reactions kinetically more reactive. So, superoxide or peroxide generated in an aprotic medium such as a lipid membrane may be much more reactive than it would be free in aqueous solution. [*See also general discussion, pp. 363–367.*]

Fridovich: One exception is the 'self' reaction, the spontaneous disproportionation, because protons are needed to give hydrogen peroxide and oxygen. In a non-protic solvent O_2^- is stable.

Hill: Very stable; in spite of some reports, with really clean glassware and solvents, solutions of superoxide in, say, dimethyl sulphoxide or even dimethylformamide can be kept for a long time.

Michelson: Superoxide is a reductant, and you mentioned that it can reduce nucleic acids. Is this true?

Hill: The reaction of O_2^- with nucleic acids and nucleotides has been described by Van Hemmen & Neuling (1975) and Lown *et al.* (1976).

Michelson: What is reduced? Also, what is formed? A 5,6-dihydropyrimidine? A 6-hydroperoxy-5,6-dihydrothymine has been identified in irradiation studies (Ekert & Monier 1959). I doubt whether the action of O_2^- is the same as on cytochrome *c*, for example.

Willson: There is evidence that superoxide reacts with bromouracil (Volkert *et al.* 1967) but the reduction of normal nucleic acid derivatives by superoxide is unlikely. The product of one-electron addition to a nucleic acid base, say (thymine)$^-$ or (adenine)$^-$, reacts with oxygen (to give superoxide) at rates of 3–4 $\times$ 10^9 dm^3 mol^{-1} s^{-1} (Scholes & Willson 1967; Loman & Ebert 1970; Willson 1971). I doubt whether the reverse reaction will go to any extent.

Hill: Well, your own work on quinones (Patel & Willson 1973) is an object lesson reminding us to define the conditions because one can swing the equilibrium one way or the other by changing the quinone and by changing the solvent.

Willson: Semiquinones and oxygen do participate in equilibrium reactions but, in the case of nucleic acid derivatives, the one-electron redox potential appears to be so different from that of oxygen (at least in water and I agree that in cells the medium may not always be water) that the reverse reaction to base and O_2^- is unlikely.

Cohen: McCord & Day (1978) used iron(II)–EDTA in a Haber–Weiss-type reaction to generate hydroxyl radicals with tryptophan as the agent to be peroxidized. What is the product? Will it interfere with the spectrum you observed? Does superoxide react with tryptophan?

Michelson: Formylkynurenine is not formed (A. M. Michelson, unpublished work, 1974).

McCord: We were following the disappearance of the tryptophan absorption. I doubt that superoxide reacts with tryptophan directly; we saw a secondary reaction with hydroxyl radical or some iron complex equivalent to that which gives many products. Armstrong & Swallow (1969) have investigated these reactions.

Fridovich: Amongst those products do you expect hydroxylated derivatives of tryptophan?

McCord: Yes.

Bielski: Adams & Wardman (1977) have summarized studies on the interaction of hydroxyl radicals with tryptophan. Hydroxyl radicals add to the heterocyclic ring at the C-2 and C-3 positions, giving optical absorption maxima at 345 and 325 nm. Addition of hydroxyl radicals to the benzene ring of tryptophan results in a maximum at 310 nm.

Willson: The radical cation can also be formed; instead of radical addition, an electron can be transferred to an attacking radical (Posener *et al.* 1976). Walrant *et al.* (1975) have obtained evidence for the reaction of O_2^- with tryptophan to give *N*-formylkynurenine both electrochemically and radiochemically.

Bielski: The rate constant for the reaction of O_2^- with tryptophan is slow, less than 24 $dm^3\ mol^{-1}\ s^{-1}$ (see pp. 43–48).

Goldstein: The evidence comes from electrochemical and radiation-chemical experiments: the reaction depends on the tryptophan concentration.

References

ADAMS, G. E. & WARDMAN, P. (1977) Free radicals and biology: the pulse radiolysis approach, in *Free Radicals in Biology* vol. 3 (Pryor, W. A., ed.), pp. 53–91, Academic Press, London & New York

ARMSTRONG, R. C. & SWALLOW, A. J. (1969) Pulse- and gamma-radiolysis of aqueous solutions of tryptophan. *Radiat. Res. 40*, 563–579

BIELSKI, B. H. J. & CHAN, P. C. (1978) Products of reaction of superoxide and hydroxyl radicals with Mn^{2+} cation. *J. Am. Chem. Soc. 100*, 1920–1921

CURNUTTE, J. T., KARNOVSKY, M. L. & BABIOR, B. M. (1976) Manganese-dependent NADPH oxidation by granulocyte particles. *J. Clin. Invest. 57*, 1059–1067

EKERT, B. & MONIER, R. (1959) Structure of thymine hydroperoxide produced by X-irradiation. *Nature (Lond.) 184*, 58–59

FOOTE, C. S., ABAKERLI, R. B. & CLOUGH, R. L. (1978) Stalking singlet oxygen, in *Abstracts of the International Conference on Chemi and Bioenergized Processes*, p. 25, Guarujà, Brazil

HENRY, J. P. & MICHELSON, A. M. (1977) Superoxide and chemiluminescence, in *Superoxide and Superoxide Dismutases* (Michelson, A.M., McCord, J. M. & Fridovich, I., eds.), pp. 283–290, Academic Press, London & New York

KELLOGG, E. W. III & FRIDOVICH, I. (1975) Superoxide, hydrogen peroxide and singlet oxygen in lipid peroxidation by a xanthine oxidase system. *J. Biol. Chem. 250*, 8812–8817

LAND, E. J. & SWALLOW, A. J. (1971) One-electron reactions in biochemical systems as studied by pulse radiolysis. V. Cytochrome *c*. *Arch. Biochem. Biophys. 145*, 365–372

LEE-RUFF, E. (1977) The organic chemistry of superoxide. *Chem. Soc. Rev. 6*, 195–214

LOMAN, H. & EBERT, M. (1970) The radiation chemistry of thymine in aqueous solution. Some reactions at the thymine-electron adduct. *Int. J. Radiat. Biol. Relat. Stud. Phys. Chem. Med. 18*, 369–379

LOWN, J. W., BEGLEITER, A., JOHNSON, D. & MORGAN, A. R. (1976) Studies related to antitumor antibiotics. Part V. Reactions of mitomycin C with DNA examined by ethidium fluorescence assay. *Can. J. Biochem. 54*, 110–119

MCCORD, J. M. & DAY, E. D. JR. (1978) Superoxide-dependent production of hydroxyl radical catalyzed by iron–EDTA complex. *FEBS (Fed. Eur. Biochem. Soc.) Lett. 86*, 139–142

PATEL, K. B. & WILLSON, R. L. (1973) Semiquinone free radicals and oxygen: pulse radiolysis study of one-electron transfer equilibria. *J. Chem. Soc. Faraday Trans. I 69*, 814–825

POSENER, M. L., ADAMS, G. E. & WARDMAN, P. (1976) Mechanism of tryptophan oxidation by some inorganic radical-anions; pulse radiolysis study. *J. Chem. Soc. Faraday Trans. I 72*, 2231–2239

SCHOLES, G. & WILLSON, R. L. (1967) Radiolysis at aqueous thymine solutions. *Faraday Soc. Trans. 63*, 298–299

SCHWARZ, H. A. (1962). *J. Phys. Chem. 66*, 255–262

VAN HEMMEN, J. J. & NEULING, W. S. A. (1975) Inactivation of biologically active DNA by gamma-ray-induced superoxide radicals and their dismutation products singlet molecular oxygen and hydrogen peroxide. *Biochim. Biophys. Acta 402*, 133–141

VOLKERT, O., BORS, W. & SCHULTE-FROHILINDE, D. (1967) Strahlenchemie wässriger, sauerstoffhaltiger Lösungen von 5-Bromuracil (11). Reaktion des OH-Radikals mit Bromuracil. *Z. Naturforsch. (B) 22*, 480–485

WALLING, C. (1975) Fenton's reagent revisited. *Acc. Chem. Res. 8*, 125–131

WALRANT, P., SANTUS, R., REDPATH, J. L. & LEXA, D. (1975) Role of diatomic oxygen ions in the formation of *N'*-formylkynurenine from tryptophan. *C. R. Acad. Sci. (D) (Paris) 5* (280), 1425–1428

WEBER, E., STEIGEMANN, W., JONES, T. A. & HUBER, R. (1978) The structure of oxy-erythrocruorin at 1.4 Å. *J. Mol. Biol. 120*, 327–336

WILLSON, P. L. (1971) Pulse radiolysis studies on reaction of triacetoneamine-*N*-oxyl with radiation-induced free radicals. *Faraday Soc. Trans. 67*, 3008–3019

Hydroxyl radicals and biological damage *in vitro*: what relevance *in vivo*?

ROBIN L. WILLSON

Department of Biochemistry, Brunel University, Uxbridge, Middlesex

Abstract Hydroxyl radicals have been implicated in various forms of tissue injury ranging from radiation-induced cell death to alloxan-induced diabetes in animals.

Although hydroxyl radicals can be readily generated *in vitro*, for example by the action of radiation or from the reaction of ferrous (Fe^{II}) ions with peroxides, they have not been observed directly *in vivo*. Their involvement in tissue damage has generally only been inferred from studies with so-called selective radical scavengers.

In this paper the validity of such inferences will be briefly discussed in the light of current knowledge about the mechanisms and rates of reactions of hydroxyl radicals with biological compounds.

Suggestions that hydroxyl radicals ($OH^{\bullet}$) are important intermediates in chemistry and biology have been made for half a century. In 1929 Risse proposed that these highly oxidizing species were formed in water exposed to X-rays and in the same year Urey *et al.* described how they might also be formed in hydrogen peroxide exposed to u.v. light. Since then, there have been many suggestions that hydroxyl radicals are also important in biological phenomena including the initiation and progression of some diseases (Table 1). I say 'suggestions' in the Table because in most instances the evidence remains circumstantial.

Although our knowledge of hydroxyl radical reactions *in vitro* has increased enormously in recent years, their relevance, particularly in normal enzyme processes *in vivo*, is still widely questioned. Walling (1975) summed up the present situation succinctly: 'Finally and much more speculatively, biochemists have been reluctant to consider hydroxyl radicals as intermediates in enzymic hydroxylation because of their high and indiscriminate reactivity.' He then

TABLE 1

'Suggestions' for $OH^{\bullet}$ involvement in biological phenomena

Phenomenon	*Reference*
Biological oxidation	Haber & Willstätter (1931)
Fenton's reagent	Haber & Weiss (1932)
Radiation damage	Weiss (1944)
Oxygen toxicity	Gerschman *et al.* (1954)
Ageing, mutation and cancer	Harman (1962)
Xanthine oxidase activity	Beauchamp & Fridovich (1970)
Phagocytic activity	Johnston *et al.* (1973)
Lysosome peroxidation	Fong *et al.* (1973)
Inflammation	McCord (1974)
Hydroxydopamine toxicity	Cohen & Heikkila (1974)
Alloxan-induced diabetes	Heikkila *et al.* (1976)
Prostaglandin synthesis	Panganamala *et al.* (1976)
Bleomycin toxicity	Lown & Sim (1976)
Ozone toxicity	Willson (1977*a*)
Microsomal ethanol oxidation	Cederbaum *et al.* (1977)

added the footnote: 'Since such formulations imply a high hydroxyl radical flux in one's liver, their reluctance has been understandable.'

This reluctance is not new. In 1932 J. Haldane wrote: 'Thus Haber and Willstätter postulate free OH radicals not only in the catalase reaction but also in the actions of acetaldehyde oxidase... I think that the majority of biochemists will demand very strong experimental evidence before they accept the chain theory of enzyme action.'

Haber & Willstätter (1931) had discussed the action of catalase, peroxidase and the enzymic oxidation of alcohol and acetaldehyde in terms of chain reactions in which the enzyme produced the initiating radical. They proposed reactions (1)–(3) for the case of catalase.

$$\text{Katalase} + H_2O_2 = \text{Desoxy-Katalase} + O_2H^{\bullet} \quad (1)$$

$$O_2H^{\bullet} + H_2O_2 = O_2 + H_2O + OH^{\bullet} \quad (2)$$

$$OH^{\bullet} + H_2O_2 = H_2O + O_2H^{\bullet} \quad (3)$$

They depicted peroxidase action on pyrogallol by reactions (4)–(6) (in which the arrow indicates the unpaired electron). Their scheme for the oxidation of alcohol consisted of reactions (7)–(10). Although some of these chain-propagating reactions such as (9) are now considered unlikely and reactions (2) and (5) are probably very slow, the very description of reactions (2), (3), (5), (6), (8) and (10) with the hydroxyl radical in reaction (10) acting at the carbon atom adja-

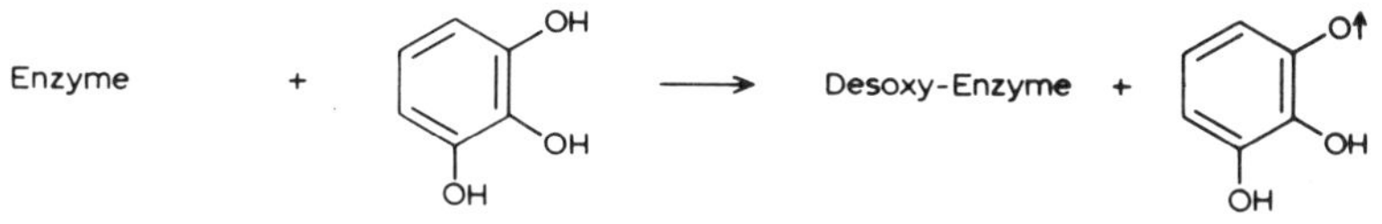

Enzyme + (OH, OH, OH) → Desoxy-Enzyme + ($O\uparrow$, OH, OH) (4)

($O\uparrow$, OH, OH) + H_2O_2 → (O, O, OH) + $\downarrow OH$ + H_2O (5)

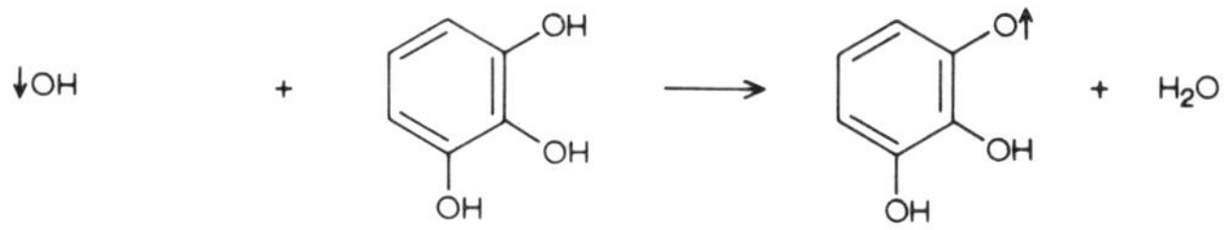

$\downarrow OH$ + (OH, OH, OH) → ($O\uparrow$, OH, OH) + H_2O (6)

$$CH_3CH_2OH + \text{Enzyme} \rightarrow CH_3\underset{\downarrow}{C}H(OH) + \text{Monodesoxy-Enzyme} + H^{\bullet} \quad (7)$$

$$CH_3\underset{\downarrow}{C}H(OH) + O_2 \rightarrow CH_3CH(OH){-}O{-}O\downarrow \quad (8)$$

$$CH_3CH(OH){-}O{-}O\downarrow + CH_3CH_2OH \rightarrow 2CH_3CHO + \downarrow OH + H_2O \quad (9)$$

$$\downarrow OH + CH_3CH_2OH \rightarrow CH_3\underset{\downarrow}{C}H(OH) + H_2O \quad (10)$$

cent to the hydroxyl group is remarkable, considering the techniques available. In the following year Haber & Weiss proposed that hydroxyl radicals were formed by Fenton's reagent (iron(II) ion and hydrogen peroxide) and later Weiss elaborated on the manner in which the different reactions of peroxidase and catalase could be attributed to free-radical processes (Haber & Weiss 1932, 1934; Weiss 1935, 1937). Reactions (11)–(17), amongst others, were discussed in these papers. (Dots, the current convention, have been inserted for clarity to indicate unpaired electrons.) Now, 40 years later, most of these reactions are common knowledge. Indeed, reactions (11) and (13) have both been referred to as the Haber–Weiss reaction, and have attracted considerable attention. In particular, the importance of reaction (13), with the formation of

$$Fe^{2+} + H_2O_2 \rightarrow Fe^{3+} + OH^{\bullet} + OH^- \quad \text{(11) (Fenton reaction)}$$

$$Fe^{2+} + OH^{\bullet} \rightarrow Fe^{3+} + OH^- \quad (12)$$

$$O_2^- + H_2O_2 \rightarrow O_2 + H_2O + OH^{\bullet} \quad \text{(13) (Haber–Weiss reaction)}$$

$$HCOOH + OH^{\bullet} \rightarrow {}^{\bullet}COOH + H_2O \quad (14)$$

$$H_2O_2 + {}^{\bullet}COOH \rightarrow H_2O + CO_2 + OH^{\bullet} \quad (15)$$

$$I^- + OH^{\bullet} \rightarrow I^{\bullet} + OH^- \quad (16)$$

$$Fe^{3+} + I^- \rightleftharpoons Fe^{2+} + I^{\bullet} \quad (17)$$

$OH^{\bullet}$, in systems which generate O_2^- has been the subject of much debate. In the proceedings of a recent symposium, one paper was even entitled 'In defense of Haber–Weiss' (Cohen 1977)!

Since 1970 many biochemically oriented papers have been published in which participation of $OH^{\bullet}$ has been invoked to explain the results. The summary of one of the latest of these (Cederbaum *et al.* 1977) ends: 'The results suggest that ethanol oxidation by microsomes can be dissociated from drug metabolism and that the mechanism of ethanol oxidation may involve in part, the interaction of ethanol with hydroxyl radicals that are generated by microsomes during the oxidation of NADPH.'

Fifty years after their formulation, hydroxyl radicals are at last becoming socially acceptable amongst biochemists. One might rightly ask two questions: Why? For how long? The first question is the easier to answer. There are perhaps two principal reasons. First, interest in free radicals generally has been stimulated by the finding by McCord & Fridovich (1969) that the well known protein erythrocuprein could catalyse the dismutation of the one-electron adduct of oxygen, O_2^- — the superoxide radical anion. Second and more fundamentally, the advent of the fast-reaction technique of pulse radiolysis has resulted in the accumulation of a mass of chemical information about free radicals in solution. This has enabled somewhat previously considered ephemeral reactions to be described with confidence — at least *in vitro.*

REACTIONS OF $OH^{\bullet}$ *IN VITRO*

The technique of pulse radiolysis was developed to enable radiation chemists to learn more about the manner in which ionizing radiation damages molecules. It was hoped that such information might be useful in radiation protection and radiation therapy. The technique has since become the method of choice for the generation and direct study of free radicals in solution with widespread applications in chemistry and biochemistry generally. Unfortunately, it is still often thought that when something is irradiated a mass of unspecific and ill-defined reactions take place, and a host of products results. This is indeed true when high radiation doses are used. However, if experimental systems are carefully designed, many free-radical reactions can be studied individually with considerable precision (see Willson 1977*a* and references therein).

It was long believed that the chemical effects of ionizing radiation in aqueous solution were due to 'activated water', although what this was nobody knew. In 1944 Weiss proposed that irradiation of water produced hydrogen atoms and hydroxyl radicals (reaction 18) and that these products could react with

$$H_2O \rightsquigarrow H^{\bullet} + OH^{\bullet} \quad (18)$$

themselves or with substrates present in solution. Whether a radical reacted with one solute or another would depend on the concentrations of the individual solutes and their respective rates of reaction with the radical in question. Although it is now known that in neutral solution the yield of $H^{\bullet}$ atoms is only small and that the reducing species present is its conjugate base, the solvated electron e_{aq}^{-}, Weiss's proposals laid the foundation for the use of 'selective radical scavengers' and free radical protection:

$$OH^{\bullet} + \text{target} \rightarrow \text{product} \qquad \text{DAMAGE}$$

$$OH^{\bullet} + \text{scavenger} \rightarrow \text{products} \qquad \text{PROTECTION}$$

Experimental support for the formation of $OH^{\bullet}$ in irradiated solutions quickly followed. In 1947 Dainton reported that radiation could, like Fenton's reagent, initiate the polymerization of acrylonitrile in aqueous solution. Stein & Weiss (1948) described how phenol could be formed from benzene and salicylic acid from benzoic acid in a similar way. They again raised the possibility that $OH^{\bullet}$ was involved in biological oxidations. Later studies showed that many amino acids and nucleic acid derivatives also reacted with $OH^{\bullet}$ (Scholes *et al.* 1949, 1956; Scholes & Weiss 1950, 1954). By the mid 1950s it was realized that $OH^{\bullet}$ could react rapidly with most organic compounds as well as several inorganic ions (Fig. 1). Although considerable progress was made in characterizing reactions of $OH^{\bullet}$ with different solutes it was not until the advent of the pulse radiolysis technique in the early 1960s that the reactions could be described with confidence.

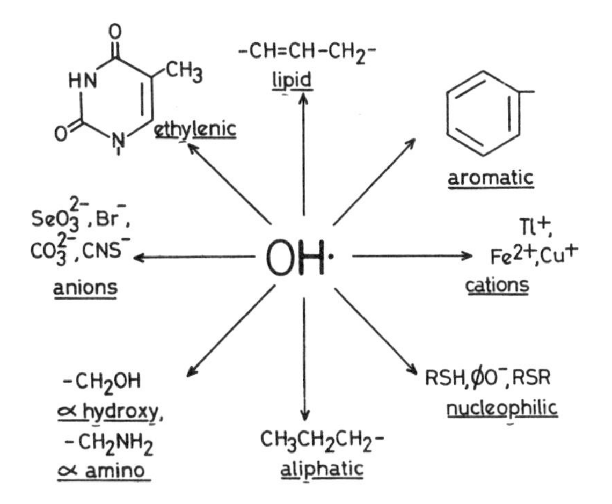

FIG. 1. Some reactions of hydroxyl radical with organic and inorganic compounds.

For example, the yield of destruction of thymine in conditions where only OH' radicals were thought to be active was found to be twice that of adenine. It was initially thought that this was because their rates of reaction with OH' were different. Quantitative analysis of the yields of base destruction of mixed solutions indicated, however, that the rates of reaction were similar. This conclusion was later confirmed when the actual absolute rate constants were measured by pulse radiolysis (Scholes *et al.* 1965).

OH' reaction rate constants

Many of the presently available OH' rate constants have come from one of three types of experiment:

(*a*) pulse-radiolysis studies in which the reactions have been observed directly by following the growth in absorption of a reaction product or the loss of absorption of a reactant (e.g. phenylalanine, benzoate or thymine);

(*b*) by pulse-radiolysis competition experiments with a reference solute which yields a strongly absorbing species on reaction with OH' (e.g. thiocyanate ion). Knowing the rate of reaction of OH' with the reference solute, one can obtain the rate of reaction of OH' with the unknown solute (Adams *et al.* 1965);

(*c*) from stationary-state radiation competition experiments with thymine as a reference solute. The extent of reaction of OH' with thymine can be determined by measurement of the loss in absorption at 265 nm, provided that the solute

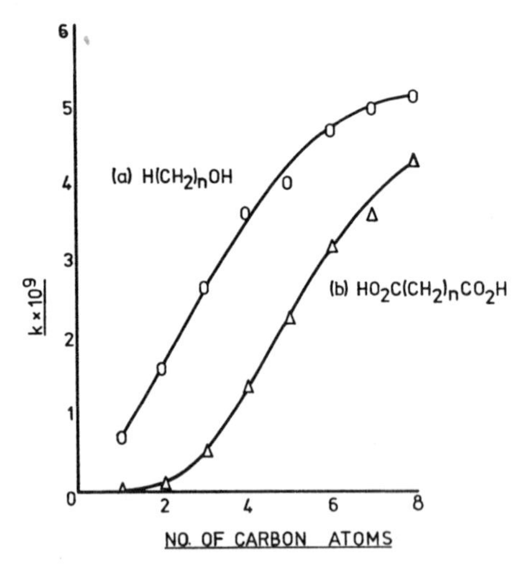

FIG. 2. Rate of reaction of n-alkanols and alkanedioic acids with hydroxyl radical as a function of chain length; the units of the absolute rate, k, are $l\ mol^{-1}\ s^{-1}$.

does not absorb at this wavelength or yield products which interact with thymine or have similar absorption spectra. Fig. 2 illustrates typical data obtained with the thymine system and plotted on an absolute scale by taking the rate of reaction of $OH^{\bullet}$ with thymine (obtained directly by pulse radiolysis) as 4.5×10^9 $l\ mol^{-1}\ s^{-1}$ (Scholes & Willson 1967; Willson *et al.* 1971).

The rates of reaction of $OH^{\bullet}$ with many substances have now been measured (see Dorfman & Adams 1973). If $OH^{\bullet}$ is generated in a biological medium, its lifetime will be very short (probably less than 1 μs) and it will react very close to where it is formed.

Types of $OH^{\bullet}$ reaction

Reactions of $OH^{\bullet}$ can be classified into three main types: (*a*) hydrogen abstraction; (*b*) addition; and (*c*) electron transfer. Which type of reaction takes place depends on the substance present in the medium. Although the rates of reaction of some solutes may be similar, the product radicals may have widely different properties. For example, methanol yields principally the radical $^{\bullet}CH_2OH$, which has strong reducing properties:

$$OH^{\bullet} + CH_3OH \rightarrow {}^{\bullet}CH_2OH + H_2O \qquad (19)$$

and $^{\bullet}CH_2OH$ can react with oxygen to form a peroxy radical which can then decay in the presence of base to form O_2^- (Adams & Willson 1969; Asmus *et al.* 1973):

$$^{\bullet}CH_2OH + O_2 \longrightarrow HOCH_2O_2^{\bullet} \xrightarrow{OH^-} {}^-OCH_2O_2^{\bullet} \longrightarrow CH_2O + O_2^-$$

With higher alcohols more radicals are formed which have their unpaired electron located on carbon atoms distant from the hydroxy group (e.g. reactions 20):

$$CH_3CH_2CH_2OH \longrightarrow \begin{cases} CH_3CH_2\dot{C}HOH \text{ (readily oxidized)} \\ \left.\begin{array}{l} CH_3\dot{C}HCH_2OH \\ {}^{\bullet}CH_2CH_2CH_2OH \\ CH_3CH_2CH_2O^{\bullet} \end{array}\right\} \text{(resistant to oxidation)} \end{cases} \qquad (20)$$

These radicals are relatively resistant to oxidation (Adams *et al.* 1968; Adams & Willson 1969).

The related peroxy radicals do not decay rapidly to yield O_2^-; they may even have oxidizing properties (e.g. Packer *et al.* unpublished results) (reaction 21).

$$Me_2C(OH)CH_2O_2^{\bullet} + RO^- \rightarrow Me_2C(OH)CH_2O_2^- + RO^{\bullet} \qquad (21)$$

Clearly, if alcohols are used as $OH^{\bullet}$ scavengers, factors other than chain length must be considered when interpreting the data obtained.

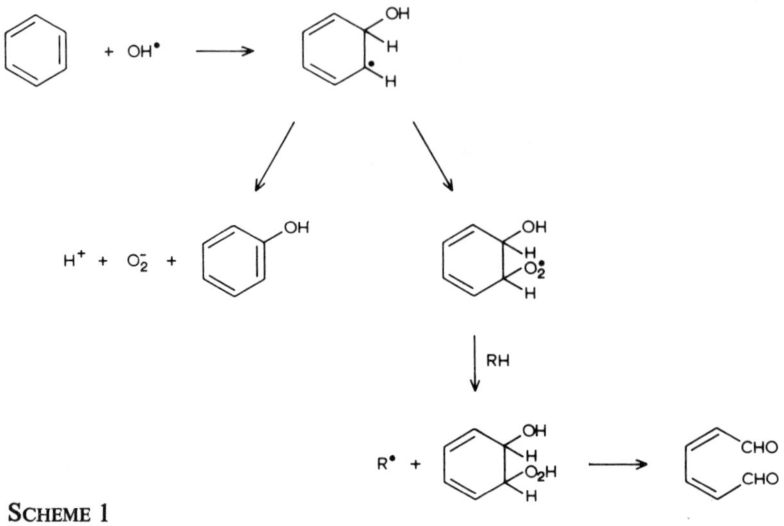

SCHEME 1

With some substances addition rather than abstraction may occur and, with biological compounds, this may have a considerable bearing on the resulting damage. As long ago as 1956 Daniels *et al.* showed that attack of OH˙ on benzene could lead to the production of muconaldehyde as well as phenol (Scheme 1). The dialdehyde may be involved in benzene toxicity. It is widely

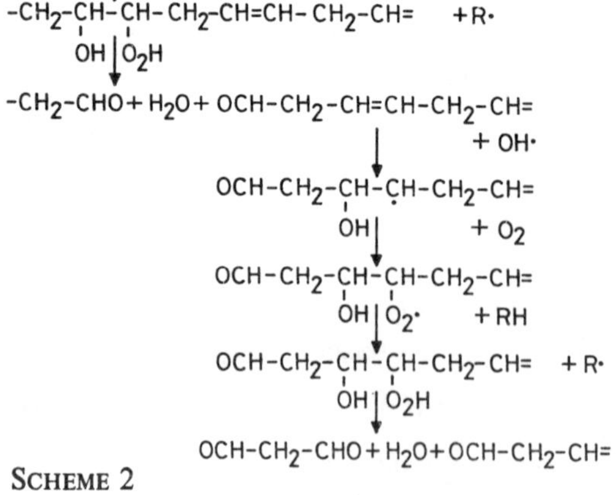

-CH_2-CH=CH-CH_2-CH=CH-CH_2-CH=
↓ + OH·
-CH_2-CH(OH)-ĊH-CH_2-CH=CH-CH_2-CH=
↓ + O_2
-CH_2-CH(OH)-CH(O_2·)-CH_2-CH=CH-CH_2-CH=
↓ + RH
-CH_2-CH(OH)-CH(O_2H)-CH_2-CH=CH-CH_2-CH= + R·
↓
-CH_2-CHO + H_2O + OCH-CH_2-CH=CH-CH_2-CH=
↓ + OH·
OCH-CH_2-CH(OH)-ĊH-CH_2-CH=
↓ + O_2
OCH-CH_2-CH(OH)-CH(O_2·)-CH_2-CH=
↓ + RH
OCH-CH_2-CH(OH)-CH(O_2H)-CH_2-CH= + R·
↓
OCH-CH_2-CHO + H_2O + OCH-CH_2-CH=

SCHEME 2

accepted that in lipids hydrogen abstraction from an activated methylene group is the site of electrophilic attack. Although hydroxyl radicals probably react principally at this site, the possibility of some minor addition to the double bond with subsequent production of aldehydes, as with benzene, deserves consideration (Scheme 2).

OH˙ GENERATION *IN VIVO?*

Hydroxyl radicals may be generated in several ways: (i) effect of ionizing radiation on water; (ii) photolysis of hydrogen peroxide; (iii) Fenton's reaction; (iv) electron transfer to ozone; (v) radical–peroxide reactions: (*a*) superoxide–peroxide (iron-catalysed?); (*b*) organic radical–peroxide (high concentration of peroxide — low concentration of oxygen).

Exposure to ionizing radiation leads inevitably to the generation of OH˙ as a result of the ionization of water. The photolysis of hydrogen peroxide is unlikely *in vivo*.

$$H_2O \xrightarrow{h\nu} H_2O^+ + e^-$$
$$H_2O^+ + H_2O \rightarrow OH^\bullet + H_3O^+$$

Hydroxyl radicals might also be formed from ozone or from hydrogen peroxide either by a Fenton-type reaction or by electron donation from free radicals.

Ozone and OH˙

At pH > 13 the hydroxyl radical in solution exists in its basic form, $O^{-\bullet}$. With oxygen the ozone–electron adduct (the ozonide ion) is rapidly formed. This species might also be formed in neutral solution by one-electron transfer from a suitable donor (reaction 22).

$$O_3 + \text{electron donor} \rightarrow O_3^- \quad (22)$$

Pulse-radiolysis studies show that, if it were formed, it could rapidly protonate or decay to give OH˙ (Buxton 1969; Gall & Dorfman 1969) (reactions 23–26).

$$O_3^- \rightleftharpoons O_2 + O^- \quad (23)$$
$$O^- + H_2O \rightleftharpoons OH^\bullet + OH^- \quad (\text{p}K_a\ 11.9) \quad (24)$$
$$O_3^- + H_2O \rightleftharpoons HO_3 + OH^- \quad (\text{p}K_a\ 10.4) \quad (25)$$
$$HO_3 \rightarrow OH^\bullet + O_2 \quad (?) \quad (26)$$

Thus OH· may have a role in ozone toxicity (Willson 1977*a*). Its subsequent reactions may also lead to O_2^- as described above.

Fenton's reaction in vivo?

The Fenton reaction (11, p. 21) is relatively slow ($k = 62\,\text{l}\,\text{mol}^{-1}\,\text{s}^{-1}$) compared to the free-radical reactions discussed so far (Keene 1964). Nevertheless, provided that a sufficient concentration of hydrogen peroxide is formed intracellularly, the reaction may occur. It would be most damaging if it took place in the vicinity of some vital molecule such as nucleic acid. I have previously suggested (1977*b*) that, if iron became decompartmentalized from some normally safe site (i.e. one in which its ability to catalyse free-radical reactions was inhibited), it might be transported to a sensitive site and there catalyse reactions leading to the production of $OH^{\bullet}$. I suggested that iron might be 'decompartmentalized' through the action of an agent such as a drug, a toxic chemical or even a virus complexing with the metal, a process previously proposed for the toxic effect of hydroxyquinoline (Fig. 3). Lown & Sim (1977) have since suggested, independently, a similar mechanism for the chemotherapeutic action of the cancer drug, bleomycin (BLM) (reactions 27–30). The drug was particularly damaging to DNA *in vitro* when iron(II) ions were present. The activity was diminished by superoxide dismutase and by catalase or by the presence of isopropanol.

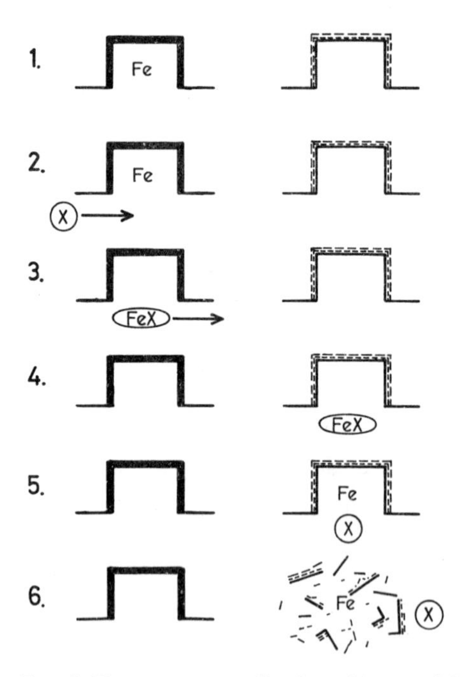

FIG. 3. Decompartmentalization of iron and its possible consequences: the site on the left is a safe site for iron; that on the right, an unsafe site (1). X is an agent that can bind iron and transport it to an unstable site (steps 2–5). There it can catalyse free-radical reactions and lead to the break up of the site (step 6).

$$BLM + Fe^{2+} \rightleftharpoons (BLM)Fe^{2+} \quad (27)$$
$$(BLM)Fe^{2+} + O_2 \rightarrow (BLM)Fe^{3+} + O_2^- \quad (28)$$
$$(BLM)Fe^{2+} + O_2^- \xrightarrow{H^+} (BLM)Fe^{3+} + H_2O_2 \quad (29)$$
$$(BLM)Fe^{2+} + H_2O_2 \rightarrow (BLM)Fe^{3+} + OH^{\bullet} + OH^- \quad (30)$$

Haber–Weiss and radical–peroxide reactions

The possibility that $OH^{\bullet}$ is formed by the Haber–Weiss reaction (13) has aroused much interest. Recent results indicate that the reaction is very slow in neutral solution and will not compete with the dismutation between the acidic and basic forms of O_2^- and the formation of H_2O_2 (reactions 31–33):

$$O_2^- + H_2O_2 \rightarrow O_2 + OH^- + OH^{\bullet} \quad (13)$$
$$O_2^- + H_2O \rightleftharpoons HO_2 + OH^- \quad (pK_a\ 4.9) \quad (31)$$
$$O_2^- + HO_2 \rightleftharpoons O_2 + HO_2^- \quad (32)$$
$$HO_2^- + H_2O \rightleftharpoons H_2O_2 + OH^- \quad (33)$$

The Haber–Weiss reaction may be catalysed by metal ions, both the following contributing to the overall reaction (13) (Van Hemmen & Meuling 1977; Ilan & Czapski 1977; McCord & Day 1977; Halliwell 1978):

$$Fe(chelate)^{3+} + O_2^- \rightleftharpoons Fe(chelate)^{2+} + O_2$$
$$Fe(chelate)^{2+} + H_2O_2 \rightarrow Fe(chelate)^{3+} + OH^- + OH^{\bullet}$$

However, other radical reactions should be considered, particularly when the concentration of hydrogen peroxide (compared with that of oxygen) is high. It has long been known from radiation studies that radicals from ethanol, isopropanol and formic acid react with H_2O_2 yielding $OH^{\bullet}$ and this can result in chain reactions (34–37).

$$OH^{\bullet} + HCO_2H \rightarrow {}^{\bullet}CO_2H + H_2O \quad (34)$$
$${}^{\bullet}CO_2H + H_2O_2 \rightarrow CO_2 + H_2O + OH^{\bullet} \quad (35)$$
$$OH^{\bullet} + (CH_3)_2CHOH \rightarrow (CH_3)_2\dot{C}OH + H_2O \quad (36)$$
$$(CH_3)_2\dot{C}OH + H_2O_2 \rightarrow (CH_3)_2CO + H_2O + OH^{\bullet} \quad (37)$$

Pulse-radiolysis studies have shown that the rate of one-electron transfer between related solutes often increases with increasing difference in the one-electron oxidation potentials (E_7^1) (Meisel & Neta 1975) of the couples involved (Patel & Willson 1973; Rao & Hayon 1973; Wardman & Clark 1976). Measurements using the pulse-radiolysis semiquinone-equilibrium method have shown that the E_7^1 (O_2/O_2^-) is considerably greater than that of related couples containing several quinones (Q), NAD^+, reduced riboflavin (FH_2), lipoic acid or CO_2.

It seems possible that other radical reactions with H_2O_2, such as (38)–(41), may occur in preference to that with O_2^- in some systems in which the oxygen and radical concentrations are low.

$$RSSR^{-\cdot} + H_2O_2 \rightarrow RSSR + OH^- + OH^\cdot \quad (38)$$

$$Q^{\cdot -} + H_2O_2 \rightarrow Q + OH^- + OH^\cdot \quad (39)$$

$$FH^\cdot + H_2O_2 \rightarrow F^{+\cdot} + OH^- + OH^\cdot \quad (40)$$

$$CO_2^- + H_2O_2 \rightarrow CO_2 + OH^- + OH^\cdot \quad (41)$$

ENZYME SUICIDE AND SELF-PROTECTION

If, as is sometimes postulated, $OH^\cdot$ radicals are produced in normal enzyme reactions, some mechanism must exist to prevent the enzyme destroying itself (Fig. 4). It is unlikely that any enzyme will be resistant *per se* to $OH^\cdot$ attack,

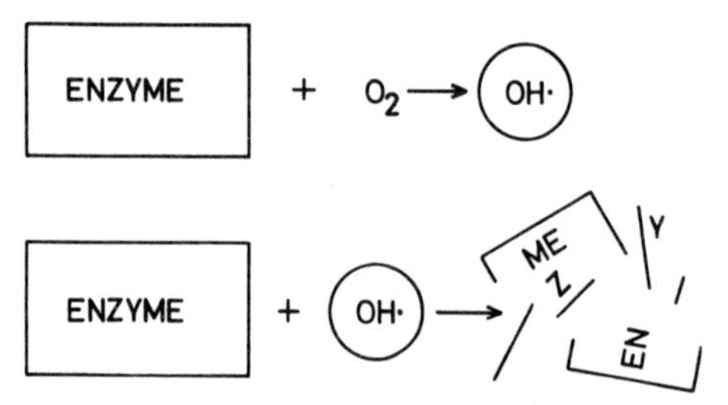

FIG. 4. Enzyme suicide as a result of the generation of hydroxyl radicals.

although the radical may be formed at a relatively insensitive site close to a more reactive species which immediately traps it. Another mechanism, however, may apply to systems in which high concentrations of easily-oxidizable ions such as CO_3^{2-}, CNS^- or I^- are also present. Hydroxyl radicals react rapidly with these ions to yield less-oxidizing radicals which are more selective in their action. For example, although lysozyme is inactivated by hydroxyl radicals, an excess of thiocyanate ions (which react rapidly with $OH^\cdot$) does not confer protection (Aldrich *et al.* 1969; Adams *et al.* 1972*a*, *b*); in contrast, ribonuclease is protected to a considerable extent (Fig. 5). The reason for this difference is that the radical $(CNS)_2^-$, formed according to reaction (42), reacts rapidly in

$$OH^\cdot + 2CNS^- \rightarrow (CNS)_2^- + OH^- \quad (42)$$

neutral solution with only two of the natural amino acids: cysteine and tryptophan. Neither enzyme contains cysteine. Lysozyme contains tryptophan (and at least one of these residues is vital for its activity) but ribonuclease does not. Other radicals such as CO_3^-, SeO_3^-, I_2^- and Br_2^- (formed similarly from $OH^\cdot$) are also selective in their action. Thus, if a system generating $OH^\cdot$ also contains an anion which reacts with $OH^\cdot$ to yield a radical which does not react with the enzyme, this could provide a safe means of radical production suitable, for example, for bacterial killing. In some systems, generation of radicals more

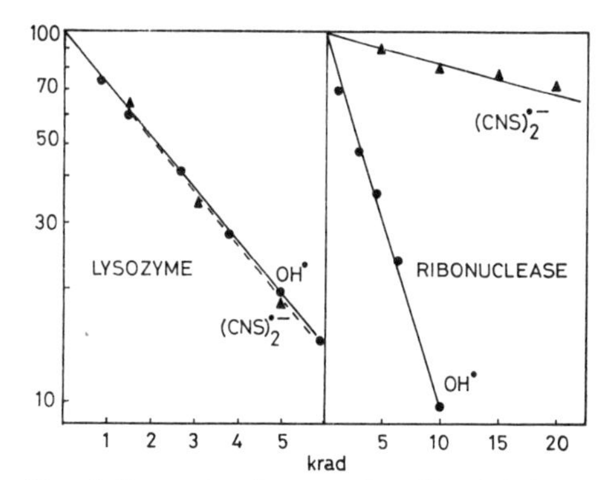

FIG. 5. Protection (measured as % activity) of ribonuclease against damage due to radiation by thiocyanate ions; lysozyme is not protected.

specific than $OH^•$ may lead to additional killing. Hydroxyl radicals might react at relatively resistant, non-vital sites as well as more sensitive vital sites and thus a fraction of the radicals will essentially be wasted, in terms of critical damage. Conversion into more specific radicals may lead to a greater attack at the more sensitive vital sites. This may partly explain the reported increase in erythrocytic haemolysis in the presence of carbonate ions (Michelson & Durosay, in press).

CELLULAR DAMAGE BY $OH^•$ RADICALS

Many biological effects may result from $OH^•$ reactions; membranes, enzymes, nucleic acids and polysaccharides may be affected. Damage may result from the direct action of the radicals themselves or from the action of toxic products. If damage is extensive, the cell may die. If damage is slight, the cell may survive but may undergo some inheritable change which leads to uncontrolled proliferation; hence, cancer. Damage to a nucleic acid base may be direct or indirect or through the covalent 'grafting' of a foreign chemical which has also undergone free-radical attack (e.g. reactions 43–45).

$$\text{DNA} + \text{radical} \rightarrow (\text{DNA radical}) \quad (43)$$

$$\text{carcinogen} + \text{radical} \rightarrow (\text{carcinogen radical}) \quad (44)$$

$$(\text{carcinogen radical}) + (\text{DNA radical}) \rightarrow \text{DNA-carcinogen covalent adduct} \quad (45)$$

Many compounds have been added to cellular and subcellular systems to scavenge $OH^•$ in the hope of changing a particular effect, which can then be attributed to the reaction of this radical. If the magnitude of the change for different scavengers increases with their increasing rate of reaction with $OH^•$, this is considered additional evidence. As we have seen, however, different solutes can lead to different types of radicals and care must be taken in this sort of interpretation. The scavenger must be specific for $OH^{\bullet}$ and must not yield another radical or stable products which can either enter into subsequent reactions with other solutes or interfere with other concurrent free-radical reactions. Finally, it must not affect the biochemistry of the system in such a way that its sensitivity is changed.

The assignment of a particular biological effect to a particular radical species has taxed radiation chemists and radiation biologists for decades. Radiobiologists have still not overcome the problem. In radiation chemistry the assignment of an effect with any confidence has only come from a careful qualitative and quantitative analysis of the radiation products: a reaction scheme is constructed in which the rate constants of the individual reactions and the concentrations of reactants and products correspond.

From *in vitro* studies we can say that certain reactions can occur *in vitro*. Whether these reactions are relevant *in vivo*, however, remains an open question. The characterization of the role of oxygen free radicals in cellular or subcellular systems will, I feel, as in radiobiology, continue to tax the ingenuity of biochemists for many years to come.

At a previous Ciba Foundation Symposium in 1956 on the subject of *Ionizing Radiations and Cell Metabolism* the Chairman, Professor Haddow, said in his opening remarks:

> 'My colleague J. A. V. Butler has pointed out that the basic puzzle of radiobiology, one which has been stressed especially by L. H. Gray is still unsolved — namely that comparatively small doses of radiation produce marked biological changes, although in general rather large doses are required to produce easily observable chemical changes...
>
> Again to quote Butler, we are at the moment in the position of a man who tries to elucidate the mechanism of a telephone exchange by throwing bricks into it and observing some of the results...
>
> From the study of the influence of ionizing radiations on cell metabolism may, however, flow the most profound consequences for the theory of ageing, for the theory of carcinogenesis, and for the theory of heredity.'

Twenty-two years have passed since that meeting; Butler, Gray, Haddow and Weiss are no longer with us. Yet if we substitute 'oxygen free radicals' for 'radiobiology', 'irradiation' and 'ionizing radiations', these statements still ring very true today.

ACKNOWLEDGEMENT

Financial support from the Cancer Research Campaign and the International Atomic Energy Agency is gratefully acknowledged.

References

ADAMS, G. E. & WILLSON, R. L. (1969) Pulse radiolysis studies on the oxidation of organic radicals in aqueous solution. *Trans. Farad. Soc. 65*, 2981–2987

ADAMS, G. E., BOAG, J. W., CURRANT, J. & MICHAEL, B. D. (1965) Absolute rate constants for the reaction of the hydroxyl radical with organic compounds in *Pulse Radiolysis*, pp. 131–143 (Ebert, M., Keene, J. P., Swallow, A. J. & Baxendale, J. H., eds.), Academic Press, London & New York

ADAMS, G. E., MICHAEL, B. D. & WILLSON, R. L. (1968) Electron transfer studies by pulse radiolysis. *Adv. Chem. Ser. 81*, 289–308

ADAMS, G. E., ALDRICH, J. E., BISBY, R. H., CUNDALL, R. B., REDPATH, J. L. & WILLSON, R. L. (1972*a*) Selective free radical reactions with proteins and enzymes: reactions of inorganic radical anion with amino acids. *Radiat. Res. 49*, 278–289

ADAMS, G. E., BISBY, R. H., CUNDALL, R. B., REDPATH, J. L. & WILLSON, R. L. (1972*b*) Selective free radical reactions with proteins and enzymes: the inactivation of ribonuclease. *Radiat. Res. 49*, 290–299

ALDRICH, J. E., CUNDALL, R. B., ADAMS, G. E. & WILLSON, R. L. (1969) Identification of essential residues in lysozyme. A pulse radiolysis method. *Nature (Lond.) 221*, 1049–1050

ASMUS, K. D., MÖCKEL, H. & HENGLEIN, A. (1973) Pulse radiolytic study of the site of $OH^{\bullet}$ attack on aliphatic alcohols in aqueous solution. *J. Phys. Chem. 77*, 1218–1221

BEAUCHAMP, C. & FRIDOVICH, I. (1970) A mechanism for the production of ethylene from methional. *J. Biol. Chem. 245*, 4641–4646

BUXTON, G. V. (1969) Pulse radiolysis of aqueous solutions: some rates of reaction of $OH^{\bullet}$ and $O^{-\bullet}$ and pH dependence of the yield at $O_3^{-\bullet}$. *Trans. Farad. Soc. 65*, 2150–2158

CEDERBAUM, A. I., DICKER, E., RUBIN, E. & COHEN, G. (1977) The effect of dimethylsulfoxide and other hydroxyl radical scavengers on the oxidation of ethanol by rat liver microsomes. *Biochem. Biophys. Res. Commun. 78*, 1254–1262

COHEN, G. (1977) In defense of Haber–Weiss, in *Superoxide and Superoxide Dismutases* (Michelson, A. M., McCord, J. M. & Fridovich, I., eds.), pp. 317–321, Academic Press, London & New York

COHEN, G. & HEIKKILA, R. E. (1974) The generation of hydrogen peroxide, superoxide radical and hydroxyl radical by 6-hydroxydopamine, dialuric acid and related cytotoxic agents. *J. Biol. Chem. 249*, 2447–2452

DAINTON, F. S. (1947) Effect of gamma and X-rays on dilute aqueous solutions of acrylonitrile. *Nature (Lond.) 160*, 268–269

DANIELS, M., SCHOLES, G. & WEISS, J. (1956) Chemical action of ionizing radiations in solution, Part XV. Effect of molecular oxygen in the irradiation of aqueous benzene solutions with X-ray. *J. Chem. Soc.*, 832–834

DORFMAN, L. M. & ADAMS, G. E. (1973) *Reactivity of the Hydroxyl Radical in Aqueous Solutions*, NSRDS-NBS No. 46, US Department of Commerce, National Bureau of Standards, Bethesda

FONG, K., MCCOY, P. B., POYER, J. L., KEELE, B. B. & MISRA, H. (1973) Evidence that peroxidation of lysosomal membranes is initiated by hydroxyl free radicals produced during flavin enzyme activity. *J. Biol. Chem. 248*, 7792–7797

GALL, B. L. & DORFMAN, L. M. (1969) Pulse radiolysis studies. XV. Reaction of the oxide radical ion and of the ozonide ion in aqueous solution. *J. Am. Chem. Soc. 91*, 2199

GERSCHMAN, R., GILBERT, D. L., NYE, S. W., DWYER, P. & FENN, W. O. (1954) Oxygen poisoning and X-irradiation. A mechanism in common. *Science (Wash. D.C.) 119*, 623–626

HABER, F. & WEISS, J. (1932) Uber die Katalyse des Hydroperoxydes. *Naturwissenschaften 20*, 948–950

HABER, F. & WEISS, J. (1934) The catalytic decomposition of hydrogen peroxide by iron salts. *Proc. R. Soc. Lond. A 147*, 332–351

HABER, F. & WILLSTÄTTER, R. (1931) Unpaarigkeit und Radikalketten im Reaktionsmechanismus organischer und enzymatischer Vorgänge. *Berichte 64*, 2844–2856

HADDOW, A. (1956) Chairman's opening remarks, in *Ionizing Radiations and Cell Metabolism (Ciba Found. Symp.)*, pp. 1–2, Churchill, London [now Edinburgh]

HALDANE, J. B. S. (1932) Chain reactions in enzymatic catalysis. *Nature (Lond.) 130*, 61

HALLIWELL, B. (1978) Superoxide-dependent formation of hydroxyl radicals in the presence of iron chelates. *FEBS (Fed. Eur. Biochem. Soc.) Lett. 92*, 321–326

HARMAN, D. (1962) Role of free radicals in mutation, cancer, aging and the maintenance of life. *Radiat. Res. 16*, 753–763

HEIKKILA, R. E., WINSTON, B., COHEN, G. & BARDEN, H. (1976) Alloxan-induced diabetes — evidence for the hydroxyl radical as a cytotoxic intermediate. *Biochem. Pharmacol. 25*, 1085–1092

ILAN, Y. A. & CZAPSKI, G. (1977) The reaction of superoxide radical with iron complexes of EDTA studied by pulse radiolysis. *Biochim. Biophys. Acta 498*, 386–394

JOHNSTON, R. B., KEELE, B., WEBB, L., KESSLER, D. & RAJAGOPALAN, K. V. (1973) Inhibition of phagocytic bactericidal activity by superoxide dismutase: a possible role for superoxide anion in the killing of phagocytized bacteria. *J. Clin. Invest. 52*, 44a (abstr.)

KEENE, J. P. (1964) Pulse radiolysis of ferrous sulphate solution. *Radiat. Res. 22*, 14–20

LOWN, J. W. & SIM, S. (1977) The mechanism of the bleomycin-induced cleavage of DNA. *Biochem. Biophys. Res. Commun. 77*, 1150–1157

MCCORD, J. M. (1974) Free radicals and inflammation: protection of synovial fluid by superoxide dismutase. *Science (Wash. D.C.) 185*, 529–531

MCCORD, J. M. & DAY, E. M. (1978) Superoxide-dependent production of hydroxyl radical catalyzed by iron-EDTA complex. *FEBS (Fed. Eur. Biochem. Soc.) Lett. 86*, 139–144

MCCORD, J. M. & FRIDOVICH, I. (1969) Superoxide dismutase: an enzyme function for erythrocuprein (hemocuprein). *J. Biol. Chem. 244*, 6049–6055

MEISEL, D. & NETA, P. (1975) Oxidation reduction potential of riboflavin studied by pulse radiolysis. *J. Phys. Chem. 79*, 2459–2461

PACKER, J., SLATER, T. F. & WILLSON, R. L. Unpublished.

PANGANAMALA, R. V., SHARMA, H. M., HEIKKILA, R. E., GEER, J. C. & CORNWELL, D. G. (1976) Role of hydroxyl radical scavengers dimethyl sulfoxide, alcohols and methional in the inhibition of prostaglandin synthesis. *Prostaglandins 11*, 599–607

PATEL, K. & WILLSON, R. L. (1973) Semiquinone free radicals and oxygen. Pulse radiolysis study of one electron transfer equilibria. *J. Chem. Soc. Faraday Trans. I 69*, 814–825

RAO, P. S. & HAYON, E. (1973) Rate constants of electron transfer processes in solution: dependence on the redox potential of the acceptor. *Nature (Lond.) 243*, 344–346

SCHOLES, G. & WEISS, J. (1950) Chemical action of ionizing radiations on nucleic acids in aqueous systems. *Nature (Lond.) 166*, 640

SCHOLES, G., STEIN, G. & WEISS, J. (1949) Action of X-rays on nucleic acid. *Nature (Lond.) 164*, 709–710

SCHOLES, G. & WEISS, J. (1954) Chemical action of X-rays on nucleic acids and related substances in aqueous solutions. *Biochem. J. 56*, 65–72

SCHOLES, G. & WILLSON, R. L. (1967) γ-Radiolysis of aqueous thymine solutions. *Trans. Farad. Soc. 63*, 2983–2993

SCHOLES, G., WEISS, J. & WHEELER, C. M. (1956) Formation of hydroperoxides from nucleic acid by irradiation with X-rays in aqueous systems. *Nature (Lond.) 178*, 157

SCHOLES, G., SHAW, P., WILLSON, R. L. & EBERT, M. (1965) Pulse radiolysis studies of aqueous solutions of nucleic acid, in *Pulse Radiolysis* (Ebert, M., Keene, J. P., Swallow, A. J. & Baxendale, J. H., eds.), pp. 151–164, Academic Press, London & New York

STEIN, G. & WEISS, J. (1948) Chemical effects of ionizing radiations. *Nature (Lond.) 161*, 650

UREY, H. C., DAWSEY, L. H. & RICE, F. O. (1929) The absorption spectrum and decomposition of hydrogen peroxide by light. *Trans. Farad. Soc. 51*, 1371–1383

VAN HEMMEN, J. J. & MEULING, W. J. A. (1977) Inactivation of *Escherichia coli* by superoxide radicals and their dismutation products. *Arch. Biochem. Biophys. 182*, 743–748

WALLING, C. (1975) Fenton's reagent revisited. *Acc. Chem. Res. 8*, 125–131
WARDMAN, P. & CLARKE, E. D. (1976) Oxygen inhibition of nitro-reductase: electron transfer from nitro radical-anions to oxygen. *Biochem. Biophys. Res. Commun. 69*, 942–949
WEISS, J. (1935) Elektronenübergangsprozesse im Mechanismus von Oxydations- und Reduktionsreaktionen in Lösungen. *Naturwissenschaften 23*, 64–69
WEISS, J. (1937) Reaction mechanism of the enzymes catalase and peroxidase in the light of the theory of chain reactions. *Trans. Farad. Soc.*, 1107–1116
WEISS, J. (1944) Radiochemistry of aqueous solutions. *Nature (Lond.) 153*, 748–750
WILLSON, R. L. (1977*a*) Free radicals and electron transfer in biology and medicine. *Chem. Ind.*, 183–193
WILLSON, R. L. (1977*b*) Iron, zinc, free radicals and oxygen in tissue disorders and cancer control, in *Iron Metabolism (Ciba Found. Symp. 51)*, pp. 331–354, Elsevier/Excerpta Medica/North-Holland, Amsterdam
WILLSON, R. L., GREENSTOCK, C. L., ADAMS, G. E., WAGEMAN, R. & DORFMAN, L. M. (1971) The standardisation of hydroxyl rate data from radiation chemistry. *Int. J. Radiat. Phys. Chem. 3*, 211–220

Discussion

Reiter: With regard to the protection against hydroxyl radicals by thiocyanate ions, Morrison & Steel (1968) reported that thiocyanate stabilizes lactoperoxidase. How does this compare with the data you obtained with lysozyme and ribonuclease?

Willson: If $OH^•$ radicals are generated in the lactoperoxidase system, a mechanism is needed to protect the generating enzyme from their damaging action. The reaction with CNS^- may suffice. We ought to purify the lactoperoxidase to see whether it reacts with thiocyanate radicals.

Fridovich: One need not consider free hydroxyl radicals; any strong oxidant generated at the active site may react with some component of the active site or with some exogenous small molecule. For example, the copper-zinc superoxide dismutase is gradually inactivated by hydrogen peroxide, especially at high pH. One can protect the enzyme against this inactivation with various electron donors, such as dianisidine, imidazole or polyunsaturated fatty acids. The interaction of the copper at the active site with the H_2O_2 apparently generates a powerful oxidant which can inactivate the enzyme by attacking one of the imidazole ligands to the copper atom. The exogenous electron donor may preferentially attack the oxidant and prevent inactivation of the enzyme. Maybe a similar situation pertains with lactoperoxidase.

Reiter: Is the inactivation irreversible?

Fridovich: Yes.

Hill: All the active-site imidazole groups are attacked (A. E. G. Cass & H. A. O. Hill, unpublished results). At high concentrations of hydrogen peroxide the protein denatures.

Michelson: Dr Willson, you question the existence of hydroxyl radicals in biology but you have quoted the reaction of semiquinones with hydrogen peroxide to give hydroxyl radicals; therefore, hydroxyl radicals must exist biologically. For example, glucose oxidase, a flavoprotein which produces hydrogen peroxide, will automatically release hydroxyl radicals by reaction of H_2O_2 with a semiquinone-type intermediate.

Willson: There is good evidence for the formation of $OH^{\bullet}$ from the reaction of the radical $^{\bullet}CO_2H$ with hydrogen peroxide (Hart 1951). Although the one-electron redox potentials of many quinone–semiquinone couples, $E_7^1(Q/Q^-)$ are greater than that of the CO_2–formate radical-ion couple $E_7^1(CO_2/CO_2^-)$, they are often less than that of the oxygen–superoxide radical ion couple $E_7^1(O_2/O_2^-)$. I mentioned earlier (p. 29) that the rate constants of several related electron-transfer reactions increase with increasing difference in the one-electron potential of the redox couples involved: the reactions were reversible. In these reactions involving hydrogen peroxide dissociative electron capture occurs and the reactions are irreversible. The formation of $OH^{\bullet}$ from some semiquinone–hydrogen peroxide reactions may occur at least as readily, if not more readily, than the related Haber–Weiss reaction involving the superoxide radical anion.

The reactions (1) and (2) are envisaged where Q^- is a semiquinone or semi-reduced flavin. The net reaction is again the Haber–Weiss reaction (3).

$$Q + O_2^- \rightleftharpoons Q^- + O_2 \quad (1)$$

$$Q^- + H_2O_2 \rightarrow OH^{\bullet} + OH^- + Q \quad (2)$$

$$O_2^- + H_2O_2 \rightarrow OH^{\bullet} + OH^- + O_2 \quad (3)$$

Michelson: That will depend very much on local concentrations.

Willson: Yes; that is why there is still a large question mark concerning the relevance of such reactions *in vivo*.

Michelson: We were interested in the relative reactivities of O_2^-, formate radical anion, carbonate radical anion and hydroxyl radicals (Maral *et al.* 1979). We used a very simple system, γ-irradiation with the right buffer conditions to produce the different radicals. We could draw no simple generalization from the results which represent loss of *enzymic activity*, that is attack at a particular site in the protein, rather than global attack. Obviously, superoxide dismutase is very resistant to O_2^- but much less so to hydroxyl radicals. Table 1 shows the data for hydroxyl-radical attack on different kinds of enzymes in the absence and presence of carbonate. One may suppose that there is an induced specificity; as hydroxyl radicals will attack the protein anywhere many are lost without touching the active site. Carbonate radicals live much longer but are much less reactive and must react specifically with the metal or some part of the active site to increase the loss of enzymic activity.

TABLE 1 (Michelson)
Inactivation of enzymes by various oxygen-containing radicals, measured as half-life of enzymic activity

Enzyme	*Half-life of enzymic activity, $t_{1/2}$/min in the absence of carbonate*			*with carbonate*		
	O_2^-	CO_2^-	HO$^\cdot$	O_2^-	CO_2^-	HO$^\cdot$
Bovine CuSOD	835	491	49	13	33	9
Human CuSOD	867	794	32	22	19	7
Bacterial CuSOD	1665	> 300	79	280	180	26
Bacterial FeSOD	> 300	> 300	50	7	5	10
Human MnSOD	2040	2053	19	18	26	1.54
Bovine MnSOD	> 720	> 720	55	21	38.5	1.82
Pancreatic RNAse	193	68	13	158	90	43.6
Glucose oxidase	> 300	> 300	30	> 300	> 300	42.5
Catalase	> 500	> 500	52	> 500	> 500	61
Glutathione peroxidase	25	10.5	92	24	12	121

Exposure to γ-rays (^{60}Co): 680 rad/min. Since the yield of production of the different radicals under various buffer conditions (presence or absence of formate, N_2O, anaerobic or saturated with O_2) is not identical, the above figures for half-lives of enzymic activity represent the kinetics of loss of activity for a given energy input.

Willson: I agree; but equally well the converse may sometimes be true: a scavenger which produces a more specific active free radical might increase inactivation.

Michelson: We have those results also: the rate of inactivation (half-life) of, for instance, ribonuclease by hydroxyl radicals is 13 min but 43.6 min when carbonate is present (see Table 1). Glucose oxidase, a flavoprotein, is protected by carbonate; catalase is slightly protected — the difference reflects the chemical difference between the haematinic iron and the ordinary iron in an iron enzyme. Glutathione peroxidase, a selenium enzyme, is also protected by carbonate. In each system one has to look at both partners in the social exchange (Table 1).

Hill: There is an important difference in the case of the dismutase: the substrate is an anion. So the possibility of the reactive radical getting into the active site is much higher. You have to have a targetted radical. In all the other examples you chose, the substrates are neutral.

Michelson: No; the formal radical cation is also charged and behaves differently (Table 1). We were surprised, however, to find that glutathione peroxidase lost activity more rapidly with O_2^- than with hydroxyl radicals (25 min and 92 min, respectively).

Willson: One has got to be extremely careful in this particular system where you irradiate materials for a long time with a ^{60}Co γ-source.

Michelson: But we used only 680 rad/min for no more than 10–20 min to determine the initial exponentials.

Willson: The value of the data would be greatly enhanced if associated pulse-radiolysis experiments were undertaken. Then we should be able to establish more firmly the initial reaction mechanisms.

Michelson: That, of course, requires a fairly heavy investment in technology and does not infallibly lead to a mechanism.

Stern: Returning to the idea of enzyme suicide, we found in an investigation of superoxide dismutase activity in red cells that removal of enzymic scavenging sources of hydrogen peroxide in the red cells (i.e. catalase and glutathione peroxidase) in the presence of 1,4-naphthoquinone-2-sulphonic acid, a substance which generates O_2^- and H_2O_2 when incubated with red cells, results in loss of superoxide dismutase activity (S. E. M. McMahon & A. Stern, to be published).

Willson: The superoxide dismutase from erythrocytes contains no methionine or tryptophan. This may be significant because tryptophan and methionine are amino acids that are relatively more sensitive to free-radical attack. In contrast, aspartic acid is relatively abundant in the enzyme: this amino acid tends to be relatively resistant to free radicals. Nature seems to have designed these enzymes to be resistant to free-radical damage.

Fridovich: The enzyme is incredibly stable.

Cohen: Analogies sometimes suffer from being not quite correct. I take issue with the analogy of throwing bricks into a telephone exchange to learn how a telephone works being likened to using hydroxyl radical scavengers in biological systems to learn whether or not hydroxyl radicals play a role. As a result of the early studies of Beauchamp & Fridovich (1970) we became interested in hydroxyl radicals and in the Haber–Weiss reaction. Thinking in terms of hydroxyl radicals provided a useful guide for designing experiments in biological systems. Our results were dramatic — we did not feel we were throwing bricks. For example, in attempts to prevent the diabetic state induced by alloxan, or to stop the sympathectomy induced by 6-hydroxydopamine, the use of hydroxyl radical scavengers in complex biological systems (that is, in living animals) leads to the complete blocking of the phenomenon in question (Heikkila *et al.* 1976; Cohen *et al.* 1976; Cohen 1978). So, thinking in terms of hydroxyl radical scavengers has been successful. I agree with you that there is no 'hard evidence' for hydroxyl radicals *in vivo*; the supportive evidence is mainly inferential. However, it is also possible to side with Professor Michelson's view that undoubtedly they are formed. But, we will need ways of distinguishing between $OH^{\cdot}$ or $OR^{\cdot}$ in biological systems.

One question that you raised is, what happens to the scavengers? They may themselves be converted into free radicals that may exhibit toxicity in biological systems. Swallow's data (1953) on X-irradiation of ethanolic solutions may provide an answer to why giving an animal ethanol is successful in some experiments. Apparently, the free radicals formed from ethanol can transfer electrons to NAD^+ and NADH is formed; in other words what may have been a damaging radical is transformed into something that the biological system can handle.

Fridovich: Maybe we are throwing bricks but our aim is good!

Slater: As you have indicated, Dr Willson, hydroxyl radicals may be produced in the microsomal ethanol-oxidizing system and this may in part explain the protective action of some antioxidants (DiLuzio & Hartman 1967) against ethanol-induced liver disturbances. However, the occurrence, as well as the physiological significance, of this oxidizing system is still an area of rich controversy (see Slater 1972; also Comporti 1978).

Willson: Dr Cohen and his colleagues have suggested (Cederbaum *et al.* 1977) 'that ethanol oxidation by microsomes can be dissociated from drug metabolism and that the mechanism of ethanol oxidation may involve, in part, the interaction of ethanol with hydroxyl radicals that are generated by microsomes during the oxidation of NADPH'.

Cohen: The microsomal system that oxidizes ethanol is of great interest to those working on alcohol metabolism. The system involves cytochrome P450 and the generation of peroxide. We picked it up as a potential biological example of the Haber–Weiss reaction. Hydroxyl radical scavengers, such as benzoate, mannitol, dimethyl sulphoxide and thiourea, inhibited the rate of oxidation of ethanol by microsomes. In more recent work (G. Cohen & A. L. Cederbaum, unpublished results) we used methional and the corresponding keto acid (2-oxo-4-methylthiobutyric acid): these agents effectively blocked the oxidation of ethanol and also gave rise to ethylene. Because ethylene production from these agents has been used to detect hydroxyl radical production we may infer that the microsomes produce hydroxyl radicals.

Willson: Methional may also react with alkoxy radicals, $RO^\bullet$ (Pryor & Tang 1978). This may occur in some of the scavenging systems used (but not those such as Professor Fridovich uses).

Klebanoff: I shall present some data which are at least compatible with another mechanism for the formation of ethylene from methional, namely, the initiation of ethylene formation by electron abstraction by singlet oxygen (see pp. 263–284; also Klebanoff & Rosen 1978).

Cohen: O'Brien suggested that also (Rahimtula *et al.* 1978).

Michelson: Let me go out on a limb. So far we have considered the Haber–Weiss reaction as more or less a chemical accident in the cell. If hydroxyl radicals play an important part in phagocytic killing, it is not a chemical accident. So, we have to postulate a hydroxyl-producing enzyme. If we do that, we must have a second postulate: that there is a super-superoxide dismutase which eliminates hydroxyl radicals. The cell must provide an automatic defence.

People have been looking for a 'Haber–Weiss enzyme' but has anybody looked for a hydroxyl-protective enzyme?

Bielski: To destroy $OH^•$ radicals in a biological system one does not need an enzyme. It is well documented that $OH^•$ radicals attack biological molecules indiscriminately at rates ranging from 10^6 to $10^{10}\,l\,mol^{-1}\,s^{-1}$. Most large molecules (e.g. enzymes) have many sites with which $OH^•$ radicals can react but most of them have no relevance to the process of inactivation. This was confirmed by radiation studies where it was observed that g values of inactivation were always smaller than the g value of the radical responsible for inactivation.

Fridovich: Dorfman & Adams (1973) have put together a useful tabulation of rate constants for the reactions of hydroxyl radicals. To underline Dr Bielski's point, one cannot find a rate constant slower than $10^6\,l\,mol^{-1}\,s^{-1}$; some rates are as fast as $10^{10}\,l\,mol^{-1}\,s^{-1}$. Many substances that abound in cells — glutathione, ascorbate and so on — react at about that rate, i.e. by diffusion-limited reactions. Perhaps we would be going too far out on a limb to look for such an enzyme.

McCord: Nature seems to exemplify the old adage that an ounce of prevention is worth a pound of cure by putting superoxide dismutases, catalases and glutathione peroxidases almost everywhere. In the specialized case of phagocytic cells there would seem to be no advantage in having a scavenging enzyme for hydroxyl radicals since the intention is destruction.

Fridovich: And even if the preventive mechanisms are not perfectly efficient, there are antioxidants such as tocopherol which can minimize the extent of damage due to those surviving radicals.

Willson: Furthermore, repair enzymes exist. However, I suggest that protection is better undertaken at the previous step: by prevention of the formation of superoxide radicals. For example, zinc ions bind to sensitive reducing species which would otherwise combine with iron and be autooxidized. In this way zinc could suppress the formation of superoxide radicals.

Fridovich: The production of superoxide is avoided generally by the use of enzymes specially designed for the multivalent reduction of oxygen; most of the oxygen consumption by animal cells is accounted for by cytochrome oxidase which makes water, without intermediates.

McCord: I disagree that hydroxyl radicals are produced *in vivo* by Fenton's reaction. Day and I (McCord & Day 1978) have proposed that with physiologically attainable concentrations superoxide is the reductant of iron(III) to iron(II), whereas Fenton's reaction requires very high concentrations of hydrogen peroxide for this role because it is a poor reductant. In our system we observed production of hydroxyl radicals at micromolar concentrations of hydrogen peroxide (at which Fenton's reaction does not proceed).

Willson: I accept that Fenton's reaction is slow. However, this type of reaction may be important with regard to tissue damage induced by xenobiotic substances which complex iron, for instance damage induced by the action of drugs such as bleomycin (Lown & Sim 1977).

Michelson: Another point is that in Dr McCord's system the catalytic centre is alternately oxidized and reduced the whole time whereas in Fenton's reaction hydrogen peroxide is present in massive excess.

Keberle: In the metabolism of drugs oxidative reactions take place at selected sites in the molecule: e.g. at the ω or $\omega - 1$ carbon atom of an aliphatic side-chain, or in the α position relative to a double bond or a hetero-atom. Although these oxidative processes occur by way of hydroxyl radicals, they are highly stereoselective. Take, for instance, the optical isomers of α-phenyl-α-ethyl-glutarimide: the dextrorotatory form is only hydroxylated in the glutarimide ring, whereas the laevorotatory form is hydroxylated in the side-chain (Kerbele *et al.* 1963).

The question is whether this stereoselectivity is due to cytochrome P450 or, perhaps, to the possibility that hydroxyl radicals can react stereoselectively in solution and in the presence of organic materials, such as proteins or membranes, serving as a symmetric carrier. Has this problem been studied?

Willson: Not to my knowledge. I find it hard to accept that the hydroxyl radical would be stereoselective when reacting in simple aqueous solution.

References

BEAUCHAMP, C. & FRIDOVICH, I. (1970) A mechanism for the production of ethylene from methional. *J. Biol. Chem. 245*, 4641–4646

CEDERBAUM, A. L., DICKER, E., RUBIN, E. & COHEN, G. (1977) The effect of dimethylsulfoxide and other hydroxyl radical scavengers on the oxidation of ethanol by rat liver microsomes. *Biochem. Biophys. Res. Commun. 78*, 1254–1262

COHEN, G. (1978) The generation of hydroxyl radicals in biological systems. *Photochem. Photobiol. 28*, 669–676

COHEN, G., HEIKKILA, R. E., ALLIS, B., CABBAT, F., DEMBIEC, D., MACNAMEE, D., MYTILINEOU, C. & WINSTON, B. (1976) Destruction of sympathetic nerve terminals by 6-hydroxydopamine: protection by 1-phenyl-3-(2-thiazolyl)-2-thiourea, diethyldithiocarbamate, methimazole, cysteamine, ethanol and n-butanol. *J. Pharmacol. Exp. Ther. 199*, 336–352

COMPORTI, M. (1978) in *Biochemical Mechanisms of Liver Injury* (Slater, T. F., ed.), pp. 475–479, Academic Press, London

DILUZIO, N. R. & HARTMAN, A. D. (1967) Role of lipid peroxidation in the pathogenesis of the ethanol-induced fatty liver. *Fed. Proc. 26*, 1436–1442

DORFMAN, L. & ADAMS, G. (1973) *Reactivity of the Hydroxyl Radical in Aqueous Solutions*, National Standard Reference Data System No. 46, Bethesda

HART, E. J. (1951) Mechanism of the γ-ray induced oxidation of formic acid in aqueous solution. *J. Am. Chem. Soc. 73*, 68–73

HEIKKILA, R. E., WINSTON, B., COHEN, G. & BARDEN H. (1976) Alloxan-induced diabetes: evidence for the hydroxyl radical as a cytotoxic intermediate. *Biochem. Pharmacol. 25*, 1085–1092

KEBERLE, H., RIESS, W. & HOFFMAN, K. (1963) Über den stereospezifischen Metabolismus der optischen Antipoden von α-phenyl-α-äthyl-Glutarimid (Doriden[R]). *Arch. Int. Pharmacodyn. Ther. 163*, 142, 117–124

KLEBANOFF, S. L. & ROSEN, H. (1978) Ethylene formation by polymorphonuclear leukocytes: role of myeloperoxidase. *J. Exp. Med. 148*, 490–506

LOWN, J. W. & SIM, S. (1977) The mechanism of bleomycin-induced cleavage of DNA. *Biochem. Biophys. Res. Commun. 77*, 1150–1157

MARAL, J., MICHELSON, A. M. & MONNY, C. (1979), in preparation

MCCORD, J. M. & DAY, E. D. JR. (1978) Superoxide-dependent production of hydroxyl radical catalyzed by iron-EDTA complex. *FEBS (Fed. Eur. Biochem. Soc) Lett. 86*, 139–142

MORRISON, M. & STEEL, W. F. (1968) Lactoperoxidase, the peroxidase in the salivary gland, in *Biology of the Mouth* (Parson, P. H., ed.), pp. 89–110, American Association for the Advancement of Science, Washington, D.C.

PRYOR, W. A. & TANG, R. H. (1978). Ethylene formation from methional. *Biochem. Biophys. Res. Commun. 81*, 498–503

RAHIMTULA, A. D., HAWCO, F. J. & O'BRIEN P. J. (1978) The involvement of $^1O_2$ in the inactivation of mixed function oxidase and peroxidation of membrane lipids during the photosensitized oxidation of liver microsomes. *Photochem. Photobiol. 28*, 811–817

SLATER, T. F. (1972) in *Free Radical Mechanisms in Tissue Injury*, pp. 180–182, Pion, London

SWALLOW, A. J. (1953) The radiation chemistry of ethanol and diphosphopyridine nucleotide and its bearing on dehydrogenase action. *Biochem. J. 54*, 253–257

WILLSON, R. L. (1977) 'Free?' radicals and electron transfer in biology and medicine. *Chem. Ind.*, 183–193

Reaction rates of superoxide radicals with the essential amino acids

BENON H. J. BIELSKI and GRACE G. SHIUE

Chemistry Department, Brookhaven National Laboratory, Upton, New York

Abstract Upper limits for the rates of reaction of amino acids with superoxide radicals have been determined spectrophotometrically by the stopped flow method. Rate measurements at 23 °C for the reaction of HO_2 with amino acids were made in the pH range between 1 and 2; similar measurements for O_2^- were taken near pH 10.

The results show that, overall, amino acids are relatively unreactive toward both HO_2 and O_2^-. Computed second-order rate constants for their interaction with HO_2 range from $10\ l\ mol^{-1}\ s^{-1}$ for aliphatic amino acids to about $600\ l\ mol^{-1}\ s^{-1}$. The second-order rate constants for the interaction of amino acids with O_2^- are smaller and range from 0.1 to about $20\ l\ mol^{-1}\ s^{-1}$.

The study of the rates of reaction of superoxide radicals with the essential amino acids is part of an overall program aimed at the understanding of the basic chemistry of HO_2 and O_2^-. Since amino acids are building blocks for a variety of larger molecules, in particular proteins, they constitute on a quantitative basis a large fraction of the chemicals found in a mammalian cell. Judging from studies on the hydroxyl radical studies, which is six to nine orders of magnitude more reactive, one can safely assume that the reactivity of HO_2 and O_2^- will not change drastically toward amino acids which are integral parts of larger molecules. Hence information obtained from the study of free amino acids could be extrapolated to a relatively large number of biological compounds. Although the rate constants under discussion are small, experimental verification of their order of magnitude appears essential for the understanding of the overall role of superoxide radicals in biological systems.

In the present study the superoxide free radicals were generated in oxygen-saturated formate solutions at pH 11.5 by ^{60}Co γ-rays (Bielski & Richter 1977) or by vacuum u.v. photolysis (Holroyd & Bielski 1978). In the first case the irradiated formate solutions containing superoxide radicals with a half-life of about 20–30 min were transferred for study to a fast-kinetics spectrophotometer (Durrum Instrument Co., Model D 110). In the second instance the superoxide radical was generated inside a plasma lamp, which is an integral part of the kinetic spectrophotometer.

Depending on the experimental conditions, the rates of reaction of HO_2 and O_2^- with the amino acids were studied by three different methods:

(*A*) Direct observation of the disappearance of HO_2 or O_2^- is possible only in the presence of those amino acids (AAs) which are transparent in the u.v. range between 230 and 280 nm. The corresponding rate constants for reactions (1) and (2) were computed with equation (4). Reaction (3) is written as a

$$O_2^- + AA \xrightarrow{k_1} \text{Product(s)} \tag{1}$$

$$HO_2 + AA \xrightarrow{k_2} \text{Product(s)} \tag{2}$$

$$O_2^- \xrightarrow{k_3} \tfrac{1}{2}O_2 + \tfrac{1}{2}H_2O_2 \tag{3}$$

$$k_{1\text{ or }2} = \frac{0.693}{[AA]}\,[1/(t_{\frac{1}{2}})_{1\text{ or }2} - 1/(t_{\frac{1}{2}})_3] \tag{4}$$

first-order reaction in spite of the fact that mixed kinetics are observed for O_2^- decay in the presence of large quantities of amino acid. This is an approximation, but the error introduced is small and the complexity of using mixed first- and second-order kinetics is not warranted by the data.

(*B*) Aromatic amino acids with high absorbance in the u.v. were studied in the alkaline pH range by pseudo-first-order competition kinetics with nitroblue tetrazolium at 560 nm (Rapp *et al.* 1973).

(*C*) In cases where neither direct monitoring of HO_2/O_2^- nor kinetic competition studies could be used because of too strong absorbance of the amino acid in the u.v. range or the lack of an appropriate scavenger for competition kinetics, a method based on oxygen consumption, that had been developed earlier (Bielski & Richter 1977), was used. The latter method also lends itself to studies of induced chain reactions; such a reaction scheme had been postulated for cysteine (Owen & Brown 1969; Barton & Packer 1970).

The amino acids (purchased from Sigma Chemical Co.) were used without further purification. To minimize the effect of possible metal impurities, which when chelated by amino acids could act as efficient dismutation catalysts (Klug-Roth & Rabani 1976), we carried out all experiments in the presence of 50μM-ethylenediaminetetraacetic acid (EDTA).

The rate studies, at 23 °C, covered a 10- to 100-fold concentration range of the given amino acid. Column 2 of Tables 1 and 2 lists the upper limit of the concentration range studied.

The results in Tables 1 and 2 show that the rate constants for the interaction

TABLE 1

Rates of reaction of O_2^- with amino acids

Amino acids	$[AA]/mol\ l^{-1}$	*pH*	*Method*	$k_1/l\ mol^{-1}\ s^{-1}$
DL-Alanine	0.100	10.0	*A*	$<0.06 \pm 0.02$
L-Arginine	0.150	10.1	*A*	$<0.13 \pm 0.03$
DL-Asparagine	0.100	10.1	*A*	$<0.16 \pm 0.02$
DL-Aspartic acid	0.100	10.0	*A*	$<0.18 \pm 0.04$
L-Cysteine	0.050	10.9	*A*	$<15.00 \pm 2.00$
L-Cystine	0.0005	10.0	*A*	$<0.40 \pm 0.07$
Glycine	0.100	8.8	*A*	$<0.42 \pm 0.12$
L-Glutamic acid	0.025	8.7	*A*	$<0.39 \pm 0.07$
L-Glutamine	0.100	10.0	*A*	$<0.25 \pm 0.05$
L-Histidine	0.150	10.0	*A*	$<1.00 \pm 0.21$
DL-Isoleucine	0.100	8.0	*A*	$<2.00 \pm 0.40$
L-Leucine	0.100	9.9	*A*	$<0.21 \pm 0.02$
DL-Lysine	0.100	8.5	*A*	$<3.30 \pm 0.03$
DL-Methionine	0.100	8.3	*A*	$<0.33 \pm 0.05$
L-Phenylalanine	0.043	10.1	*B*	$<0.36 \pm 0.05$
L-Proline	0.100	10.0	*A*	$<0.16 \pm 0.05$
DL-Serine	0.100	9.0	*A*	$<0.53 \pm 0.04$
DL-Threonine	0.150	10.1	*A*	$<0.21 \pm 0.05$
L-Tryptophan	0.020	10.6	*B*	$<24.00 \pm 3.00$
L-Tyrosine	0.005	10.8	*B*	$<10.00 \pm 2.00$
DL-Valine	0.150	10.1	*A*	$<0.18 \pm 0.02$

TABLE 2

Rates of reaction of HO_2 with amino acids

Amino acids	$[AA]/mol\ l^{-1}$	*pH*	*Method*	$k_2/l\ mol^{-1}\ s^{-1}$
DL-Alanine	0.100	1.6	*A*	$<44.0 \pm 11.0$
L-Arginine	0.100	1.6	*A*	$<63.0 \pm 14.0$
DL-Asparagine	0.100	1.4	*A*	$<53.8 \pm 10.0$
DL-Aspartic acid	0.100	1.5	*A*	$<12.0 \pm 4.0$
L-Cysteine	0.100	1.4	*A,C*	$\leqslant 601.0 \pm 85.0$
Glycine	0.100	1.5	*A*	$<48.6 + 4.0$
L-Glutamic acid	0.100	1.6	*A*	$<30.0 \pm 6.0$
L-Glutamine	0.100	1.5	*A*	$<23.0 \pm 6.0$
L-Histidine	0.050	1.8	*A*	$\leqslant 95.0 \pm 14.0$
DL-Isoleucine	0.100	1.4	*A*	$<38.9 \pm 5.0$
L-Leucine	0.100	1.4	*A*	$<23.0 \pm 4.0$
DL-Lysine	0.100	1.4	*A*	$<13.3 \pm 3.0$
DL-Methionine	0.100	1.5	*A*	$<48.8 \pm 15.0$
L-Phenylalanine	0.005	1.3	*A*	$<180.0 \pm 50.0$
D-Proline	0.100	1.4	*A*	$<17.3 \pm 3.0$
DL-Serine	0.100	1.2	*A*	$<54.6 \pm 8.0$
DL-Threonine	0.100	1.4	*A*	$<12.5 \pm 4.0$
DL-Valine	0.100	1.5	*A*	$<10.5 \pm 1.3$

of amino acids with HO_2 are larger by one to two orders of magnitude than the corresponding values for the reaction with O_2^-. We should caution that, although the difference in the rates is true for cysteine and histidine, for which a good linear relationship could be established between observed pseudo-first-order rate constant and scavenger concentration, the same is not true for the other amino acids studied. High rate values can arise from inherent experimental errors. For example, in alkaline solutions the absorbance of O_2^- is high but its spontaneous decay is low; a favourable situation for high signal-to-noise ratio. In the acid pH range the situation is the reverse. The absorbance due to HO_2 is at best one-half that of O_2^- in the alkaline range and its spontaneous decay rate is 10^3 times faster. Such a situation gives relatively large experimental errors which because of the pseudo-first-order conditions have a profound effect on the computed second-order rate constant. The lack of appropriate HO_2 scavengers has made it impossible at this time to evaluate the rate constants for the aromatic amino acids by competition kinetics. Overall, the reported second-order rate constants in Tables 1 and 2 should be considered only as upper limits of the true rates of reaction.

Of the essential amino acids the most reactive toward HO_2 and O_2^- is cysteine. The second-order rate constants $k_{RSH+HO_2} \leqslant 601.0 \pm 85.0\ l\ mol^{-1}\ s^{-1}$ (measured at pH 1.4) and $k_{RS^-+O_2^-} = 15.0 \pm 2.0\ l\ mol^{-1}\ s^{-1}$ (measured at pH 10.9) were determined for a cysteine concentration range from 0.01 to 0.1 mol/l. All cysteine solutions were prepared with oxygen-free water. Spontaneous oxidation due to mixing with O_2-saturated formate solutions was measured and corrected for in the kinetic O_2^- experiments. Since it had been shown that cysteine may undergo a chain oxidation (Owen & Brown 1969; Barton & Packer 1970), the chain length in the present experimental conditions was established as 1.3, which was negligible and did not interfere with the determination of the rate constant of interest.

The rate constants for cysteine determined in the present study are much smaller than the earlier reported values of $k_{O_2^-+RSH} > 5 \times 10^4\ l\ mol^{-1}\ s^{-1}$ (Barton & Packer 1970) and $k_{O_2^-+RSH} = 1.8 \times 10^4\ l\ mol^{-1}\ s^{-1}$ (Al-Thannon *et al.* 1974). The earlier values, results of radiation studies, had been computed from *G* values for observed hydrogen peroxide yields. The determination of accurate rate constants from *G* values is often difficult since a small change in *G* can have a large effect on the computed rate constant. Contrary to the conclusions of the earlier investigations (that HO_2 does not react with cysteine), the present study shows as expected that HO_2 as an oxidizing agent is more reactive than O_2^-.

Of the other sulphur-containing amino acids both cystine and methionine are unreactive. In the case of cystine the generation of O_2^- from the negatively charged amino acid radical and molecular oxygen (reaction 5) is rapid: $k_{RSSR^- + O_2} = 4.3 \pm 0.3 \times 10^8$ l mol^{-1} s^{-1} (Barton & Packer 1970). The present

$$RSSR^- + O_2 \rightarrow RSSR + O_2^- \quad (5)$$

results suggest that reaction (5) is unidirectional and does not involve an equilibrium.

The heterocyclic and aromatic amino acids were studied in the alkaline pH range by competition kinetics with nitroblue tetrazolium. As is evident from the rate constants, histidine ($k < 1.0 \pm 0.2$ l mol^{-1} s^{-1}) and phenylalanine ($k < 0.36 \pm 0.05$ l mol^{-1} s^{-1}) are the least reactive of this group of compounds, and tyrosine ($k < 10.00 \pm 2.0$ l mol^{-1} s^{-1}) and tryptophan ($k < 24.0 \pm 3.0$ l mol^{-1} s^{-1}) are modestly reactive. Since these amino acids were studied by the same method and over similar concentration ranges, one can assume that the relative degree of increased reactivity is correct. Some recent photolysis studies of oxygenated solutions of tryptophan suggests that the reaction product of tryptophan and O_2^- is N'-formylkynurenine (Walrant *et al.* 1975).

In acid solution histidine has a convenient u.v. window and hence was studied by method *A*. The rate constant for its reaction with HO_2 ($k < 95.0 \pm 14.0$ l mol^{-1} s^{-1}) was linear in the range from 12.5 to 50 mmol/l. Linearity of the pseudo-first-order rate constant with scavenger concentration was not reached for phenylalanine, for which the limit in present experimental conditions was 5 mmol/l. The corresponding computed rate constant ($k < 180.0 \pm 50.0$ l mol^{-1} s^{-1}) is high and should be used with reserve. The low rate value obtained at pH 10.1 ($k < 0.36 \pm 0.05$ l mol^{-1} s^{-1}) from competition studies with nitroblue tetrazolium indicates that phenylalanine does not react with O_2^-.

The least reactive are the aliphatic amino acids, which for practical purposes should be considered as 'non-reactive' since one would not expect either HO_2 or O_2^- to be able to break any of the chemical bonds encountered in these compounds. If one takes the average value of the rate constants for these 'non-reactive' amino acids ($k \approx 0.6$ l mol^{-1} s^{-1} for O_2^- and $k \approx 30.0$ l mol^{-1} s^{-1} for HO_2) and uses it as a yardstick for the sensitivity of the analytical methods used, it becomes apparent that of all the essential amino acids the only ones that react at all with HO_2 and O_2^- are cysteine, histidine, tyrosine, tryptophan and possibly phenylalanine.

ACKNOWLEDGEMENTS

We thank Dr C. A. Long for constructive criticism of this manuscript and for the many stimulating discussions and helpful suggestions. This research was carried out at Brookhaven National Laboratory under contract with the US Department of Energy and supported in part by its Division of Basic Energy Sciences and NIH Grant 1 RO1 GM 23656-01.

References

AL-THANNON, A. A., BARTON, J. P., PACKER, J. E., SIMS, R. J., TRUMBORE, C. N. & WINCHESTER, R. V. (1974) Radiolysis of aqueous solutions of cysteine in the presence of oxygen. *Int. J. Radiat. Phys. Chem. 6*, 233–248

BARTON, J. P. & PACKER, J. E. (1970) The radiolysis of oxygenated cysteine solutions at neutral pH. The role of RSSR and O_2^-. *Int. J. Radiat. Phys. Chem. 2*, 159–166

BIELSKI, B. H. J. & RICHTER, H. W. (1977) A study of the superoxide radical chemistry by stopped-flow radiolysis and radiation induced oxygen consumption. *J. Am. Chem. Soc. 99*, 3019–3023

HOLROYD, R. A. & BIELSKI, B. H. J. (1978) The photochemical generation of superoxide radicals in aqueous solutions. *J. Am. Chem. Soc. 100*, 5796–5800

KLUG-ROTH, D. & RABANI, J. (1976) Pulse radiolytic studies on reactions of aqueous superoxide radicals with copper(II) complexes. *J. Phys. Chem. 80*, 588–591

OWEN, T. C. & BROWN, M. T. (1969) The radiolytic oxidation of cysteine. *J. Org. Chem. 34*, 1161–1162

RAPP, U., ADAMS, W. C. & MILLER, R. W. (1973) Purification of superoxide dismutase from fungi and characterization of the reaction of the enzyme with catechols by electron spin resonance spectroscopy. *Can. J. Biochem. 51*, 158–171

WALRANT, P., SANTUS, R., REDPATH, J. L. & LEXA, D. (1975) Role of diatomic oxygen ions in the formation of *N'*-formylkynurenine from tryptophan. *C. R. Hebd. Séances Acad. Sci., Sér. D 280*, 1425–1428

Discussion

Fridovich: I understand that during photolysis the yield (or concentration) of superoxide reaches a maximum and then declines, simultaneously with an increase in concentration of peroxide. How does this happen?

Bielski: As can be seen from reactions (2) to (5) the mechanism of O_2^- formation during photolysis is very similar to the radiation-induced mechanism:

$$H_2O \xrightarrow{h\nu} OH + H \quad (1)$$

$$H + O_2 \rightarrow HO_2 \rightleftharpoons O_2^- + H^+ \quad (2)$$

$$H + HCOO^- \rightarrow H_2 + COO^- \quad (3)$$

$$OH + HCOO^- \rightarrow H_2O + COO^- \quad (4)$$

$$CO_2^- + O_2 \rightarrow CO_2 + O_2^- \quad (5)$$

Initially there is little hydrogen peroxide present since it is a secondary product, but as oxygen becomes depleted, especially near the window where the light intensity is the highest, O_2^- is converted into hydrogen peroxide by reaction (6)

$$H + O_2^- \rightarrow HO_2^- \quad (6)$$

and the ratio of $[O_2^-]/[H_2O_2]$ decreases monotonically with time.

Fridovich: Is the accumulated peroxide being photolysed to give $OH^\bullet$ which reacts with the O_2^-?

Bielski: Only after a relatively high concentration of hydrogen peroxide has been formed. In our experimental conditions the ratio $[H_2O]/[H_2O_2]$ is of the order of 10^6–10^7 and hence direct photolysis of hydrogen peroxide would be negligible.

Stern: These measurements were made at the extremes of pH at which the amino acids are obviously not typically in a structural form seen in physiological conditions. Do these results reflect the behaviour at physiological pH?

Bielski: We used the extremes of the pH scale because in those conditions HO_2 and O_2^- can be studied as separate species and their respective spontaneous decay rates are more favourable for stopped-flow experimentation. Assuming that the observed rates of reaction of the amino acid with HO_2 in the acid range and with O_2^- in the alkaline range represent the limiting rate values for the given system, we can compute rate values for any other pH by equation (7) where [AA] is the concentration of the amino acid, x is the ratio

$$k_{obs} = [AA]\left(\frac{k_{HO_2} + k_{O_2^-}x}{1 + x}\right) \tag{7}$$

$K_{HO_2}/[H^+]$, and k_{HO_2} and $k_{O_2^-}$ are the respective rate constants for HO_2 and O_2^- with the amino acid.

Stern: Might cysteine, for instance, be more reactive at physiological pH than at pH 10?

Bielski: I do not think so.

Fridovich: The ionized form of cysteine is more readily oxidizable, certainly under the influence of catalysis by trace metals. One might expect more rapid oxidation at higher pH by an oxidant.

Willson: Results with selective free radicals, such as those from carbonate and thiocyanate ion, and with hydroxyl radicals lead me to question your assumption that the rates with the amino acids alone closely reflect their reactivity in proteins. For example, the reactivity of *N*-benzoyl- or *N*-formylmethionine with selective free radicals is much less than that of free methionine (Adams *et al.* 1973). On the other hand, the rate constant for the reaction of $OH^{\cdot}$ with glycylglycine is 10 times rather than twice that for glycine at pH 7 (Scholes *et al.* 1965).

Bielski: Your observation represents most likely more than one type of chemical reaction; for instance, at low pH an OH radical will abstract a H atom, but at high pH it may preferentially remove only an electron — hence the difference in rates. In contrast the HO_2 and O_2^- are unreactive species which at best undergo a mild oxidation-reduction reaction.

As to polypeptides we have been unable to make a strong enough solution to bring about an observable reaction. Considering that the concentrations of such compounds are often only 10 times higher than the concentration of the added O_2^- they would have to react at rates comparable to the dismutation rate of O_2^- at the given pH.

Fridovich: You could do it if you changed the method of assay and followed the loss of activity of a polypeptide which had catalytic activity.

Bielski: That could work. What is badly needed for such competition studies are indicators which react with O_2^- and HO_2 at relatively slow rates. They are essential for future work.

Goldstein: You and Dr Willson (p. 16) mentioned the reaction of O_2^- with tryptophan to give *N*-formylkynurenine but the reaction is highly dependent on the concentration of tryptophan. Tryptophan is a good interceptor of solvated electrons with an efficiency comparable with that of oxygen. This reduces the concentration of O_2^-. Do you observe a concentration effect?

Bielski: Although we studied a 100-fold concentration range (up to the limit of solubility), we did not observe a concentration effect.

Hill: Are you satisfied with just adding EDTA to remove the metals from amino acids? How do metal–EDTA complexes react?

Bielski: We have studied (J. Weinstein & B. Bielski, unpublished results) the effect of 15 metal–EDTA complexes on the dismutation rate of O_2^-. So far we have found only three complexes that had an effect: iron(III), iron(II) and cobalt(II). With all other metals addition of EDTA eliminated the catalytic effect of the cation present.

Fridovich: Would they catalyse or otherwise influence the rate of reaction with the amino acids?

Michelson: Even more important would be to know whether a complex of iron(II) or iron(III) with the amino acid will complex with O_2^-.

Hill: One could have an amino acid–EDTA mixed complex.

Bielski: To find an answer to these questions we are rebuilding our stopped-flow spectrophotometer to accommodate two mixing chambers which should in principle allow us to study the following reactions:

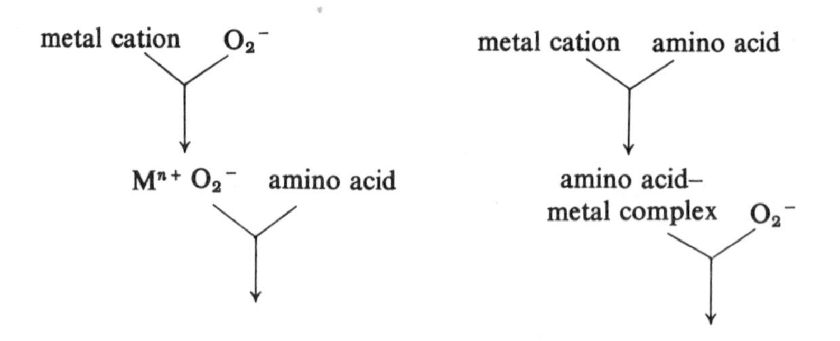

Willson: What is the mechanism of the reaction of nitroblue tetrazolium (NBT) with O_2^-?

Bielski: We are in the midst of studying the reduction of NBT by O_2^-. The first objective of this investigation is the determination of the molar extinction coefficient for the stable blue reduction product generally referred to as NBT^-. Although we do not know the true nature of this blue product, we can generate it quantitatively by reducing NBT with either CO_2^- or O_2^-. A preliminary value for its molar absorbance at 560 nm is in the range 18 000–20 000 l mol^{-1} cm^{-1}.

Slater: Is the reduction that you studied to the monoformazan or to the diformazan?

Bielski: It appears to be a single-electron product because in pulse-radiolysis experiments we interact 10μM-O_2^- with 2mM-NBT and one would not expect one NBT molecule to encounter two O_2^- radicals. If NBT^{2-} were formed, we would expect an absorbance change, which so far we have not observed. We have since found that with short pulses (0.5 μs) there is at least one short-lived transient formed which decays by second-order kinetics to a more strongly absorbing species at 530–550 nm. This new finding suggests that the final blue product could be NBT^{2-}.

Willson: Does NBT^- react with oxygen?

Bielski: So far we have seen no back reaction but we shall check that out.

Willson: Anclair *et al.* (1978) have suggested an equilibrium.

Bielski: When too much of the blue end-product is formed it rapidly precipitates, thus interfering with any possible equilibrium measurements.

Hill: Is an equilibrium established?

Bielski: If one compares the quantitative reduction of NBT by O_2^- to the reduction by CO_2^- in the absence of oxygen, one finds that the observed absorbance change at 560 nm is mole-for-mole the same. This suggests that in the presence of O_2 there is no back reaction and if an equilibrium should exist it would be far to one side.

Slater: Have you determined the absorption spectrum of the NBT reaction product that you follow at 560 nm?

Bielski: Yes.

Slater: The absorption spectrum of the NBT monoformazan is different from that seen for the diformazan (Pearse 1972); the latter has a maximum absorption of 560 nm and its formation involves more than a single-electron reduction.

Bielski: The absorbance of the radiation-formed reduction products of NBT varies with pH. There are strong indications that this variation is due mainly to a change in the mechanism of reduction with pH.

Fridovich: If the NBT^- radical were formed, it should be detectable by e.s.r. spectroscopy. The mono- and di-formazans are not radicals and would be excluded.

What would happen if the NBT^- absorbed weakly and, after its formation, dismutated into the monoformazan or even the diformazan and that was the species that you could observe?

Bielski: Our recent results indicate that what you propose is the most likely pathway.

Fridovich: What was the shortest time of observation?

Bielski: A few microseconds.

FORMATION OF COPPER-SUPEROXIDE COMPLEXES

Bielski: We have recently submitted for publication (Bielski & Long 1979) a note on the formation of complexes between copper(II) and HO_2/O_2^-. Because of its relevance to the general subject matter of this meeting I shall summarize our findings.

When a 10–100mM-$Cu(ClO_4)_2$ solution is mixed with a 10–15μM-solution of O_2^-, a transient is observed which has an absorption maximum in the u.v. range, and this maximum shifts with pH. At pH 2.58 the peak is at 225 nm and at pH 5.56 the maximum is at 265 nm. At the higher pH there is an additional absorption maximum at 360 nm of much lower absorbance. Since we do not yet know if O_2^- and HO_2 in these complexes are associated with one or two copper atoms, for the time being we shall refer to them as CuO_2^+ and $Cu(OOH)^{2+}$, respectively. In general the spectrum of the copper-superoxide complex at 5.56 strongly resembles the spectrum of the manganese(II)-superoxide complex which also has two absorption maxima, one at 270 and a weaker one at 420 nm (Bielski & Chan 1977).

The mechanism of the formation and decay of the copper-superoxide complexes can be described by the following set of equations (1)–(5):

$$Cu^{2+} + HO_2 \rightarrow Cu(OOH)^{2+} \quad (1)$$

$$Cu^{2+} + O_2^- \rightarrow CuO_2^+ \quad (2)$$

$$2Cu(OOH)^{2+} \rightarrow 2Cu^{2+} + H_2O_2 + O_2 \quad (3)$$

$$Cu(OOH)^{2+} + CuO_2^+ + H_2O \rightarrow 2Cu^{2+} + H_2O_2 + O_2 + OH^- \quad (4)$$

$$2CuO_2^+ + 2H_2O \rightarrow 2Cu^{2+} + H_2O_2 + O_2 + 2OH^- \quad (5)$$

The second-order rate constants for the decay were measured over 3–4 half-lives. The preliminary approximate values are: at pH 0.6 $k_3 = 4.0 \times 10^3$ l mol^{-1} s^{-1} and at pH 5.36, $k_5 = 1.5 \times 10^4$ l mol^{-1} s^{-1}. This decay mechanism is analogous to the decay mechanism of the superoxide free radicals (equations 6–7).

$$2HO_2 \xrightarrow{10^6\,l\,mol^{-1}s^{-1}} H_2O_2 + O_2 \quad (6)$$
$$O_2^- + HO_2 + H_2O \xrightarrow{10^8\,l\,mol^{-1}s^{-1}} H_2O_2 + O_2 + OH^- \quad (7)$$
$$2O_2^- + 2H_2O \xrightarrow{<0.3\,l\,mol^{-1}s^{-1}} H_2O_2 + O_2 + 2OH^- \quad (8)$$

The observation of the Cu^{2+}–superoxide complexes is possible because in present experimental conditions all superoxide radicals HO_2/O_2^- are consumed in a single step reaction (reaction 1 and/or 2) by the large excess of Cu^{2+}. If all superoxide radicals are not consumed in the first step, they will react rapidly with the formed Cu^{2+}–superoxide complex in a dismutation step (see reactions 11 and 12). This was confirmed experimentally where it was observed that the signal of the Cu^{2+}–superoxide complex decreased with a decrease in the initial concentration ratio of Cu^{2+}/O_2^-, the O_2^- concentration being kept constant. These findings suggest that HO_2 and O_2^- most likely do not reduce Cu^{2+} to Cu^+ as has been suggested by the generally-accepted old mechanism for the copper-catalysed dismutation of superoxide radicals (reactions 9 and 10).

Old mechanism:

$$Cu^{2+} + O_2^- \rightarrow Cu^+ + O_2 \quad (9)$$
$$Cu^+ + O_2^- + 2H^+ \rightarrow Cu^{2+} + H_2O_2 \quad (10)$$

Our results suggest that the superoxide radical first forms a copper complex, in which the Cu^{2+} cation plays the same role as the proton in HO_2. As in the spontaneous disproportionation of superoxide radicals, where the controlling step above pH 2.5–3.0 is the reaction between HO_2 and O_2^- (reaction 7), we propose an analogous step between the copper–superoxide complex and O_2^- reactions 11 and 12).

New mechanism:

$$Cu^{2+} + O_2^- \rightarrow CuO_2^+ \quad (1)$$
$$Cu^{2+} + HO_2 \rightarrow Cu(OOH)^{2+} \quad (2)$$
$$Cu(OOH)^+ + O_2^- + H_2O \rightarrow Cu^{2+} + H_2O_2 + O_2 + OH^- \quad (11)$$
$$CuO_2^+ + O_2^- + 2H_2O \rightarrow Cu^{2+} + H_2O_2 + O_2 + 2OH^- \quad (12)$$

We do not know yet if this mechanism applies also to enzymic reactions. If it does it may explain why superoxide dismutase is specific for O_2^-, as the enzyme would have to be activated by O_2^-.

A spectrum similar but not identical to the one obtained from Cu^{2+} and superoxide radicals can be obtained when one mixes an oxygen-free 10–20 μM–CuCl solution with a dilute solution of perchloric acid containing 250 μM–O_2. The spectrum has an absorption maximum at 240 nm. Preliminary studies indicate that this complex has different properties from the complexes formed from Cu^{2+} and superoxide radicals. The reason for this difference may be that the former is a monocopper complex while the latter are dicopper complexes.

Reiter: Could this mechanism apply to the reaction of copper with ascorbic acid?

Bielski: Possibly. I do not know.

Reiter: Copper and ascorbic acid together will kill bacteria and inactivate certain toxins. This reaction is therefore of great interest, in particular against Gram-negative organisms (see Kligler *et al.* 1938; Drath & Karnovsky 1974).

Bielski: Ascorbic acid may chelate copper in such a fashion, that oxygen may form a ternary complex with it, for example (**1**).

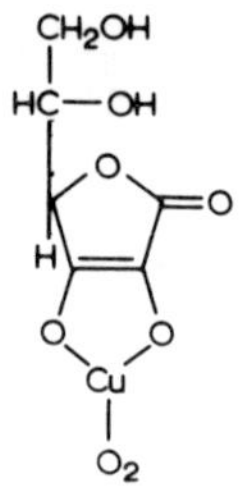

(1)

Michelson: Copper(III) has been implicated in the action of galactose oxidase (Dyrkacz *et al.* 1976). Is this likely?

Bielski: Copper(III) would only be formed with a strong oxidizing agent, for example with the hydroxyl radical. In our scheme we do not know whether copper undergoes hydrolysis. If it were to, it would provide a significant driving force.

Hill: This mechanism contains several attractive features because the previous mechanism invoked production of copper(I) which implied a change in the ligand environment of the copper. Such a change would introduce a slow step in the catalytic cycle: your mechanism obviates this.

I should add that, for convenience, we write $Cu^{2+} + O_2^-$. But what is the role of protons in the dismutase reaction? It would probably be less misleading to write the reaction as $Cu^{2+} + O_2^-(H_2O)_n$. Although you bring water into the last steps, it is more likely to be present initially: there is no O_2^- in water, only solvated O_2^-.

Bielski: We have a problem how to describe the complex as we know so little about it.

Dormandy: We now need a third presentation from you, Dr Bielski, to tie the amino acid, metal and oxygen reactions together since biological systems contain all those components.

Bielski: We hope to study such ternary systems with the syringe system I described earlier.

Willson: Artificial superoxide dismutases, such as copper salicylate complexes, have been shown to react rapidly with O_2^-. Such a mechanism has been proposed for the anti-inflammatory action of aspirin (de Alvare *et al.* 1976; Younes & Weser 1977; Younes *et al.* 1978).

Fridovich: Copper by itself will efficiently catalyse the dismutation of O_2^-. Also, one can construct metal buffers (the terminology was introduced by G. Schwarzenbach many years ago), which consist of a complex with a definite dissociation constant so that it can maintain a constant concentration of the free metal in solution. In complexes of copper(II) with tyrosine or salicylate free copper probably acts as the catalyst. Addition of a strong complexing agent like EDTA eliminates all the activity due to these copper complexes but has no effect on the activity of superoxide dismutase enzymes. The copper(II) complexes seem to be active not because of the specific nature of the complex but rather because the chelating agents are sufficiently weak to allow some free copper in solution. How much free copper is needed for catalysis?

Bielski: Very little.

Hill: One does not have to go so far as free copper. Three things can happen with these copper complexes. First, some free copper may be released. Secondly, the monocomplexes, e.g. $Cu(Tyr)(H_2O)_2$ could be formed, with a nice open site. Thirdly, even the complex itself is relatively open — it is not dissimilar from the structure of the copper site in the enzyme. In the EDTA complex the copper is enveloped and inactivated. We do not know which species is responsible for the dismutation when the complex is added. I guess that all those three possibilities could contribute to different degrees.

Fridovich: Biologically, the enzyme accounts for all the dismutase activity: one can separate the protein components of a crude extract of, say, *E. coli* on a gel, find a few bands of dismutase activity, each of which can be identified with a superoxide dismutase which has been isolated, and the sum of which accounts fully for the total activity of the crude extract.

References

ADAMS, G. E., REDPATH, J. L., BESBY, R. H. & CUNDALL, R. B. (1973) Selective free radical reactions with proteins and enzymes: reactions of inorganic radical anions with trypsin. *J. Chem. Soc. Faraday Trans. I 69*, 1608–1617

ANCLAIR, C., TORRES, M. & HAKIM, J. (1978) Superoxide anion involvement in NBT reduction catalyzed by NADPH-cytochrome P-450 reductase: a pitfall. *FEBS (Fed. Eur. Biochem. Soc.) Lett. 89*, 26–38

BIELSKI, B. H. J. & CHAN, P. C. (1977) Enzyme-catalysed chain oxidation of nicotinamide adenine dinucleotide by superoxide radicals, in *Superoxide and Superoxide Dismutases* (Michelson, A. M., McCord, J. M. & Fridovich, I., eds.), pp. 409–416, Academic Press, London & New York

BIELSKI, B. H. J. & LONG, C. A. (1979) *J. Am. Chem. Soc.*, in press

DE ALVARE, L. R., GODA, K. & KIMURA, T. (1976) Mechanism of superoxide anion scavenging reaction by bis-(salicylato)-copper(II) complex. *Biochem. Biophys. Res. Commun. 69*, 687–694

DRATH, D. B. & KARNOVSKY, M. L. (1974) Bactericidal activity of metal-mediated peroxide-ascorbate system. *Infect. Immun. 10*, 1077–1083

DYRKACZ, G. R., LIBBY, R. D. & HAMILTON, G. A. (1976) Trivalent copper as a probable intermediate in the reaction catalyzed by galactose oxidase. *J. Am. Chem. Soc. 98*, 626–628

KLIGLER, I. J., GUGGENHEIM, K. & WARBURY, F. M. (1938) Influence of ascorbic acid on the growth and toxin production of *Cl. tetani* and the detoxification of tetanus toxin. *J. Pathol. Bacteriol. 46*, 619–629

PEARSE, A. G. E. (1972) in *Histochemistry: Theoretical and Applied*, vol. 2, p. 893, Churchill Livingstone, London [now Edinburgh]

SCHOLES, G., SHAW, P., WILLSON, R. L. & EBERT, M. (1965) Pulse radiolysis studies of aqueous solutions of nucleic acid and related substances, in *Pulse Radiolysis* (Ebert, M. *et al.*, eds.), Academic Press, New York

YOUNES, M. & WESER, V. (1977) Superoxide dismutase activity of copper-penicillamine: possible involvement of Cu(I) stabilized sulphur radical. *Biochem. Biophys. Res. Commun. 78*, 1247–1253

YOUNES, M., LENGFELDER, E., ZIENEAU, S. & WESER, V. (1978) Pulse radiolytically generated superoxide and Cu(II)- salicylates. *Biochem. Biophys. Res. Commun. 81*, 576–580

Interactions between iron metabolism and oxygen activation

R. R. CRICHTON

Unité de Biochimie, Université Catholique de Louvain, Louvain-la-Neuve, Belgium

Abstract On account of its easy access in aqueous solution to the two states ferrous (Fe^{II}) and ferric (Fe^{III}), iron is ideally suited for the activation of molecular oxygen. It is, therefore, logical to seek links between the normal and pathological metabolism of iron and oxygen activation. The pathways of intracellular iron metabolism require changes in the oxidation state of iron both in its deposition in the storage form, ferritin, and in its mobilization from the storage form and use in the cell. Evidence is presented which shows that iron oxidation and deposition in ferritin involves activation of molecular oxygen with formation of a stable peroxo-complex as an intermediate in which the oxygen is bound between two iron atoms attached to adjacent polypeptide chains. The release of iron from ferritin is thought to involve reduction by a flavin, which is associated with the protein, and serves as a cofactor being alternately reduced by NADH or NADPH and oxidized by iron(III). The nature of the low-molecular-weight iron complex which serves to transfer storage iron to transferrin and to supply iron for intracellular use remains to be established.

The consequence of excessive iron overload can be rationalized on the basis of oxidative free-radical reactions which provoke lesions typical of deregulated oxygen activation. In some cases these pathological defects can be reversed by iron chelators. Progress in the development of chelation therapy for iron overload are reviewed.

Of the elements which could potentially be involved in the activation of molecular oxygen, iron and copper are the two which, on account of their abundance in biological systems, seem to be the most likely candidates. Perhaps the best illustration of their combined role in oxygen activation is cytochrome oxidase, the terminal enzyme in the electron-transport chain which reduces molecular oxygen to water using haem iron and enzyme-bound copper. It is my object in this paper to show that activation of oxygen by iron is a normal physiological mechanism, subsequently to trace the intracellular pathways of iron metabolism, and to point out that in pathological conditions associated with primary or secondary iron overload, many of the effects observed are due to oxidative free-radical reactions which result from the deregulation of oxygen activation.

By way of introduction it is perhaps best to discuss briefly the chemistry of iron in aqueous solution. This task is made easier by the fact that the subject has been recently discussed (Aisen 1977). Iron in aqueous solution has access to two oxidation states, ferrous Fe(II) and ferric Fe(III). At physiological pH values the autoxidation of Fe(II) is favoured, and the physiologically stable form would, in theory, be Fe(III). However, the chemistry of free Fe(III) at physiological pH values is dominated by the hydrolysis of Fe(III) to yield insoluble ferric hydroxides and oxyhydroxides. In view of the enormous importance of iron as a component of many enzymes (see below) and of the toxicity of iron due to its potential for free-radical formation, biological systems have evolved mechanisms for storing and transporting iron in its physiologically-stable trivalent state; by use of complexing agents of suitable capacity, they have succeeded in overcoming the tendency to hydrolysis to an insoluble form which would otherwise have rendered iron merely a rust-coating on all organic matter.

IRON IN THE ACTIVATION OF MOLECULAR OXYGEN

Iron is involved in the transport of molecular oxygen by haemoglobin and haemerythrin, and there appears to be good reason to believe that in both of these transport systems the oxygen is already partially activated. For oxy-haemoglobin the formulation as a superoxo-iron(III) complex does not seem unreasonable (Wallace & Caughey 1975) but in oxyhaemerythrin the X-ray crystallographic study establishes that the oxygen is liganded end-on to two iron atoms; here the best description would seem to be peroxo-bisiron(III) complex (Hendrickson *et al.* 1975). The importance of iron as a component of the terminal oxidase of the mitochondrial electron-transfer pathway, cytochrome oxidase, has already been mentioned. Iron is also involved as an essential component of several other enzymes which activate molecular oxygen, such as the microsomal cytochrome P450 system involved in the hydroxylation of organic substrates, both haem- and non-haem-containing dioxygenases, pteridine-linked monooxygenases, such as the hydroxylases of the aromatic amino acids phenylalanine, tyrosine and tryptophan, several iron containing monooxygenases and molybdenum iron-sulphur flavin hydroxylases such as xanthine oxidase (Boyer 1975). Iron is also a component of enzymes involved in defence against O_2^- and O_2^{2-}, namely certain superoxide dismutases, catalase and peroxidase. The potential for radical production through redox reactions between iron(II) ions and peroxides *in vivo* is considerable and, as we shall see later, many of the symptoms of iron toxicity are similar to those produced by the decomposition of lipid hydroperoxides, a reaction that has often been suggested for the initiation of destructive radical reactions *in vivo* (Pryor 1976).

IRON TRANSFER WITHIN CELLS

I shall briefly trace the path of iron from the exterior of the cell, i.e. from the plasma membrane, to the interior of the cell, and see what are the essential principles that determine the way in which iron can fulfill its essential roles within the cell. First we must concern ourselves with the mechanisms whereby transport iron in the extracellular space, essentially bound to transferrin, is accumulated by the cell. Transferrin, the principal iron-binding protein of plasma can bind a maximum of 2 g atoms of iron/molecule. Our current concepts of iron uptake from transferrin have not greatly changed from the pioneering studies of Jandl *et al.* (1959) who showed that the iron atoms of transferrin are taken up by membrane receptors on immature erythroid cells, and presumably on other tissues (for reviews see Aisen & Brown 1975; Morgan 1974).

Transferrin, containing one or two g atoms of iron/molecule of molecular weight 80 000, binds to specific receptors on the plasma membrane of cells. The next step in the pathway of iron assimilation is debatable, but may be explained by the interiorization of a portion of the plasma membrane, containing the bound transferrin, by endocytosis (Hemmaplardh & Morgan 1977) followed by fusion of the endocytic vacuole with a lysosome to give a phagosome (Trump & Berezesky 1977). The acidic pH (around 4.5) in the interior of the phagosome should facilitate the release of iron from transferrin, and the iron could be assimilated by a low-molecular-weight chelator (such as citrate) or by lysosomal ferritin. The apotransferrin would subsequently be returned to the plasma membrane associated with its receptor and released for subsequent re-use as has been proposed for other plasma membrane components (Schneider *et al.* 1977; Tulkens *et al.* 1977). Experiments to control this hypothesis have been initiated in rat fibroblast cells in culture (J. N. Octave, Y.-J. Schneider, A. Trouet & R. R. Crichton, unpublished results). We have found that iron-loaded transferrin is taken up by rat fibroblast cells in culture, that the transferrin is initially associated with the plasma membrane and that subsequently the iron is released to cytosol ferritin. Iron release from transferrin to rat fibroblasts could occur in phagolysosomes and the iron could be either directly released to the cytosol via a low-molecular-weight chelator (Jacobs 1977) or else pass through lysosomal ferritin before being released to the cytosol.

That intracellular iron is found almost exclusively in ferritin (and to a small extent in a number of iron-containing enzymes) is already well established. However, the possibility that there exists a small amount of intracellular iron in a mobile, chelatable pool (Jacobs 1977) seems highly likely, if not absolutely proven. This pool of iron would have as its principal role to enable exchange between storage iron (ferritin) and intracellular iron-containing proteins, and to enable intracellular storage iron to exchange with extracellular (transferrin-bound) iron. As will be discussed later, there is good reason to believe that this pool of chelatable iron may be the site of iron mobilization for several iron chelators.

The bulk of intracellular iron is either in ferritin, found in the cytosol and in lysosomes, or in intracellular iron-containing proteins. In non-erythroid cells, ferritin iron represents the most important repository of iron (in the average human some 25% of the total body iron is in ferritin or in haemosiderin, a poorly defined, predominantly lysosomal form of iron that is assumed to be formed by breakdown of ferritin). The origin of lysosomal ferritin is not clear; however, it seems most likely that either it is due to the transport of ferritin within the cell as a secretory protein, or else ferritin is accumulated in lysosomes by autophagocytosis. We assume that in most cells ferritin serves a dual function, as a means of maintaining a reserve of storage iron within the cell for use in the synthesis of iron-containing enzymes, and as a detoxification mechanism which enables iron to be maintained in a soluble form within the cell.

IRON DEPOSITION IN FERRITIN

Before discussing the mechanisms by which ferritin accumulates iron, I shall briefly summarize the structure of ferritin. The best characterized ferritin, that of horse spleen, consists of a nearly-spherical hollow protein shell, apoferritin, of outside diameter 13.0 nm, composed of 24 polypeptide chains each of molecular weight 18 500. The protein shell encloses an internal cavity of diameter 7.5 nm in which a variable amount of iron is deposited, largely as iron(III) oxyhydroxide together with some phosphate (for reviews of this and subsequent aspects of the structure see Harrison *et al.* 1977; Harrison 1977; Crichton 1973). The capacity of ferritin to store iron(III) in the hydrolysed state in a soluble form is perhaps best illustrated by the fact that aqueous solutions of ferritin are readily prepared which contain a concentration of 0.3–0.5 mol/l in iron! The iron content of ferritin can vary from zero to a maximum of around 4500 atoms/molecule; the mean value for horse spleen ferritin is around 2500 atoms/molecule. Since the apoferritin protein shell has a molecular weight of 440 000, this means that the molecular weight of ferritin can vary from 440 000 up to 900 000. The full iron core has a lobed appearance imposed by the apoferritin protein shell which presumably explains why four 'subunits' are sometimes seen in electron microscopy of ferritin iron cores (Farrant 1954). In partially filled shells, the iron atoms are located in microcrystalline particles attached to the inner surface of the protein (Massover & Cowley 1973).

The structure of the apoferritin molecule has been determined by X-ray analysis at a resolution of 0.28 nm (Banyard *et al.* 1978). The crystals used were cubic of space group F432. The tertiary structure of the molecule can best be described as a series of four nearly parallel regions of α-helix 3.4–4.2 nm long with a fifth shorter helical region disposed roughly at right angles to the others, accounting for some 60–70% of the polypeptide chain. These five helices are joined by non-helical sections which are mostly disposed on the exterior of the molecule and for which the electron density is less well defined (Fig. 1). The high helix content is in good agreement with c.d. and o.r.d. results (for review see Crichton 1973). The quaternary structure is characterized by the presence of channels along the four-fold axes of symmetry which have a diameter of around 1.3 nm. The localization of certain heavy-atom sites close to two-fold axes have important implications for the mechanism of iron deposition and are discussed later.

Apoferritin from horse spleen has an isoelectric point of 4.4 and has a relatively high content of non-polar amino acids (45%). One quarter of the amino acids is Glx or Asx. The NH_2-terminus is blocked by *N*-acetylation of the terminal serine residue. We are presently engaged in the determination of

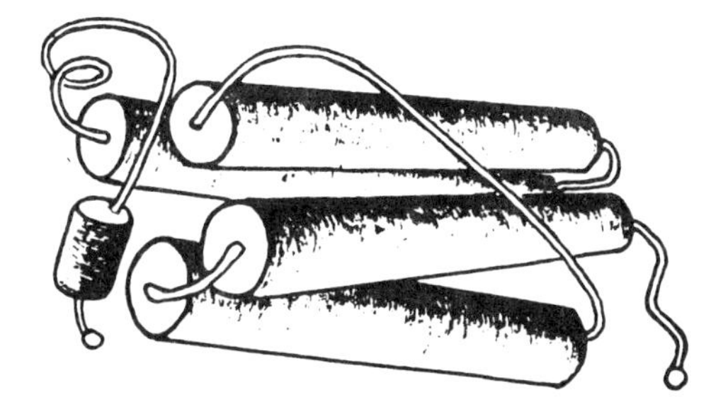

FIG. 1. A model of the apoferritin subunit (from Banyard *et al.* 1978): the subunit is composed of four nearly parallel helices 3.4–4.2 nm long and a short helix at right angles to them. The helical regions are connected by non-helical segments; their assignment is less certain than for the helices. Four of the short helices line the channels between the subunits in the quaternary structure, the long helices lying along the molecular surfaces.

the amino acid sequence of the horse-spleen apoferritin molecule, and to date, more than 80% of the sequence is established.

Ferritin is present in all mammalian cells with the greatest concentrations found in liver, spleen and bone marrow (for review see Crichton 1973). It appears that apoferritins from different tissues of the same species have different primary structures as indicated by amino acid compositions, immunological reactivity, peptide fingerprints, isoelectric focusing and subunit molecular weights (for reviews see Harrison 1977; Drysdale 1977).

Iron deposition in ferritin seems likely to involve an oxidative mechanism. The earliest studies on iron deposition indicated that Fe(II) was rapidly taken up by apoferritin in the presence of a suitable oxidant, and converted into the oxyhydroxide (hydrolysed) form of Fe(III). The nature of the product formed was determined by the apoferritin, and it was adduced that apoferritin plays a catalytic role in iron oxidation and deposition to form ferritin (reviewed in Crichton 1973). On the basis of these experiments several models were proposed for iron deposition in ferritin (Fig. 2). Of the three models presented I should like to direct attention to the first and the last. The first model supposes that Fe(II) is oxidized to Fe(III) on catalytic sites on the protein, subsequently migrates to the interior of the protein where it is hydrolysed and precipitates,

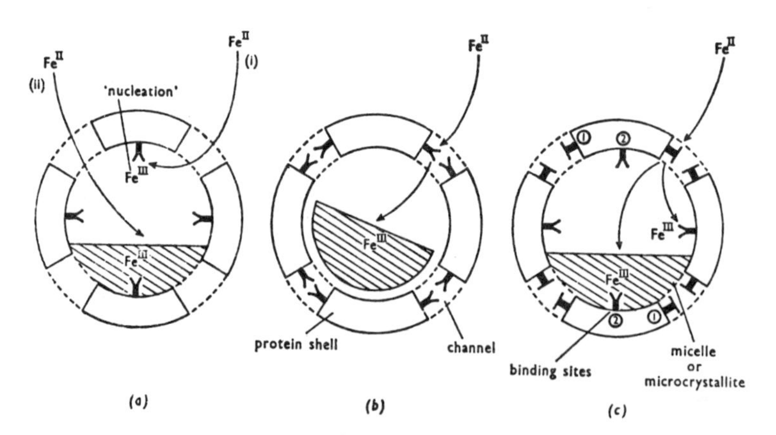

FIG. 2. Schematic models for the formation of ferritin (from Macara *et al.* 1972): the drawings show sections through the protein shell with subunits and intersubunit channels. In model (*a*) Fe(II) is bound at sites inside the protein shell where it is oxidized to Fe(III) which acts as a nucleus for the formation of a microcrystal of ferric oxyhydroxide. The micelle grows by addition and oxidation of further iron atoms on its surface. In model (*b*) all Fe(II) atoms which enter the molecule are bound and oxidized at sites located in the subunit channels. It is assumed that Fe(III) is less firmly bound to these sites than Fe(II), and forms ferric oxyhydroxide crystals within the protein shell. Model (*c*) is similar to model (*b*) except that the protein has specific sites for both oxidation and nucleation.

forming an iron(III) oxyhydroxide micelle, on the surface of which subsequent iron oxidation, hydrolysis and deposition occur. The third model assumes that there are catalytic sites which bind and oxidize Fe(II). The Fe(III) formed migrates to the interior of the protein where it is hydrolysed and is bound by heteronucleation sites which act as foci for the subsequent formation of the iron oxyhydroxide micelle. The essential difference between this mechanism and the 'crystal growth' model proposed above is that the apoferritin protein shell is involved throughout the process of iron deposition independent of the amount of iron deposited in the interior of the protein shell. In contrast the first mechanism predicts that, once the iron micelle has begun to be deposited, all subsequent iron deposition occurs on the surface of the growing micelle.

I now propose a molecular mechanism for iron oxidation by ferritin, which involves appropriate activation of molecular oxygen, and subsequently I shall speculate on the way in which the Fe(III) is hydrolysed and finds its way into the interior of the protein shell, where, both from neutron low-angle scattering studies on ferritin (Stuhrmann *et al.* 1976) and from high resolution electron microscopy studies on degraded ferritin fractions (Massover 1978), it is reasonable to assume that the iron micelle is firmly fixed.

When we began to look at iron deposition in ferritin we were struck by a number of apparently conflicting observations. In the first place, it is clear that if we assume that molecular oxygen is the electron acceptor in iron oxidation, then we must envisage a mechanism which ultimately reduces molecular oxygen to the oxidation level of water, i.e. we must assume that four iron atoms are oxidized for each molecule of dioxygen consumed. This prediction is confirmed by analysis of the consumption of molecular oxygen associated with iron deposition in ferritin. Both Melino *et al.* (1978) and Wauters *et al.* (1978) find that iron deposition in ferritin is accompanied by the consumption of 0.24 mol of dioxygen/g atom of iron deposited in ferritin. The second conflicting observation, apparently unrelated to iron deposition was that Fe(II) was released from ferritin after incubation with α,α'-bipyridyl (Dognin & Crichton 1975). Finally, there were several anomalies in the type of kinetics observed for iron deposition in several different buffers.

The model which we have proposed for iron deposition in ferritin can be summarized in four steps (Fig. 3). In the first step Fe(II) is bound to specific binding sites on adjacent polypeptide chains. We assume that these sites have a much greater affinity for Fe(II) than for Fe(III). Subsequent to iron binding a molecule of dioxygen is bound between the two atoms of Fe(II) and is reduced to form an end-on peroxo-complex in which the oxygen is coordinated to the two iron atoms. The formal valence state of the iron atoms is now Fe(III) and the peroxo-complex is assumed to be stable, or at least to be in equilibrium

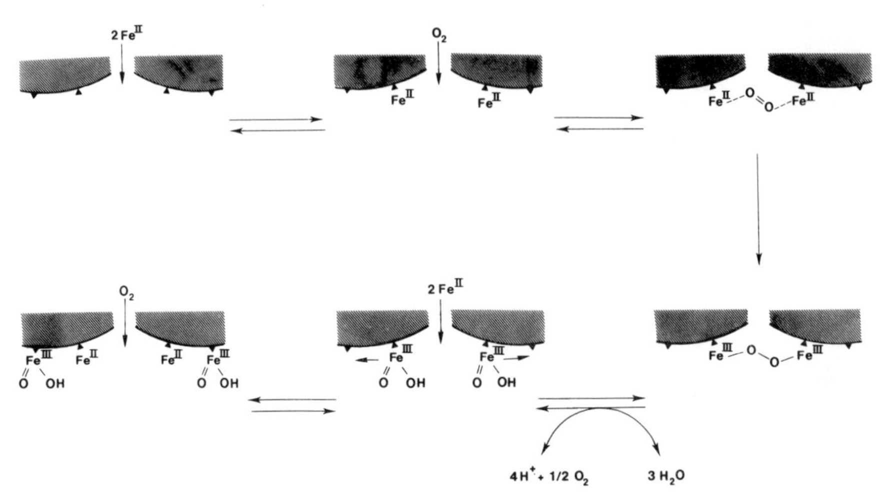

FIG. 3. A model for ferritin iron oxidation and deposition (from Crichton & Roman 1978). The iron-binding sites involved in iron oxidation are located close to a two-fold axis of symmetry of the molecule. The sites of heteronucleation to which the iron migrates to form the oxyhydroxide micelle are in the interior of the molecule.

with the subsequent hydrolysed intermediates. In the final steps the peroxo-complex is assumed to be hydrolysed; we have given a formal stoichiometry in the figure; it is clear that release of a single atom of oxygen does not occur. One possible mechanism would be that two incoming Fe(II) atoms are oxidized by the peroxo-intermediate, and then hydrolysed as outlined in reaction (1).

$$Fe^{III}—O—O—Fe^{III} + 2Fe^{2+} + 6H_2O \rightarrow 4FeO.OH + 8H^+ \quad (1)$$

The evidence in support of this mechanism is now summarized. First, we have observed (Wauters *et al.* 1978) that the rapid kinetics of iron deposition in ferritin (measured by stopped-flow techniques) in iron-complexing buffers at pH 5.5 can best be linearized by assuming that iron enters into the rate equation to the second power at least, consistent with the mechanism proposed. More recently, we have established (Pâques *et al.* 1979) that, in non-complexing buffers, the rate-limiting step in ferritin iron deposition is first order and corresponds to the deprotonation of a group or groups on the protein with a pK of around 6.00. Free superoxide was not detected as an intermediate in iron deposition (Wauters *et al.* 1978) and the stoichiometry of iron oxidized/oxygen consumed showed that four iron atoms were oxidized per molecule of dioxygen reduced (Melino *et al.* 1978; Wauters *et al.* 1978). That two sites for iron binding located on adjacent polypeptide chains are involved in iron oxidation and deposition is supported by the X-ray crystallographic results of Banyard *et al.* (1978) who reported that the major sites of Tb^{3+} fixation are located close to a two-fold axis of symmetry of the apoferritin molecule, and at a distance of 0.43 nm from its symmetry-related pair. Tb^{3+} inhibits iron deposition in ferritin and is assumed to bind to carboxyl groups (Treffry *et al.* 1977). We have confirmed the results of Treffry *et al.* (1977) that the site of uranyl oxide is close to a two-fold axis of symmetry. Uranyl ions are also known to have a high affinity for carboxyl groups. We also found that Cr^{3+} binds on a two-fold axis of the protein, crystallized in cubic form (symmetry F432) and blocks the subsequent fixation of uranyl oxide. We conclude that 12 g atoms of Cr^{3+} are bound per apoferritin molecule and have confirmed this value by atomic absorption (E. Pâques, R. Huber & R. R. Crichton, unpublished results). Since we have previously shown that the binding of 12 g atoms of Cr^{3+}/molecule of apoferritin results in 85–90% inhibition of iron deposition (Wauters *et al.* 1978), it seems reasonable to assume that the catalytic site of apoferritin is composed of two adjacent polypeptide chains, each of which binds an atom of Fe^{2+} as proposed in the model.

Finally we have evidence that the peroxo-intermediate is formed, and is sufficiently stable to be able to oxidize several organic substrates. The release of Fe(II) from ferritin as the pink-coloured complex with α,α'-bipyridyl (Dognin & Crichton 1975; Crichton & Roman 1978) can be explained (Fig. 4) by the formation of bipyridyl *N*-oxide. A similar result is found with triphenylphosphine, which is oxidized by ferritin to the corresponding oxide (A. Crutzen, K. Böneman & R. R. Crichton, unpublished results).

Further indications that activated oxygen exists at the catalytic site of ferritin is summarized below in the section on iron release from ferritin by low-molecular-weight chelating agents.

The molecular mechanisms involved in Fe(III) hydrolysis and its subsequent

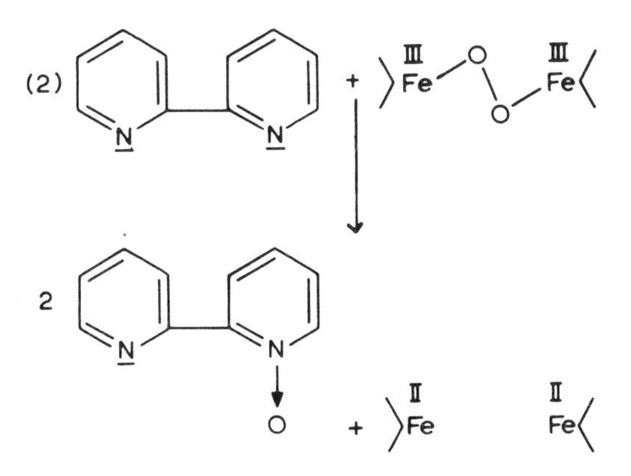

FIG. 4. Mechanism for Fe(III) reduction by α,α'-bipyridyl (from Crichton & Roman 1978).

migration to subsequent sites of heteronucleation in the interior of the protein shell will not be discussed here. However, the capacity of apoferritin to oxidize and deposit iron is inhibited both by Zn^{2+} (Macara *et al.* 1973) and by chemical modification of carboxyl groups (Wetz & Crichton 1976) and of cysteine and histidine residues in apoferritin (Bryce & Crichton 1973). On the basis of the model proposed for iron oxidation (Fig. 3) we may assume that the most likely ligands of iron are imidazole nitrogen and carboxyl oxygen atoms.

IRON RELEASE FROM FERRITIN

I shall not insist here on the various mechanisms that have been proposed for iron mobilization from ferritin but shall concentrate my attention on the most effective methods of iron release, and the role of low-molecular-weight chelators. A considerable body of evidence suggests that iron release from ferritin involves reduction of the iron to Fe(II) and the subsequent complexation of the Fe(II) by an appropriate, and as yet unidentified, complexing agent. The most rapid release of iron from ferritin is observed in anaerobic conditions by the reduced form of FMN in the presence of a suitable Fe(II) chelator such as α,α′-bipyridyl (Sirivech *et al.* 1974), according to reaction (2). Since $FMNH_2$ can be

$$FMNH_2 + 2Fe(III) \rightarrow FMN + 2Fe(II) \xrightarrow{\alpha,\alpha'-\text{bipyridyl}} Fe(II)-\text{bipyridyl complex} \quad (2)$$

readily obtained from FMN and reduced pyridine nucleotides (NADPH and NADH), we have studied the release of iron from ferritin using the system described in reactions (3) + (2). In aerobic conditions (i.e. around 250μM-

$$NAD(P)H + H^+ + FMN \rightarrow NAD(P)^+ + FMNH_2 \quad (3)$$

concentration of O_2) the iron of ferritin is released by this system after a lag phase which depends on the concentration of reactants used (Crichton *et al.* 1975). The lag phase corresponds to the autoxidation of $FMNH_2$ by the dissolved O_2; as soon as the concentration of O_2 diminishes below 2–3 μmol/l iron reduction and release is observed and eventually attains a plateau which corresponds to the mobilization of 70% of the ferritin iron present. In principle the same effect could be obtained by use of a suitable $NAD(P)^+$-dependent dehydrogenase together with its substrate to generate NAD(P)H (reaction 4) together with FMN, ferritin and a chelator of Fe(II) such as α,α′-bipyridyl (reactions 3 and 2): in effect this system can readily mobilize ferritin iron

$$AH_2 + NAD(P)^+ \rightleftharpoons A + NAD(P)H + H^+ \quad (4)$$

TABLE 1
Iron release from ferritin by several systems which generate $FMNH_2$ in the presence of α,α'-bipyridyl

Source of $FMNH_2$	*Velocity of ferritin iron release to bipyridyl (g atom* mol^{-1} min^{-1}*)*	*% of iron released after 24 h at 20 °C*
FMN/NADH	212.0	68.9
FMN/NAD^+ + ethanol/ alcohol dehydrogenase	71.7	79.5
FMN/$NADP^+$/glucose-6-phosphate/glucose-6-phosphate dehydrogenase	150.4	76.4

In all cases the final concentration of reagents was: 2.5mM-FMN, 5mM-α,α'-bipyridyl, 3.75mM-NADH, -NAD^+ and -$NADP^+$. The ferritin solutions were in 100mM-MOPS (morpholinoethanesulphonate) buffer, pH 7.0. The final concentrations of glucose-6-phosphate and of glucose-6-phosphate dehydrogenase were 7.5 mmol/l and 10 μg/ml, respectively, and of ethanol and alcohol dehydrogenase 7.5 mmol/l and 10 μg/ml, respectively. In the experiments with alcohol dehydrogenase a final concentration of 50mM-semicarbazide was used to trap the acetaldehyde formed (R. R. Crichton, F. Roman & M. Clesse, unpublished observations).

(Table 1) using as electron donor either ethanol + NAD^+ with alcohol dehydrogenase or glucose-6-phosphate + $NADP^+$ with glucose-6-phosphate dehydrogenase (R. R. Crichton, F. Roman & M. Clesse, in preparation).

It is clear from a closer examination of the dependence of iron release from ferritin on the concentration of NADH and FMN (Roman 1975; Wauters 1977) that the flavin essentially plays the role of a coenzyme, being alternately reduced and oxidized. It is, therefore, not unreasonable to suppose that the flavin is associated with the protein. The conventional procedures for ferritin isolation produce a protein which, after reduction to form apoferritin, gives a difference spectrum on denaturation against undenatured protein characteristic of a reduced flavin. We have recently isolated ferritin by a procedure which removes the flavin (R. R. Crichton *et al.*, unpublished observations). These results suggest, therefore, that iron mobilization from ferritin may involve a reductive mechanism in which a flavin cofactor mediates electron transfer from reduced pyridine nucleotides to iron, with concomitant reduction of the iron to Fe(II) followed by its release from the protein by an appropriate low-molecular-weight chelator.

In view of the interest in the development of iron chelators for the treatment of iron overload, particularly in the case of thalassaemia (discussed below) we have examined the effects of several chelators on iron release from ferritin *in vitro*, using ferritin that was prepared by the modified isolation procedure, and thus did not contain any associated flavin. The results are presented in Table 2.

TABLE 2

Release of iron from ferritin by several chelating agents in the presence and absence of riboflavin mononucleotide (FMN)

Chelator	*Final concentration (mmol/l)*	*Release of iron (g atom/mol)* without FMN 4 h	without FMN 24 h	with FMN 4 h	with FMN 24 h
Desferrioxamine B	1.0	67.7	225.3	11.4	11.4
α,α′-Bipyridyl	20.0	15.2	41.8	78.1	284.3
Rhodotorulic acid	1	116.0	249.5	20.9	22.1
2,3-Dihydroxybenzoate	1	31.4	41.7	26.1	26.8
Pyridine-2-aldehyde 2-pyridyl-hydrazone	1	37.3	85.4	73.2	107.2

All experiments were carried out with 1μM-ferritin in 200mM-MOPS buffer, pH 7.4.

We can see that in the absence of the cofactor FMN desferrioxamine B and rhodotorulic acid, which are both hydroxamic acid derivatives (Nielands 1977), are the most effective. However, when we add FMN, iron release by desferrioxamine B and rhodotorulic acid is almost completely blocked. Iron release by 2,3-dihydroxybenzoate is also inhibited. In contrast, iron release by α,α′-bipyridyl and pyridine-2-aldehyde 2-pyridylhydrazone is markedly increased; FMN does not affect iron release in the absence of light, a fact which suggests that photoreduction of the flavin is necessary for the effects observed (F. Roman, F. Roland & R. R. Crichton, unpublished results, 1978). We propose that, after binding of the reduced flavin to the protein, the chelator, if it is too bulky to have access to the iron on the catalytic sites of the protein (desferrioxamine B and rhodotorulic acid) or has a complexation constant inferior to that of the

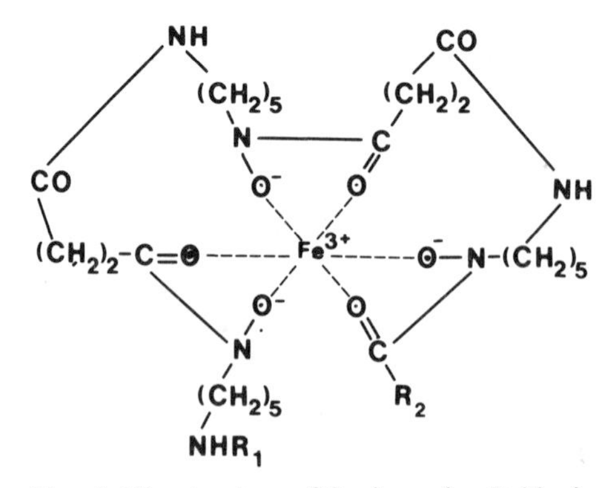

FIG. 5. The structure of ferrioxamine B (the iron complex of desferrioxamine B).

sites (dihydroxybenzoate) it will be unable to remove the iron. For the other two chelators two possible mechanisms could be invoked. The reduced flavin could generate Fe(II) on the catalytic sites, which iron is accessible to the chelator for complexation. Alternatively, the peroxo-intermediate at the catalytic sites of ferritin can transfer its peroxide to FMN with formation of a hydroperoxo-riboflavin derivative. The hydroperoxo-riboflavin derivative is assumed to be a stronger oxidant than the ferriperoxo-intermediate and can oxidize the chelator; for α,α'-bipyridyl we have shown that the corresponding *N*-oxide is formed (Crichton & Roman 1978).

IRON OVERLOAD AND OXYGEN ACTIVATION

Iron overload can result from a number of causes, all of which have in common the fact that the elimination of iron in man is limited to about 1 mg/day, and thus any increase in iron accumulation, whether due to increased dietary absorption or to transfusion (as in the secondary iron overload observed in transfused thallasaemia) cannot readily be compensated for by an increase in iron excretion. The ultrastructural and biochemical pattern in iron overload can best be summarized as an increase in hepatic ferritin in the cytoplasm and the deposition of ferritin in lysosomes, which is subsequently transformed to haemosiderin. Clearly the risk that iron can become decompartmentalized in such a situation is increased, and production of free radicals can thereafter lead to tissue damage, of which the most extensively studied is the peroxidation of membrane lipids. Iron can also increase the rate of lipid oxidation by catalysing the decomposition of lipid hydroperoxides. As Graziano (1976) pointed out, it is the Fe(II) form of iron which is the most potent free-radical generator, and thus the lipid-oxidizing potential of iron is greater when ascorbate is present, since ascorbate can regenerate Fe(II) from Fe(III). One of the terminal polymer products of lipid peroxidation, lipofuscin, is reported to be present at very high levels in the tissues of patients with thalassaemia or primary haemochromatosis (Graziano 1976).

During initial clinical trials on 2,3-dihydroxybenzoate (Peterson *et al.* 1974, 1976) as a potential iron chelator for treatment of iron overload in thalassaemia a lightening of the skin was observed in patients receiving the drug. Skin pigmentation is common in iron overload and is due to increased melanin production, which is thought to arise from a free-radical mechanism analogous to that involved in melanin formation by sunlight, in which iron in the skin catalyses free-radical formation and amplifies melanin formation especially in areas of skin exposed to the u.v. irradiation of the sun (Alexander 1960). Since the amounts of iron removed by 2,3-dihydroxybenzoate were not large in the

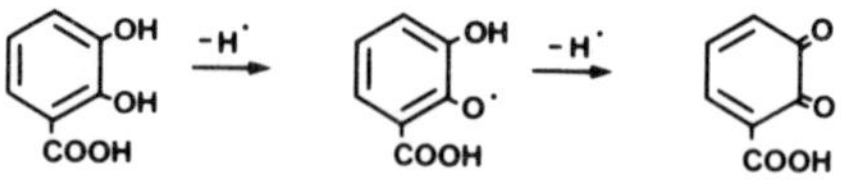

FIG. 6. A scheme whereby 2,3-dihydroxybenzoate might act as a free-radical scavenger (from Graziano *et al.* 1976).

study in question, it seems unlikely that 2,3-dihydroxybenzoate was removing sufficient iron from the skin to explain the reduction in pigmentation. It seemed likely that 2,3-dihydroxybenzoate was acting as a free-radical scavenger (Fig. 6), thereby terminating free-radical chain reactions involved in melanin synthesis. Evidence that 2,3-dihydroxybenzoate can indeed inhibit malonaldehyde formation in H_2O_2-stressed erythrocytes was obtained in support of this hypothesis, and it was further established that in erythrocytes from patients with thalassaemia major, thalassaemia intermedia, sickle-cell disease and haemoglobin Köln disease, all of which show an increased susceptibility to peroxidation, the formation of malonaldehyde was substantially inhibited by 2,3-dihydroxybenzoate (Graziano *et al.* 1976). Further evidence that 2,3-dihydroxybenzoate can act as a free-radical scavenger was found in studies of the effects of the drug on hepatotoxicity of CCl_4, an *in vivo* system in which the early appearance of lipid peroxidation in liver has been proposed as the most likely key event in the chemical pathology of CCl_4 (Graziano 1976).

It might be expected that iron chelators, if they were able to reduce tissue iron levels to more nearly normal values, would be valuable tools in preventing the free-radical chain reactions that characterize iron overload; in the case of 2,3-dihydroxybenzoate an additional function as a radical scavenger seems to be indicated. To the extent that decompartmentalization of iron can lead to free-radical production and tissue disorders, the interaction between iron metabolism and oxidative free-radical reactions is apparent. Further, iron is ideally suited to play the role of catalyst in the sequence of reactions (reactions 5 and 6) that constitute the Haber–Weiss reaction (1934) (reaction 7). To the

$$O_2^- + Fe^{3+} \rightarrow O_2 + Fe^{2+} \quad (5)$$

$$Fe^{2+} + H_2O_2 \rightarrow Fe^{3+} + OH^- + OH^{\cdot} \quad (6)$$

$$O_2^- + H_2O_2 \rightarrow O_2 + OH^- + OH^{\cdot} \quad (7)$$

extent that ferritin iron deposition and mobilization involve changes in redox states of iron, activation of oxygen to a peroxo-state, whose reactivity as an oxidant can apparently be enhanced by flavins, the links between oxygen activation and toxicity and the normal and pathological metabolism of iron are slowly beginning to take shape.

References

AISEN, P. (1977) Some physicochemical aspects of iron metabolism, in *Iron Metabolism (Ciba Found. Symp. 51)*, pp. 1–14, Elsevier/Excerpta Medica/North-Holland, Amsterdam

AISEN, P. & BROWN, E. B. (1975) Structure and function of transferrin. *Prog. Hematol. 9*, 25–56

ALEXANDER, P. (1960) Protection of macromolecules *in vitro* against damage by ionizing radiation, in *Radiation Protection and Recovery* (Hollaender, A., ed.), Pergamon Press, New York

BANYARD, S. H., STAMMERS, D. K. & HARRISON, P. M. (1978) Electron density map of apoferritin at 2.8 Å resolution. *Nature (Lond.) 271*, 282–284

BOYER, P. D. (ed.) (1975) *The Enzymes*, vol. XII: Oxidation-reduction Part B, Academic Press, London & New York

BRYCE, C. F. A. & CRICHTON, R. R. (1973) The catalytic activity of horse spleen apoferritin. *Biochem. J. 133*, 301–309

CRICHTON, R. R. (1973) Ferritin. *Struct. Bonding 17*, 67–134

CRICHTON, R. R. & ROMAN, F. (1978) A novel mechanism for ferritin iron oxidation and deposition. *J. Mol. Catal. 4*, 75–82

CRICHTON, R. R., WAUTERS, M. & ROMAN, F. (1975) Ferritin iron uptake and release, in *Proteins of Iron Storage and Transport in Biochemistry and Medicine* (Crichton, R. R., ed.), pp. 287–294, Elsevier/Excerpta Medica/North-Holland, Amsterdam

DOGNIN, J. & CRICHTON, R. R. (1975) Mobilisation of iron from ferritin fractions of defined iron content by biological reductants. *FEBS (Fed. Eur. Biochem. Soc.) Lett. 54*, 234–236

DRYSDALE, J. W. (1977) Ferritin phenotypes: structure and metabolism, in *Iron Metabolism (Ciba Found. Symp. 51)*, pp. 41–57, Elsevier/Excerpta Medica/North-Holland, Amsterdam

FARRANT, J. L. (1954) An electron microscopic study of ferritin. *Biochim. Biophys. Acta 13*, 569–576

GRAZIANO, J. H. (1976) Potential usefulness of free radical scavengers in iron overload, in *Iron Metabolism and Thalassemia* (Bergsma, D., Cerami, A., Peterson, C. M. & Graziano, J. H., eds.), Birth Defects, Vol. XII, pp. 135–143, Alan R. Liss, New York

GRAZIANO, J. H., MILLER, D. R., GRADY, R. W. & CERAMI, A. (1976) Inhibition of membrane peroxidation in thalassaemic erythrocytes by 2,3-dihydroxybenzoic acid. *Br. J. Haematol. 32*, 351–356

HABER, F. & WEISS, J. (1934) The catalytic decomposition of hydrogen peroxide by iron salts. *Proc. R. Soc. Lond. A 147*, 332–351

HARRISON, P. M. (1977) Ferritin: an iron storage molecule. *Semin. Hematol. 14*, 55–70

HARRISON, P. M., BANYARD, S. H., HOARE, R. J., RUSSELL, S. M. & TREFFRY, A. (1977) The structure and function of ferritin, in *Iron Metabolism (Ciba Found. Symp. 51)*, pp. 19–35, Elsevier/Excerpta Medica/North-Holland, Amsterdam

HEMMAPLARDH, D. & MORGAN, E. H. (1977) The role of endocytosis in transferrin uptake by reticulocytes and bone marrow cells. *Br. J. Haematol. 36*, 85–96

HENDRICKSON, W. A., KLIPPENSTEIN, G. L. & WARD, K. B. (1975) Tertiary structure of myohemerythrin at low resolution. *Proc. Natl. Acad. Sci. U.S.A. 72*, 2160–2164

JACOBS, A. (1977) An intracellular transit iron pool, in *Iron Metabolism (Ciba Found. Symp. 51)*, pp. 91–100, Elsevier/Excerpta Medica/North Holland, Amsterdam

JANDL, J. H., INMAN, J. K., SIMMONS, R. L. & ALLEN, D. W. (1959) Transfer of iron from serum iron-binding protein to human reticulocytes. *J. Clin. Invest. 38*, 161–185

MACARA, I. G., HOY, T. G. & HARRISON, P. M. (1972) The formation of ferritin from apoferritin. *Biochem. J. 126*, 151–162

MACARA, I. G., HOY, T. G. & HARRISON, P. M. (1973) The formation of ferritin from apoferritin — inhibition and metal ion-binding studies. *Biochem. J. 135*, 785–789

MASSOVER, W. H. (1978) The ultrastructure of ferritin macromolecules III. Mineralized iron in ferritin is attached to the protein shell. *J. Mol. Biol. 123*, 721–726

MASSOVER, W. H. & COWLEY, J. M. (1973) The ultrastructure of ferritin macromolecules. *Proc. Natl. Acad. Sci. U.S.A. 70*, 3847–3851

MELINO, G., STEFANINI, S., CHIANCONE, E., ANTONINI, E. & FINAZZI-AGRO, A. (1978) Stoichiometry of iron oxidation by apoferritin. *FEBS (Fed. Eur. Biochem. Soc.) Lett. 86*, 136–138

MORGAN, E. H. (1974) Transferrin and transferrin iron, in *Iron in Biochemistry and Medicine* (Jacobs, A. & Worwood, M., eds.), pp. 29–71, Academic Press, London

NIELANDS, J. B. (1977) Siderophores: diverse roles in microbial and human physiology, in *Iron Metabolism (Ciba Found. Symp. 51)*, pp. 107–119, Elsevier/Excerpta Medica/North-Holland, Amsterdam

PÂQUES, E. P., PÂQUES, A. & CRICHTON, R. R. (1979) A kinetic study of the mechanism of ferritin formation: the effects of buffers, of pH and of the iron content of the molecule. *J. Mol. Catalys.*, in press

PETERSON, C. M., GRAZIANO, J. H., GRADY, R. W., DE CIIUTIS, A., JONES, R. L. & CERAMI, A. (1974) Clinical evolution of 2,3-dihydroxybenzoic acid as an oral chelating drug. *Am. Soc. Hematol. 168*, 101 (abstr.)

PETERSON, C. M., GRAZIANO, J. H., GRADY, R. W., JONES, R. L., VLASSARA, H. V., CANALE, V. C., MILLER, D. R. & CERAMI, A. (1976) Chelation studies with 2,3-dihydroxybenzoic acid in patients with thalassaemia major. *Br. J. Haematol. 38*, 477–485

PRYOR, W. A. (1976) Free radical reactions in biological systems, in *Free Radicals in Biology*, Vol. 1 (Pryor, W. A., ed.), pp. 1–49, Academic Press, London & New York

ROMAN, F. (1975) *Mémoire de Licence*, Université Catholique de Louvain

SCHNEIDER, Y.-J., TULKENS, P. & TROUET, A. (1977) Recycling of fibroblast plasma-membrane antigens internalized during endocytosis. *Biochem. Soc. Trans. 5*, 1164–1167

SIRIVECH, S., FRIEDEN, E. & OSAKI, S. (1974) The release of iron from horse spleen by reduced flavins. *Biochem. J. 143*, 311–315

STUHRMANN, H. B., HAAS, J., IBEL, K., KOCH, M. H. J. & CRICHTON, R. R. (1976) Low angle neutron scattering of ferritin studied by contrast variation. *J. Mol. Biol. 100*, 399–413

TREFFRY, A., BANYARD, S. H., HOARE, R. J. & HARRISON, P. M. (1977) Structure and iron-binding properties of ferritin and apoferritin, in *Proteins of Iron Metabolism* (Brown, E. B., Aisen, P., Fielding, J. & Crichton, R. R., eds.), pp. 3–11, Grune and Stratton, New York

TRUMP, B. F. & BEREZESKY, I. K. (1977) A general model of intracellular iron metabolism, in *Proteins of Iron Metabolism* (Brown, E. B., Aisen, P., Fielding, J. & Crichton, R. R., eds.), pp. 359–364, Grune and Stratton, New York

TULKENS, P., SCHNEIDER, Y.-J. & TROUET, A. (1977) The fate of the plasma membrane during endocytosis. *Biochem. Soc. Trans. 5*, 1809–1815

WALLACE, W. J. & CAUGHEY, W. S. (1975) Mechanism for the autoxidation of hemoglobin by phenols, nitrite and 'oxidant' drugs. *Biochem. Biophys. Res. Commun. 62*, 561–567

WAUTERS, M. (1977) Doctoral thesis, Université Catholique de Louvain

WAUTERS, M., MICHELSON, A. M. & CRICHTON, R. R. (1978) Studies on the mechanism of ferritin formation: superoxide dismutase, rapid kinetics and Cr^{3+} inhibition. *FEBS (Fed. Eur. Biochem. Soc.) Lett., 91* 276–280

WETZ, K. & CRICHTON, R. R. (1976) Chemical modification as a probe of the topography and reactivity of horse spleen apoferritin. *Eur. J. Biochem. 61*, 545–550

Discussion

Dormandy: We have been interested in iron overload but only rarely see patients with acute iron poisoning (for example, children who take large doses of iron, such as their mother's iron tablets). A recent such case which ended fatally reinforced my belief that iron does cause its toxic effects by catalysing free-radical oxidation. Normally, serum is an antioxidant but this child's serum became a pro-oxidant when tested in our *in vitro* assay system (Stocks *et al.* 1974*a,b*). We also found TBA-reactive material in the child's liver and spleen. We have never previously found this in 'native' human tissue (though, of course, it can be generated *in vitro*).

Crichton: The chemistry of iron in aqueous solution is dominated by two essential reactions: oxidation and hydrolysis (Aisen 1977). Biological systems have developed safeguards against the precipitation of iron(III) oxyhydroxide, namely chelating agents of one kind or another for the transport and storage of iron.

Goldstein: Although the idea that iron overload — with ferritin, haemosiderin or whatever — causes toxicity by way of lipid peroxidation is generally accepted, the hard evidence is lacking. Dr Dormandy's studies, for instance, showing that thalassaemic red cells are more sensitive to lipid peroxidation are *in vitro* studies (Stocks *et al.* 1971). Evidence for this *in vivo* is minuscule. The same applies to haemoglobin Köln *in vivo* unless an oxidizing drug is present. You suggested that lipid peroxidation was due to free iron in red cells, but they contain practically no ferritin — in fact thalassaemic red cells contain less iron than normal red cells do. The ferritin story would not explain cirrhosis of the liver or cardiac disease with iron overload.

Crichton: I'm not so sure; Romslo (1977) has recently demonstrated that rat liver takes up ferritin iron extremely well and incorporates it in deuterohaem as long as two substances are supplied at the same time: FMN and an electron donor such as succinate (unpublished results).

Winterhalter: Whether haem synthetase works better on iron(II) or iron(III) and the fact that the final product needs iron(II) is unconnected with the synthetic mechanism.

Crichton: Romslo's evidence for a requirement for iron(II) is good.

Goldstein: But these are still *in vitro* studies.

Crichton: Yes, but at least ferritin iron can be used for haem synthesis, apparently by a process that requires reduced flavin. That is interesting because haem needs iron as iron(II) in mitochondria, which are certainly not the most anaerobic cell compartments.

Goldstein: It seems odd that although man suffers from iron overload there is no animal model of chronic iron overload in which pathology is evident. Maybe we are looking at the wrong things and generalizing to *in vivo* situations.

Crichton: True; it is difficult to find good animal models for iron metabolism. For example, humans, some of the higher monkeys and guinea-pigs are the only three species that have lost the capacity to synthesize ascorbic acid. (Some people even propose that man suffers from a megadeficiency of ascorbic acid.) Since ascorbate is not unconnected with iron metabolism it is extremely difficult to match the conditions in other experimental animals.

Another factor is that human serum contains small amounts of ferritin — about 150 ng/ml. In rats, the most commonly used experimental animals for studies on iron metabolism, the concentration of serum ferritin is 100–200 μg/ml of serum — about 1000 times higher. Furthermore, in rats the liver is much more important for the formation of red blood cells than in normal adult humans.

Keberle: In several experimental studies of iron intoxication, we have measured the concentrations of iron in various tissues and body fluids, including particularly the blood, plasma and plasma-water, of dogs given oral doses of radioactively labelled iron(II) sulphate. We have found that, as the concentration of iron in the plasma rises, the iron ions are taken up first by transferrin. As soon as the available transferrin is saturated, the iron is bound by plasma proteins. During this phase only relatively mild cardiovascular side-effects are observed.

Only when the binding capacity of the plasma proteins is exhausted and free iron ions appear in higher concentrations in the plasma-water does death ensue, quite suddenly. This is certainly due to the passive diffusion of the unbound iron ions into the tissues and the central nervous system. For this reason, I find it hard to believe that the formation of free radicals plays a decisive part in iron intoxication.

Winterhalter: A correlate in humans to these experiments was reported many years ago by Heilmayer who observed that some haemochromatotic patients died suddenly and that the serum of these patients contained free iron.

Crichton: In chronic iron overload, first, the transferrin in serum is saturated and then the iron is bound by albumin (Hershko & Rachmilewitz 1975).

Winterhalter: That is, unspecific binding.

Crichton: Yes; serious consequences ensue when that is overloaded.

Dormandy: Acute iron poisoning provides additional evidence in support of this: children after iron overload either remain perfectly fit or they die; there is almost no intermediate illness. The patients with symptoms die rapidly, in a matter of hours. I agree entirely with Dr Goldstein; there is still no clear-cut evidence that abnormal lipid peroxidation causes the premature disintegration of red cells in haemolytic disease, although peroxidation is an important mechanism of iron damage to liver, myocardium and so on.

Winterhalter: The strongest evidence comes from thalassaemic patients, who suffer a relative vitamin E deficiency. However, replenishing the levels of vitamin E in these people did not prolong the average life-span of the red cells.

Flohé: The question of whether iron toxicity is related to lipid peroxidation could be approached by comparing the toxic symptoms in different animals. In peroxide metabolism, for example, different tissues are crucial in different animals. The same should be true for iron poisoning. For instance, glutathione peroxidase activity decreases in mice, rats, pigs and birds on a selenium/vitamin E-deficient diet. However, liver necrosis occurs in the rats and both liver and pancreatic necrosis in the mice, whereas the pigs primarily suffer lesions in the small vessels of the heart and the birds develop exudative diathesis (Schwarz 1975; Flohé *et al.* 1976). Is there any evidence of such characteristic target tissues in iron intoxication?

Crichton: Not that I know of.

Slater: When some groups of piglets were given intramuscular iron colloid injections to overcome anaemia they quickly developed paralysis of the hind limbs (Patterson *et al.* 1971). This paralytic episode was shown to be related to lipid peroxidation in the muscles of these piglets which were later found to be vitamin E deficient. The sequence of events appeared to be the production of a vitamin E deficiency in the piglets as a result of the mother also being deficient in vitamin E; the administration of an iron overload then involved lipid peroxidation in the muscle. The resultant muscular damage prevented the piglets from being able to reach the sow to suckle and gain adequate nourishment. In this case, there was a clear relationship between iron administration, lipid peroxidation, and a serious pathological disturbance.

Crichton: I want to re-emphasize that the direct evidence relating iron overload to lipid peroxidation is conspicuous by its absence.

References

AISEN, P. (1977) Some physicochemical aspects of iron metabolism, in *Iron Metabolism (Ciba Found. Symp. 51)*, pp. 1–14, Elsevier/Excerpta Medica/North-Holland, Amsterdam

FLOHÉ, L., GÜNZLER, W. A. & LADENSTEIN, R. (1976) Glutathione peroxidase, in *Glutathione* (Arias, I. M. & Jakoby, W. B., eds.), pp. 115–138, Raven Press, New York

HERSHKO, C. & RACHMILEWITZ, E. S. (1975) Non-transferrin plasma iron in patients with transfusional iron overload, in *Proteins of Iron Storage and Transport in Biochemistry and Medicine* (Crichton, R. R., ed.), pp. 427–433, North-Holland, Amsterdam

PATTERSON, D. S. P., ALLEN, W. M., BERRETT, S., SWASEY, D. & DORE, J. T. (1971) The toxicity of parenteral iron preparations in the rabbit and pig with a comparison of the clinical and biochemical responses to iron-dextrose in 2 days old and 8 days old piglets. *Zentralbl. Veterinärmed. 18*, 453–464

ROMSLO, I. (1977) Iron delivery to haemoglobin, in *Iron Metabolism (Ciba Found. Symp. 51)*, pp. 189–190, Elsevier/Excerpta Medica/North-Holland, Amsterdam

SCHWARZ, K. (1975) Neuere Erkentnisse über den essentiellen Charakter einiger Spurenelemente, in *Spurenelemente in der Entwicklung von Mensch und Tier* (Betke, K. & Bidlingmaier, F., eds.), pp. 1–30, Urban & Schwarzenberg, Munich

STOCKS, J., KEMP, M. & DORMANDY, T. L. (1971) Increased susceptibility of red-blood-cell lipids to autoxidation in haemolytic states. *Lancet 1*, 266–269

STOCKS, J., GUTTERIDGE, J. M. C., SHARP, R. J. & DORMANDY, T. L. (1974*a*) Assay using brain homogenate for measuring the antioxidant activity of biological fluids. *Clin. Sci. Mol. Med. 47*, 215–222

STOCKS, J., GUTTERIDGE, J. M. C., SHARP, R. J. & DORMANDY, T. L. (1974*b*) The inhibition of lipid autoxidation by human serum and its relation to serum proteins and α-tocopherol. *Clin. Sci. Mol. Med. 47*, 223–232

Superoxide dismutases: defence against endogenous superoxide radical

IRWIN FRIDOVICH

Department of Biochemistry, Duke University Medical Center, Durham, North Carolina

Abstract Attempts to measure the rate of O_2^- production, in whole cells or in intact subcellular organelles, are frustrated by the endogenous superoxide dismutase (SOD). *Streptococcus faecalis* contains a single manganese-SOD which was isolated and used as an antigen in the rabbit. A precipitating and inhibiting antibody was obtained and used to suppress the SOD in crude lysates of *S. faecalis*. It allowed the demonstration that 17% of the total oxygen uptake by such lysates, in the presence of NADH, was associated with O_2^- production.

O_2^- attacks unsaturated lipids and breaches the integrity of membranes. When the membranes are free of lipid hydroperoxides, then both O_2^- and H_2O_2 are required and singlet oxygen appears to be the proximal attacking species. When the membrane contains some lipid hydroperoxide, then O_2^- is itself sufficient and seems to generate an alkoxy radical, by reacting with the lipid hydroperoxide. It appears likely that attack on membranes is one of the reasons for the cytotoxicity of O_2^-.

In *Escherichia coli* the manganese-SOD is derepressed by O_2^-. This enzyme is not made in the absence of oxygen and in aerobic conditions any change which results in enhanced production of O_2^- calls forth an increased synthesis of this enzyme. Increased levels of SOD, however achieved, correlate with greater resistance towards oxygen toxicity.

It is generally true that respiring cells contain more SOD than non-respiring cells. Among obligate anaerobes there is a correlation between SOD-content and tolerance towards oxygen. It is not known whether the SOD in obligate anaerobes is a retained primitive characteristic or one recently acquired by plasmid transfer.

There is an exception to the rule that copper-zinc-SOD is found in eukaryotes but not in prokaryotes, and that is the symbiotic bacterium *Photobacterium leiognathi*. This symbiont may have obtained the Cu-ZnSOD gene from the host fish.

A decade ago we realized that certain enzymes liberated the superoxide radical (McCord & Fridovich 1968*a*) and that other enzymes catalytically scavenged this radical (McCord & Fridovich 1968*b*, 1969). The traditional view of the basis of oxygen toxicity was thereby broadened and a contagious interest in the biology of oxygen radicals was initiated. Superoxide radical (O_2^-) was soon seen to be a commonplace product of oxygen reduction and an important agent of oxygen toxicity, and the superoxide dismutases were recognized as the essential defence against the reactivities of this radical (McCord *et al.* 1971). Superoxide dismutases of three distinct types were isolated and characterized from a wide range of sources and the pertinent literature grew exponentially. Several reviews have appeared (Fridovich 1972, 1974, 1975, 1976, 1977) and the proceedings of the first international symposium devoted to this topic have been published (Michelson *et al.* 1977). We can now profitably consider a few specific questions. (1) How much O_2^- is made in any given cell type? (2) What is the effect of O_2^- on biological membranes? (3) How is the biosynthesis of superoxide dismutase controlled to meet the varying demand for protection against O_2^-? (4) How can we interpret the distribution of the three varieties of superoxide dismutases?

HOW MUCH O_2^-?

O_2^- can reduce cytochrome *c* and superoxide dismutase (SOD) can interfere. O_2^- production can thus be estimated from the rate of SOD-inhibitable reduction of cytochrome *c*. The reaction of O_2^- with cytochrome *c* generates molecular oxygen, whereas the dismutation of O_2^- generates H_2O_2. Hence, O_2^- production can also be measured in terms of the SOD-reversed inhibition of oxygen consumption by ferricytochrome *c*. When applied to milk xanthine oxidase, these methods revealed that a substantial fraction of the total oxygen reduction, dependent on pH and pO_2, proceeded by the monovalent pathway (Fridovich 1970). Azzi *et al.* (1975) similarly used acetylated cytochrome *c*, to avoid interference by NADH–cytochrome *c* reductase and by cytochrome *c* oxidase, and noted that mitochondrial membranes liberated 0.5 nmol O_2^- min^{-1} (mg protein)$^{-1}$. Cadenas *et al.* (1977) found that electron-transporting subassemblies, isolated from mitochondria, produced O_2^- at much greater rates. The NADH–coenzyme Q reductase generated 9.8 nmol O_2^- min^{-1} (mg protein)$^{-1}$, and the ubiquinol–cytochrome *c* reductase liberated 6.5 nmol O_2^- min^{-1} (mg protein)$^{-1}$.

Attempts to measure the rate of O_2^- production in whole cells, or even in intact subcellular organelles, are frustrated by the endogenous SOD. A potent, specific and membrane-permeable inhibitor of superoxide dismutases would solve this problem, but none has yet come to light. We approached this problem by isolating the single SOD of *Streptococcus faecalis* and using it as an antigen in the rabbit (Britton *et al.* 1978). The antibody so obtained was both precipitating and inhibitory and it was used to suppress the SOD activity in crude soluble extracts of *S. faecalis*. It could then be shown that 17% of the oxygen consumption by such extracts, when acting on NADH, was due to the production of O_2^-. We cannot conclude that this number applies to whole cells, since disruption might change the percentage of monovalent oxygen reduction. It is, nonetheless, a useful estimate and is the best we can do at the moment. It will be interesting to apply this method to cell-free extracts of other types of cells to see how much variability there is in the percentage of monovalent oxygen reduction.

SUPEROXIDE AND MEMBRANES

The very existence of enzymes which catalytically scavenge O_2^- is an indication of the potential cytotoxicity of this radical. How else can one explain the necessity of enzymes to speed what is spontaneously a rapid reaction? This was the basis for the early thinking about the biological consequences of the production of O_2^-. Since then data have been amassed which indicate that O_2^- is cytotoxic, directly or indirectly, and that superoxide dismutases protect against this lethality. Thus, enhanced intracellular levels of SOD protect *E. coli* B (Gregory & Fridovich 1973*b*), *E. coli* K12 (Hassan & Fridovich 1977*a*), *Saccharomyces cerevisiae* (Gregory *et al.* 1974) and rat lung (Crapo & Tierney 1974) against oxygen toxicity. Furthermore, SOD added to the suspending medium protects bacteria against the lethality of a photochemical (Lavelle *et al.* 1973; Gregory & Fridovich 1974) or an enzymic (Babior *et al.* 1975) source of O_2^- and against the oxygen-enhancement of radiation lethality (Misra & Fridovich 1976; Oberley *et al.* 1976; Niwa *et al.* 1977). Fetal calf myoblasts, in culture, were similarly protected against attack by O_2^- (Michelson & Buckingham 1974).

The protective actions of extracellular SOD, alluded to above, suggested that membranes were a target for attack by oxygen radicals. The large proportion of polyunsaturated lipids in membranes and the propensity of such lipids to undergo oxidation by a free-radical chain mechanism (Barber & Bernheim 1967) further suggested membranes as a point of departure in studying the specifics of O_2^- damage. We began with solutions of linolenic acid in buffered 10% dimethoxyethane (Kellogg & Fridovich 1975) and we used the aerobic action of xanthine oxidase on acetaldehyde as the source of O_2^- and of H_2O_2. This enzymic source of O_2^- did cause linolenate peroxidation, as measured by the increase in absorbance at 233 nm due to the formation of conjugated diene hydroperoxides. The production of lipid hydroperoxide was also followed, in terms of the appearance of alkyl hydroperoxide spots on thin-layer chromatograms. SOD *or* catalase prevented lipid peroxidation. Hence both O_2^- and H_2O_2 were essential reactants.

A cooperative interaction of O_2^- and H_2O_2 has been noted many times (Beauchamp & Fridovich 1970; Cohen 1977). The explanation usually offered has invoked a reaction (1) proposed by Haber & Weiss (1934) as part of their scheme for the catalytic decomposition of H_2O_2 by iron salts. The reality

$$O_2^- + H_2O_2 \rightarrow O_2 + OH^- + OH^\cdot \quad (1)$$

of this reaction has been questioned (Halliwell 1976; Czapski & Ilan 1978), but an efficient catalysis by iron complexes has been demonstrated (McCord & Day 1978) and it probably explains the apparent disagreements. We might then have supposed that peroxidation of linolenate by the xanthine oxidase system was actually due to $OH^\cdot$, generated from O_2^- plus H_2O_2. However, the reaction mixture was rich in an $OH^\cdot$ scavenger, dimethoxyethane, and other $OH^\cdot$ scavengers, such as butanol or mannitol, were without effect. In contrast, scavengers of singlet oxygen ($^1\Delta_gO_2$) such as β-carotene or diazabicyclooctane did inhibit linolenate peroxidation. This led to the proposal that O_2^- and H_2O_2 could interact to produce $^1\Delta_gO_2$ as well as $OH^\cdot$. The xanthine oxidase reaction was seen to convert 2,5-dimethylfuran into the same product as was produced by exposure to a photochemical source of $^1O_2$, presumably diacetylethylene, and SOD or catalase inhibited that conversion. It thus appears that the O_2^- plus H_2O_2, generated by the xanthine oxidase reaction, can give rise to $^1O_2$ or to something very much like it and that this species initiated lipid peroxidation.

The xanthine oxidase reaction also caused the oxidation of lecithin, dispersed with cholesterol and dicetyl phosphate, into micelles (Kellogg & Fridovich 1977). Once again SOD or catalase protected. Scavengers of $^1O_2$, such as histidine or 2,5-dimethylfuran also protected, thus implicating $^1O_2$ generated from O_2^- plus H_2O_2. Inhibition of lipid peroxidation by SOD, under a variety of circumstances, has been reported by several independent groups. Petkau & Chelack (1974) induced lipid peroxidation in phospholipid membranes by X-irradiation and noted 81–100% protection by SOD. Gutteridge (1977) catalysed the peroxidation of ox-brain phospholipids with iron and with copper salts and noted 50% protection by SOD. Goldstein & Weissman (1977) applied the xanthine oxidase reaction to chromate-loaded multilamellar vesicles. Release of chromate, which is indicative of damage to the vesicle membranes, was prevented by SOD or by catalase. Takahama & Nishimura (1976) obtained lipid peroxidation in illuminated chloroplast fragments, in the presence of benzyl viologen. Once again, SOD protected. Pederson & Aust (1975) and

Tyler (1975) used the xanthine oxidase system plus chelated iron and also noted lipid peroxidation inhibited by SOD.

Washed human erythrocytes, exposed to the xanthine oxidase reaction, suffer oxidation of haemoglobin and lysis (Kellogg & Fridovich 1977). SOD prevented lysis but not haemoglobin oxidation; catalase prevented both. Histidine and 2,5-dimethylfuran, added as scavengers of $^1O_2$, also protected against lysis. Direct effects on the stroma could be observed, in the absence of the obscuring oxidation and precipitation of haemoglobin, by performing the exposures with 10% CO in the gas phase. Carbonmonoxy-haemoglobin is stable to H_2O_2 and did not change during exposure to the xanthine oxidase reaction and with CO, lysis, preventable by SOD, catalase, 2mM-histidine or 10mM-mannitol, was observed. It is clear that an enzymic source of O_2^- and H_2O_2 does damage the cell membrane and that the proximal attacking agent is $OH^\bullet$ and $^1O_2$ or something with reactivities similar to these species.

One way to observe the effects of O_2^- and H_2O_2 on cell membranes, in the absence of complications due to cell components such as catalase, SOD and glutathione peroxidase, is to produce artificial cells. Washed erythrocyte ghosts can be resealed by shearing and solutes present during the shearing will be enclosed within the stromal vesicles so produced. Xanthine oxidase was sealed into such vesicles along with [^{14}C]sucrose. Exposure to a diffusible substrate of xanthine oxidase (i.e. acetaldehyde) resulted in lysis of vesicles, signalled by leakage of [^{14}C]sucrose. SOD prevented this lysis, but catalase did not (Lynch & Fridovich 1978). Preparation of the washed stroma and the final shearing was a lengthy procedure and some lipid peroxidation would be expected during the aerobic manipulations of the membranes. Furthermore, the first step in the process (i.e. haemolysis) eliminated the glutathione peroxidase reaction, so that lipid peroxides could accumulate in the stroma. Alkyl hydroperoxides have been reported to react rapidly with O_2^- to yield an alkoxy radical product (Peters & Foote 1976). Membranes loaded with lipid hydroperoxides could thus be attacked directly by O_2^- without the need for H_2O_2. Resealed stromal vesicles, enriched with lipid hydroperoxides by prior exposure to a photochemical source of $^1O_2$, were vastly more sensitive to attack by O_2^-. The failure of catalase to prevent lysis of the stromal vesicles can thus be understood.

CONTROLLING THE BIOSYNTHESIS OF SOD

Intracellular levels of SOD are increased by exposure to oxygen, and elevated SOD confers greater resistance towards oxygen toxicity. This has been seen with *Streptococcus faecalis* (Gregory & Fridovich 1973*a*), *E. coli* B (Gregory & Fridovich 1973*b*), *E. coli* K12 (Hassan & Fridovich 1977*a*), yeast (Gregory *et* *al.* 1974) and rat liver (Crapo & Tierney 1974). Induction by O_2 leaves open the possibility that the actual inducer is O_2 or O_2^- or some compound uniquely derived from O_2^-. Any circumstance which gave induction of SOD at fixed pO_2 would eliminate the possibility that O_2 was itself the inducer. Increases in the rate of production of O_2^-, at fixed pO_2, do cause induction of the manganese-containing SOD (MnSOD) of *E. coli*. This has been seen in three very different sets of conditions.

E. coli, in glucose-limited chemostat culture, increase their rate of respiration in response to increased rates of dilution. The cellular content of SOD increased in parallel with the rate of respiration (Hassan & Fridovich 1977*b*). Abrupt dilution did not result in immediate increase in growth rate. There was rather a lag in growth rate of about one hour, during which time the level of SOD was increased to that characteristic of the greater rate of respiration and of growth made possible by the abundance of medium. Increased levels of SOD, caused by increased dilution rates, at fixed pO_2, did impart increased resistance towards oxygen toxicity.

E. coli, growing in an aerobic complex medium, will preferentially use glucose as a source of energy. Their reliance on a fermentative metabolism is signalled by a decline in pH, as organic acids accumulate. When the glucose is exhausted, they begin to use the organic acids and other components of the medium, in an oxidative metabolism. This change can be followed in terms of a progressive increase in the pH of the medium. The SOD content of these cells declines during the fermentative phase of growth and then increases again during the respiratory phase; all these events happen at fixed pO_2 (Hassan & Fridovich 1977*c*). Classical catabolite repression, mediated by cyclic adenylic acid, was demonstrably not involved.

Methyl viologen (paraquat) can be reduced by electron transport mechanisms in the cell to yield a viologen radical, which rapidly reduces O_2 to O_2^-. In this way it shunts electrons from the normal electron transport pathway and increases cyanide-insensitive respiration and O_2^- production. When added to cultures of *E. coli*, at fixed pO_2, it causes a profound induction of SOD (Hassan & Fridovich 1977*d*). When applied in the absence of oxygen, it had no effect. We can conclude that O_2^- itself, or some unique product of O_2^-, is the inducer, rather than O_2.

DISTRIBUTION OF SUPEROXIDE DISMUTASES

The superoxide theory of oxygen toxicity leads one to suspect that obligate anaerobes might do without SOD, since they do not ordinarily face the problem of O_2^- production. An early survey of facultative, microaerotolerant and anaerobic bacteria did show the absence of SOD only in the latter category (McCord *et al.* 1971). Additional work has shown that the correlation between SOD and oxygen tolerance is not perfect. Thus Hewitt & Morris (1975) found some SOD in 14 out of 16 obligate anaerobes. Some of these, i.e. *Chlorobium thiosulfatophilum* and *Clostridium perfringens*, had about one third of the activity found in aerobically-grown *E. coli*; others, i.e. *Clostridium acetobutylicum* and *Clostridium pasteurianum*, had only trace amounts of activity. Tally *et al.* (1977) addressed the question of the oxygen tolerance of obligate anaerobes and classified 22 strains on the basis of this tolerance. Strains which were very sensitive to oxygen lethality had little or no SOD, whereas oxygen tolerant anaerobes did contain SOD. Hatchikian *et al.* (1977) studied several strains of *Desulfovibrio* and found SOD and catalase in some but not in others. They considered the possibility that SOD in anaerobes might be a recent acquisition, perhaps *via* plasmid transfer, rather than an ancient retained characteristic.

Another aspect of the distribution of superoxide dismutases no longer seems to be as clear-cut as it did previously and that is the differences in the types of SOD found in prokaryotes and eukaryotes. Thus prokaryotes had been found to contain superoxide dismutases based on manganese or iron — but not on copper. The copper–zinc enzyme was considered to be characteristic of eukaryotes. This distinction was perturbed by the discovery of a copper-zinc-SOD in *Photobacter leiognathi* (Puget & Michelson 1974). This seems to be a very unusual situation. Thus, superoxide dismutases have been isolated from several bacteria and this is the only case yet noted of a copper-zinc-SOD in prokaryotes. Moreover, a wide range of bacteria was recently surveyed for their contents of superoxide dismutases, with cyanide-sensitivity as an indicator of the copper-zinc-SOD and H_2O_2-sensitivity to detect iron-SOD (Britton *et al.* 1978). Some bacterial species contained FeSOD, others had MnSOD and still others had both; but none had a Cu-ZnSOD. Since *P. leiognathi* is a symbiont (having been isolated from a special gland of the pony fish), we can suppose that it obtained the genetic information coding for the Cu-ZnSOD from its host fish. This proposal has yet to be tested.

References

AZZI, A., MONTECUCCO, C. & RICHTER, C. (1975) The use of acetylated ferricytochrome *c* for the detection of superoxide radicals produced in biological membranes. *Biochem. Biophys. Res. Commun. 65*, 597–603

BABIOR, B., CURNUTTE, J. T. & KIPNES, R. S. (1975) Biological defense mechanisms. Evidence for the participation of superoxide in bacterial killing by xanthine oxidase. *J. Lab. Clin. Med. 85*, 235–244

BARBER, A. A. & BERNHEIM, F. (1967) Lipid peroxidation: its measurement, occurrence, and significance in animal tissues. *Adv. Gerontol. Res. 2*, 355–403

BEAUCHAMP, C. & FRIDOVICH, I. (1970) A mechanism for the production of ethylene from methional: the generation of hydroxyl radical by xanthine oxidase. *J. Biol. Chem. 245*, 4641–4646

BRITTON, L., MALINOWSKI, D. P. & FRIDOVICH, I. (1978) Superoxide dismutase and oxygen metabolism in *Streptococcus faecalis* and comparisons with other organisms. *J. Bacteriol. 134*, 229–236

CADENAS, E., BOVERIS, A., RAGAN, C. I. & STOPPANI, A. O. M. (1977) Production of superoxide radicals and hydrogen peroxide by NADH–ubiquinone reductase and ubiquinol–cytochrome *c* reductase from beef heart mitochondria. *Arch. Biochem. Biophys. 180*, 248–257

COHEN, G. (1977) In defense of Haber–Weiss, in *Superoxide and Superoxide Dismutases* (Michelson, A. M., McCord, J. M. & Fridovich, I., eds.) pp. 317–321, Academic Press, London & New York

CRAPO, J. D. & TIERNEY, D. F. (1974) Superoxide dismutase and pulmonary oxygen toxicity *Am. J. Physiol. 226*, 1401–1407

CZAPSKI, G. & ILAN, Y. A. (1978) On the generation of the hydroxylation agents from superoxide radical — can the Haber–Weiss reaction be the source of the $OH^{\cdot}$ radicals? *Photochem. Photobiol. 28*, 651–653

FRIDOVICH, I. (1970) Quantitative aspects of the production of superoxide anion radical by xanthine oxidase. *J. Biol. Chem. 245*, 4053–4057

FRIDOVICH, I. (1972) Superoxide radical and superoxide dismutase. *Acc. Chem. Res. 5*, 321–326

FRIDOVICH, I. (1974) Superoxide dismutases. *Adv. Enzymol. 41*, 35–97

FRIDOVICH, I. (1975) Superoxide dismutases. *Annu. Rev. Biochem. 44*, 147–159

FRIDOVICH, I. (1976) Superoxide dismutase and the chemistry of hydrogen peroxide, in *Free Radicals in Biology*, vol. I (Pryor, W. A., ed.), pp. 239–277, Academic Press, New York

FRIDOVICH, I. (1977) Oxygen is toxic. *BioScience 27*, 462–466

GOLDSTEIN, I. M. & WEISSMAN, G. (1977) Effects of the generation of superoxide anion on permeability of liposomes. *Biochem. Biophys. Res. Commun. 75*, 604–609

GREGORY, E. M. & FRIDOVICH, I. (1973*a*) Induction of superoxide dismutase by molecular oxygen. *J. Bacteriol. 114*, 543–548

GREGORY, E. M. & FRIDOVICH, I. (1973*b*) Oxygen toxicity and the superoxide dismutase. *J. Bacteriol. 114*, 1193–1197

GREGORY, E. M. & FRIDOVICH, I. (1974) Oxygen metabolism in *Lactobacillus plantarum*. *J. Bacteriol. 117*, 166–169

GREGORY, E. M., GOSCIN, S. A. & FRIDOVICH, I. (1974) Superoxide dismutase and oxygen toxicity in a eukaryote. *J. Bacteriol. 117*, 456–460

GUTTERIDGE, J. M. C. (1977) The protective action of superoxide dismutase on metal ion catalysed peroxidation of phospholipids. *Biochem. Biophys. Res. Commun. 77*, 379–386

HABER, F. & WEISS, J. (1934) The catalytic decomposition of hydrogen peroxide by iron salts. *Proc. R. Soc. Lond. A 147*, 332–351

HALLIWELL, B. (1976) An attempt to demonstrate a reaction between superoxide and hydrogen peroxide. *FEBS (Fed. Eur. Biochem. Soc.) Lett. 72*, 8–10

HASSAN, H. M. & FRIDOVICH, I. (1977*a*) Enzymatic defenses against the toxicity of oxygen and of streptonigrin in *Escherichia coli*. *J. Bacteriol. 129*, 1574–1583

HASSAN, H. M. & FRIDOVICH, I. (1977*b*) Physiological function of superoxide dismutase in glucose-limited chemostat cultures of *Escherichia coli*. *J. Bacteriol. 130*, 805–811

HASSAN, H. M. & FRIDOVICH, I. (1977*c*) Regulation of superoxide dismutase in *Escherichia coli:* glucose effect. *J. Bacteriol 132*, 505–510

HASSAN, H. M. & FRIDOVICH, I. (1977*d*) Regulation of the synthesis of superoxide dismutase in *Escherichia coli* induction by methyl viologen. *J. Biol. Chem. 252*, 7667–7672

HATCHIKIAN, C. E., LEGALL, J. & BELL, G. R. (1977) Significance of superoxide dismutase and catalase activities in the strict anaerobes, sulfate reducing bacteria, in *Superoxide and Superoxide Dismutases* (Michelson, A. M., McCord, J. M. & Fridovich, I., eds.), pp. 159–172, Academic Press, London & New York

HEWITT, J. & MORRIS, J. G. (1975) Superoxide dismutase in some obligately anaerobic bacteria. *FEBS (Fed. Eur. Biochem. Soc.) Lett. 50*, 315–318

KELLOGG, E. W. III & FRIDOVICH, I. (1975) Superoxide, hydrogen peroxide and singlet oxygen in lipid peroxidation by a xanthine oxidase system. *J. Biol. Chem. 250*, 8812–8817

KELLOGG, E. W. III & FRIDOVICH, I. (1977) Liposome oxidation and erythrocyte lysis by enzymically-generated superoxide and hydrogen peroxide. *J. Biol. Chem. 252*, 6721–6728

LAVELLE, F., MICHELSON, A. M. & DIMITREJEVIC, L. (1973) Biological protection by superoxide dismutase. *Biochem. Biophys. Res. Commun. 55*, 350–357

LYNCH, R. E. & FRIDOVICH, I. (1978) Effects of superoxide on the erythrocyte membrane. *J. Biol. Chem. 253*, 1838–1845

MCCORD, J. M. & DAY, E. D. JR. (1978) Superoxide-dependent production of hydroxyl radicals catalyzed by iron–EDTA complex. *FEBS (Fed. Eur. Biochem. Soc.) Lett. 86*, 139–142

MCCORD, J. M. & FRIDOVICH, I. (1968*a*) The reduction of cytochrome *c* by milk xanthine oxidase. *J. Biol. Chem. 243*, 5753–5760

MCCORD, J. M. & FRIDOVICH, I. (1968*b*) Superoxide dismutase — an enzymic function for erythrocuprein. *Fed. Proc. 28*, 346

MCCORD, J. M. & FRIDOVICH, I. (1969) Superoxide dismutase: an enzymic function for erythrocuprein (hemocuprein). *J. Biol. Chem. 244*, 6049–6055

MCCORD, J. M., KEELE, B. B. JR. & FRIDOVICH, I. (1971) An enzyme-based theory of obligate anaerobiosis: the physiological function of superoxide dismutase. *Proc. Natl. Acad. Sci. U.S.A. 68*, 1024–1027

MICHELSON, A. M. & BUCKINGHAM, M. E. (1974) Effect of superoxide radicals on myoblast growth and differentiation. *Biochem. Biophys. Res. Commun. 58*, 1079–1086

MICHELSON, A. M., MCCORD, J. M. & FRIDOVICH, I. (eds.) (1977) *Superoxide and Superoxide Dismutases*, Academic Press, London & New York

MISRA, H. P. & FRIDOVICH, I. (1976) Superoxide dismutase and the oxygen enhancement of radiation lethality. *Arch. Biochem. Biophys. 176*, 577–581

NIWA, T., YAMAGUCHI, H. & YANO, K. (1977) Radioprotection by superoxide dismutase: reduction of oxygen effect, in *Biochemical and Medical Aspects of Active Oxygen* (Hayaishi, O. & Asada, K., eds.), pp. 209–225, University of Tokyo Press, Tokyo

OBERLEY, L. W., LINDGREN, A. L., BAKER, S. A. & STEVENS, R. H. (1976) Superoxide ion as the cause of oxygen effect. *Radiat. Res. 68*, 320–328

PEDERSON, T. C. & AUST, S. D. (1975) The mechanism of liver microsomal lipid peroxidation. *Biochim. Biophys. Acta 385*, 232–241

PETERS, J. W. & FOOTE, C. S. (1976) Chemistry of superoxide ion. II. Reaction with hydroperoxides. *J. Am. Chem. Soc. 98*, 873–875

PETKAU, A. & CHELACK, W. S. (1974) Radioprotection of model phospholipid membranes by superoxide dismutase. *Fed. Proc. 33*, 1505

PUGET, K. & MICHELSON, A. M. (1974) Isolation of a new copper-containing superoxide dismutase, bacteriocuprein. *Biochem. Biophys. Res. Commun. 58*, 830–838

TAKAHAMA, U. & NISHIMURA, M. (1976) Effects of electron donor and acceptors, electron transfer mediators, and superoxide dismutase on lipid peroxidation in illuminated chloroplast fragments. *Plant Cell. Physiol. 17*, 111–118

TALLY, F. P., GOLDIN, B. R., JACOBUS, N. V. & GORBACH, S. L. (1977) Superoxide dismutase in anaerobic bacteria of clinical significance. *Infect. Immun. 16*, 20–25

TYLER, D. D. (1975) Role of superoxide radicals in the lipid peroxidation of intracellular membranes. *FEBS (Fed. Eur. Biochem. Soc.) Lett. 51*, 180–183

Discussion

Reiter: Recently we investigated the SOD levels in an enteropathogenic strain of *E. coli* of human serotype (0111) and bovine serotype (0101). Culturing the organisms in TSYB increased the SOD level from 7.2 u/mg to 22.0 u/mg in 0101 but no increase occurred in 0111 (50/mg). Both organisms are susceptible to the bactericidal activity of the xanthine oxidase–xanthine system which is claimed to generate O_2^-. However, the organisms containing the elevated levels of SOD were not killed but only inhibited (bacteriostasis). The addition of catalase reversed the bactericidal effect of xanthine oxidase–xanthine (unpublished results). When the level of SOD increases, does the catalase increase too?

Fridovich: Not always; it depends on the organisms. We have induced SOD in *E. coli* by changing the rate of growth in a chemostat. In those conditions (glucose-limited chemostat culture) when the SOD level rose, neither the catalase nor the peroxidase level rose. Different strains will show profound differences. Although we have not investigated different strains of *E. coli*, we have looked at different microorganisms and bacteria: some have an iron-SOD, others a manganese-SOD and others have both; some are induced, some are not. One should use as a control an organism that does not show induction and demonstrate that it does not gain resistance, as you have done. Did you assay that organism for catalase?

Reiter: No.

Fridovich: You should, to check whether it can protect itself by virtue of eliminating the hydrogen peroxide.

Reiter: But we have good evidence that the H_2O_2 is not bactericidal. We use glucose oxidase–glucose to generate H_2O_2 which activates the lactoperoxidase–thiocyanate system. The lactoperoxidase oxidized in the presence of H_2O_2 and CNS^-, and free H_2O_2 can only be detected after all the CNS^- is oxidized. H_2O_2 combines preferentially with lactoperoxidase and neither the catalase (nor, for instance, *E. coli*) nor the catalase of the milk interferes with the lactoperoxidase–CNS^-–H_2O_2 system (Reiter *et al.* 1976).

Klebanoff: In your earlier studies, Professor Fridovich, of the distribution of SODs in *E. coli*, the manganese enzyme was suggested as protective against endogenous superoxide. Is the induction by paraquat of the manganese enzyme as protective against exogenous superoxide as against endogenous superoxide?

Fridovich: We had previously thought, on the basis of osmotic shock (Heppel's procedure), that the iron SOD was located in the periplasmic space. Britton & I (1977) have since demonstrated that this is not the case. The iron enzyme and the manganese enzyme both appear to be present in the matrix space of *E. coli*. That finding immediately forces a re-evaluation of whether the two enzymes are directed toward different pools of O_2^-. Since they are both in the matrix space one has to assume that both function against endogenously generated O_2^-. I now take the view that a facultative organism, like *E. coli*, must always maintain a minimal defence against O_2^-. The iron enzyme is constitutive and seems to be a minimal back-up defence, and the manganese enzyme is under repression control, thus allowing fine tuning to the needs of the moment.

Superoxide may cross the exterior of the cell easily. It crosses membranes of vesicles formed from erythrocyte stroma by traversing the anion channels, which function normally in the chloride–hydrogen carbonate exchange. The stilbene disulphonates (DIDS and SITS), which plug these anion channels and inhibit the Cl^-/HCO_3^- exchange, also prevent O_2^- permeation.

Reiter: I doubt whether the bacteria are so permeable because the O_2^- would have to traverse the cell wall of the Gram-positive organisms and outer membrane of the Gram-negative organisms before presumably damaging the inner membrane (Marshall & Reiter 1976).

Smith: When *E. coli* are maintained on glucose or glucose-free medium containing paraquat, what provides the reducing potential for the one-electron reduction of paraquat to its free radical? In mammalian systems *in vitro*, cytochrome *c* reductase has been implicated.

Fridovich: Evidently an enzyme is present but which, I don't know. The crude extract of cells grown anaerobically reduces paraquat extremely rapidly when NADH is present.

Smith: To what extent are you prepared to extrapolate your results obtained with *E. coli* to mammalian species? I ask because Crapo & Tierney (1974) described the protective effect of pre-exposing rats to 85% oxygen on subsequent exposure to a 100% oxygen atmosphere.

Fridovich: I have no data on paraquat in mammalian cells but expect that our results will be relevant. Dr Crapo has been looking at induction by oxygen but not yet at induction by paraquat. It should be done in tissue culture.

Winterhalter: If one progressively lowers the iron concentration but maintains the amounts of manganese and of O_2^-, will the bacterium shift from the iron enzyme, which it makes normally, to the manganese enzyme?

Fridovich: Anaerobically, the iron enzyme is the only one made in this *E. coli.*

Winterhalter: And if there were no iron present in the medium?

Fridovich: We have not done that. I guess that it would not make the manganese enzyme anaerobically because no O_2^- is then made and O_2^- is the inducer, even if less iron enzyme were being made by virtue of a nutritional limitation.

Winterhalter: But O_2^- could still be made by other sources, such as metabolic steps; sulphate-reducing bacteria make it.

Fridovich: We have not tried because such stringent iron limitation is difficult.

Michelson: What metal is present in the hybrid SOD (for details, see Dougherty *et al.* 1978)?

Fridovich: Iron.

Hill: Does *E. coli* synthesize the apo-manganese protein in manganese-deficient media?

Fridovich: I don't know.

Cohen: Can the manganese form of the enzyme in mammalian cells (the mitochondrial form) be induced by oxygen?

Fridovich: Although the manganese form is present in mitochondria in some tissues, it can also be present in the cytoplasm. This depends on the species. Thus human and baboon livers have large amounts of MnSOD in the cytosol whereas chicken livers, for example, contain little in the cytoplasm. Oxygen-induction studies on mammalian lung and lung cells and slices in culture confirm that the manganese rather than the copper-zinc enzyme is responsible for the observed increase in total activity (Stevens & Autor 1977).

Cohen: Has nervous system or muscle been investigated?

Fridovich: No. Maybe only the lung shows induction because other tissues are well insulated from changes in pO_2. The induction seen in lungs is modest compared to the 10-fold increases we noted. For instance, Crapo & Tierney (1974) refer to 50% increases as the total change in mammalian lungs.

Winterhalter: But the effect seems to be enormous.

Fridovich: Yes; and that suggests that we have just the amount of protection that is needed and that it is rather easily overwhelmed.

Michelson: The cell appears to be extremely intelligent because under oxygen stress it shifts biosynthesis to the manganese enzyme which is more active than the iron enzyme.

Fridovich: The rate constants of manganese enzyme are not much faster than those for the iron enzyme — it is a question of quantity rather than turnover of this specific isozyme.

Smith: Rats exposed to 90% oxygen for seven days induce not only superoxide dismutase but also glutathione peroxidase, glutathione reductase and glucose-6-phosphate dehydrogenase in their lungs (Kimball *et al.* 1976). These enzymes may well be important in the defence against oxygen toxicity.

We have also confirmed the observation of Brasher & DeAtley (1972) who exposed rats to 10% oxygen for eight days before returning them to 100% oxygen atmosphere and observed a marked protective effect.

Fridovich: Sjostrom & Crapo (1978) have repeated it, too, but found that superoxide dismutase was induced by 10% oxygen. The rate of production of O_2^- may be higher with 10% oxygen than with 20% oxygen; the relationship between the pO_2 and the rate of O_2^- production may not be monotonic. Such a situation, with an optimum rate of O_2^- production, has been seen in the autoxidation of FMN and reduced flavodoxin (Misra & Fridovich 1972).

Goldstein: The lung is not homogeneous; it is not *E. coli.* Almost any oxidant action on lungs causes type 1 cells to be replaced by type 2 cells. Amongst other things, type 2 cells contain less phosphatidylserine and phosphatidylethanolamine than type 1 cells do and so, maybe, they are less likely to be peroxidized. Could anything else besides superoxide dismutase be induced in *E. coli* in your experiments?

Fridovich: I have never thought that only superoxide dismutase was induced; superoxide dismutase seems to be one part of the multifaceted defence system. Obviously, glutathione peroxidase and catalase are also important. In our view the best defence against dangerous species like the hydroxyl radical is prevention of their production; to do so one must remove both O_2^- and peroxide, and enzymes which scavenge both these intermediates of oxygen reduction are important.

Winterhalter: Lipid peroxidation may not be the only reason by the lysis of vesicles. Other phenomena are occurring. W. Birchmeier & C. Richter (personal communication) have filled vesicles with a superoxide-generating system and demonstrated the formation of S–S bridges between molecules of the contractile apparatus of the cell membrane.

Fridovich: I do not mean to imply that lipid peroxidation is the only cause of that lysis; the underlying chemistry remains to be investigated.

Stern: Using dihydroxyfumaric acid to generate superoxide anions in the presence of lactoperoxidase we observed no lipid peroxidation (Goldberg & Stern 1977) at least as measured by malonaldehyde concentrations in red cells.

Flohé: Absence of malonaldehyde does not necessarily rule out lipid peroxidation. Treatment of erythrocytes with dialuric acid and divicine (2,6-diamino-4,5-pyrimidinediol) led to increased concentrations of malonaldehyde for only a few minutes, but later the cells appeared normal. Nevertheless these cells became leaky considerably earlier than untreated cells (Flohé *et al.* 1971); in other words, lipid peroxidation for only a few minutes or seconds caused long-lasting damage.

Fridovich: Similarly, we exposed whole erythrocytes, for example, to the acetaldehyde–xanthine oxidase reaction. That reaction can be quenched by adding an inhibitor of xanthine oxidase, such as allopurinol, but nevertheless the cells subsequently lyse after an hour or two. The damage having been done during brief exposure to the xanthine oxidase reaction leads to lysis some hours later.

Winterhalter: The erythrocytes that one sees in the circulation of thalassaemic patients are obviously the ones that have survived the challenge by the superoxide and the ones that have been damaged most inside the bone marrow are never observed in circulation because they die inside the marrow.

Michelson: Does cyclic AMP effect the induction of superoxide dismutase?

Fridovich: No; there is what looks like a catabolite repression, that is a glucose effect on MnSOD synthesis. But it is due to the primarily anaerobic metabolism in the presence of glucose and thus to production of less O_2^-. Cyclic AMP has no effect but it had the expected effect on an enzyme subject to catabolite repression (Hassan & Fridovich 1977).

Chvapil: Studying the mechanisms of retrolental fibroplasia we postulated that exposure of prematurely born babies to high concentrations of oxygen, by triggering bleeding into the vitreous body, will induce the formation of active forms of oxygen, lipid peroxidation and possibly depolymerization of hyaluronic acid or hyaloprotein of the vitreous (Hiramitsu *et al.* 1976). Dr Janice Burke (in our Department of Surgery, Ophthalmology section) tested this hypothesis both *in vitro* as well as *in vivo.*

In *in vitro* studies, Dr Burke generated O_2^- in the vitreous milieu from the xanthine–xanthine oxidase reaction. As shown in Fig. 1, vitreal viscosity was reduced in 15–30 min at 37 °C. The same drop in the viscosity resulted from injecting hyaluronidase into the test vitreous. Addition of SOD into this system did not, however, inhibit the decrease in the vitreous viscosity. The dismutase may not be as active in the presence of vitreous, but the depolymerization of the vitreous is reminiscent of a similar experiment (McCord 1974) with synovial fluid.

In vivo studies showed that endogenous SOD activities change dramatically

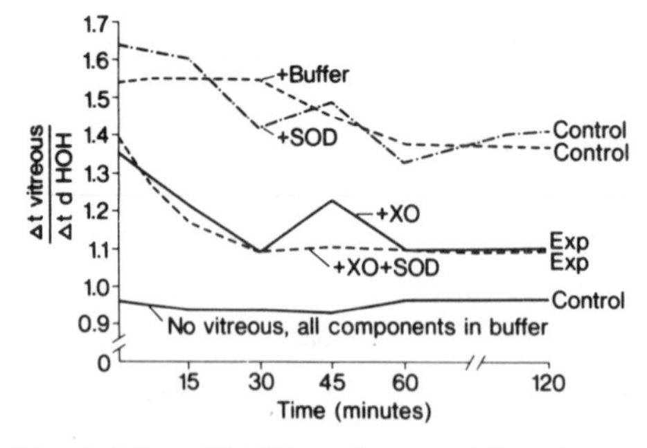

FIG. 1 (Chvapil). Effect of superoxide anion on the viscosity of bovine vitreous. Viscosity was measured as the time of flow (Δt vitreous) of 0.2 ml vitreous from a 1 ml syringe through a 22-gauge needle (silicon-treated) with respect to the rate of the flow of distilled water through the same needle (Δt d HOH).

after vitreal injury (i.e. experimental vitreous haemorrhage induced by the injection of 0.05 ml sterile homogenous haemoglobin on consecutive days 1 to 5) (Fig. 2). The change correlates with an increase in vitreal cell number predominantly due to the immigration of cells, possibly of haematogenous origin. There is a small contribution from cell proliferation, predominantly of resident vitreal cells. Although normal concentrations of vitreous SOD range between 500 and 1800 ng SOD/ml vitreous, after injection of haemoglobin the concentration of SOD increases to more than 20 000 ng/ml (see Fig. 2). When exogenous SOD is administered into the vitreous at the time of the haemoglobin injection (0.1 or 0.5 mg SOD), the cellular response (immigration) is slightly greater than with haemoglobin alone.

When SOD is administered alone, the cell response is similar to, or slightly greater than, that to saline alone. But when SOD is administered after haemoglobin (i.e. to an 'injured' or 'altered' vitreous), a pronounced inflammatory response is generated. Vitreal viscosity drops as soon as 24 h after the haemoglobin injection. The drop correlates in time with increases in endogenous vitreal concentrations of SOD. Thus, if this event is mediated by O_2^-, there is insufficient SOD present to prevent vitreal liquefaction. (It is assumed that the depolymerization of hyaluronate decreases the viscosity and that the depolymerization is mediated by free radicals.) Dr Burke could not prove, however, that SOD administered at the time of the haemoglobin injection prevents liquefaction. This may not necessarily indicate that superoxide is not involved, since the vitreous may contain an inhibitor of SOD activity. This conclusion seems to be favoured by the finding that SOD activity in the presence of vitreous is significantly lower than in the presence of buffer, as measured by the method of Beauchamp & Fridovich (1970), in which the activity is measured as the inhibition of superoxide-mediated reduction of NBT where the superoxide is generated by the xanthine–xanthine oxidase reaction.

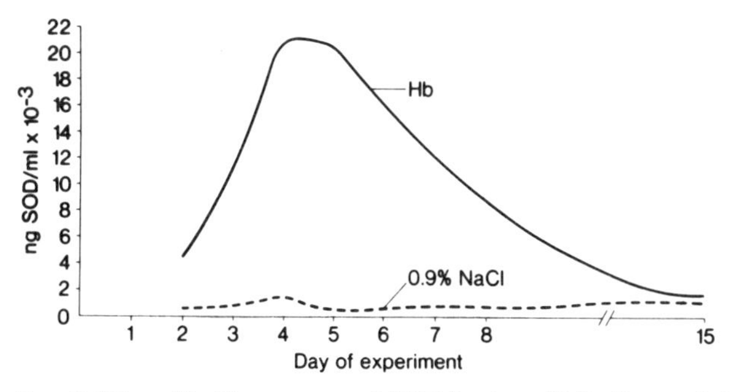

FIG. 2 (Chvapil). The amount of SOD in the rabbit vitreous injected with saline or haemoglobin.

Fridovich: There is no inhibitor as far as I know. The failure of exogenous superoxide dismutase to prevent the depolymerization directly contradicts Dr McCord's results and suggests an artifactual explanation.

McCord: I cannot explain it. *In vivo* if one is not careful to use pyrogen-free protein preparations, the injection itself may provoke the inflammation. A possible explanation for the inability of SOD to prevent liquefaction *in vivo* is that it is caused by leucocytic proteinases attacking the protein core of the hyaluronate complexes.

Chvapil: Injection of superoxide dismutase at the same time as injection of haemoglobin caused no inflammation.

Stern: Professor Fridovich, how did you regulate the concentrations of superoxide dismutase in the bacteria?

Fridovich: The lowest concentrations were obtained by growth in a glucose-minimal medium; when the bacteria use glucose primarily by the fermentative pathway the level of SOD is low. Bacteria with intermediate levels were grown aerobically in a trypticase soy-yeast extract; once they exhaust the glucose in that medium and start using the other components by oxidative pathways the level of SOD goes up substantially. The highest level was achieved by induction with pyocyanine (paraquat is only one of a family of compounds that can behave similarly and cause enhanced production of O_2^- within *E. coli*).

Klebanoff: Superoxide dismutase has been claimed as a quencher of singlet oxygen. Would you care to put that claim to rest?

Fridovich: That can be categorically stated to be nonsense! The data put forward for that claim have been refuted (Goda *et al.* 1974). For instance, Lee *et al.* used a xenon lamp directly to generate singlet oxygen by absorption of a photon. Initially they thought that SOD catalysed the quenching of $^1O_2$, but later established that a mixture of amino acids would do the same thing (Matheson *et al.* 1975). They had been using so much SOD that $^1O_2$ was reacting with the imidazole in the protein. We found that SOD had no effect on Rose-Bengal-sensitized photo-oxidation, known to be mediated by singlet oxygen (Hodgson & Fridovich 1974).

Michelson: What pleased me in the Lee *et al.* experiment was that denatured SOD performed better than native SOD — presumably because more functional groups are exposed in the denatured protein.

Fridovich: One concluding point is that in calculations on rate constants and relative concentrations the rate of quenching of singlet oxygen by water is such that the rate of reaction of $^1O_2$ with the enzyme would have to be about 10^{15} l mol^{-1} s^{-1} to allow for the enzyme to compete with water. That, of course, is orders of magnitude faster than diffusion; so not only is it wrong, it is impossible that SOD catalytically scavenges $^1O_2$.

References

BEAUCHAMP, C. & FRIDOVICH, I. (1970) A mechanism for the production of ethylene from methional. The generation of the hydroxyl radical by xanthine oxidase. *J. Biol. Chem. 245*, 4641–4646

BRASHER, R. E. & DEATLEY, R. E. (1972) Decreased pulmonary oxygen toxicity by pretreatment with hypoxia. *Arch. Environ. Health 24*, 77–81

BRITTON, L. & FRIDOVICH, I. (1977) Intracellular localization of the superoxide dismutases of *Escherichia coli*: a reevaluation. *J. Bacteriol. 131*, 815–820

CRAPO, J. & TIERNEY, D. (1974) Superoxide dismutase and pulmonary oxygen toxicity. *Am. J. Physiol. 226*, 1401–1407

DOUGHERTY, H., SADOWSKI, S. & BAKER, E. (1978). *J. Biol. Chem. 253*, 5220–5223

FLOHÉ, L., NIEBCH, G. & REIBER, H. (1971) Zur Wirkung von Divicin in menschlichen Erythrocyten. *Z. Klin. Chem. Klin. Biochem. 9*, 431–437

GODA, K., KIMURA, T., THAYER, A. L., KEES, K. & SCHAAP, A. P. (1974) Singlet molecular oxygen in biological systems: non-quenching of singlet oxygen-mediated chemiluminescence by superoxide dismutase. *Biochem. Biophys. Res. Commun. 58*, 660–666

GOLDBERG, B. & STERN, A. (1977) The role of the superoxide anion as a toxic species in the erythrocyte. *Arch. Biochem. Biophys. 178*, 218–225

HASSAN, H. M. & FRIDOVICH, I. (1977) Physiological function of superoxide dismutase in glucose-limited chemostat cultures of *Escherichia coli*. *J. Bacteriol. 130*, 805–811

HIRAMITSU, T., HASEGAWA, Y., HIRATA, K., NISHIGAKI, I. & YAGI, K. (1976) Formation of lipoperoxide in the retina of rabbit exposed to high concentration of oxygen. *Experientia 32*, 622–623

HODGSON, E. K. & FRIDOVICH, I. (1974) The production of superoxide radical during the decomposition of potassium peroxochromate(V). *Biochemistry 13*, 3811–3815

KIMBALL, R. E., REDDY, K., PEIRCE, T. H., SCHWARTZ, L. W., MUSTAFA, M. G. & CROSS, C. E. (1976) Oxygen toxicity: augmentation of antioxidant defense mechanisms in rat lung. *Am. J. Physiol 230*, 1425–1431

MARSHALL, V. M. E. & REITER, B. (1976) The effect of the lactoperoxidase-thiocyanate-hydrogen peroxide system on *Escherichia coli*. *J. Gen. Microbiol. 3*, 189

MATHESON, I. B. C., ETHERIDGE, R. D., KRATOWICH, N. R. & LEE, J. (1975) The quenching of singlet oxygen by amino acids and proteins. *Photochem. Photobiol. 21*, 165–171

MCCORD, J. M. (1974) Free radicals and inflammation: protection of synovial fluid by superoxide dismutase. *Science (Wash. D.C.) 185*, 529–531

MISRA, H. P. & FRIDOVICH, I. (1972) The role of superoxide anion in the autoxidation of epinephrine and a simple assay for superoxide dismutase. *J. Biol. Chem. 247*, 188–192

REITER, B., MARSHALL, V. M. E., BJÖRK, L. & ROSEN, C. G. (1976) Nonspecific bactericidal activity of the lactoperoxidase–thiocyanate–hydrogen peroxide system of milk against *Escherichia coli* and some Gram-negative pathogens. *Infect. Immun. 13*, 800–807

STEVENS, J. B. & AUTOR, A. P. (1977) Induction of superoxide dismutase by oxygen in neonatal rat lung. *J. Biol. Chem. 252*, 3509–3514

SJOSTROM, K. & CRAPO, J. (1978). *The Physiologist 21*, 111

Glutathione peroxidase: fact and fiction

LEOPOLD FLOHÉ

Grünenthal GmbH, Aachen, West Germany

Abstract The present knowledge of glutathione (GSH) peroxidase is briefly reviewed: GSH peroxidase has a molecular weight of about 85 000, consists of four apparently-identical subunits and contains four g atom of selenium/mol. The enzyme-bound selenium can undergo a substrate-induced redox change and is obviously essential for activity. In accordance with the assumption that a selenol group is reversibly oxidized during catalysis, ping-pong kinetics are observed. Limiting maximum velocities and Michaelis constants, indicating the formation of an enzyme–substrate complex, are not detectable. The enzyme is highly specific for GSH but reacts with many hydroperoxides.

It can be deduced from the kinetic analysis of GSH peroxidase that in physiological conditions removal of hydroperoxide is largely independent of fluctuations in the cellular concentration of GSH. However, the system will abruptly collapse if the rate of hydroperoxide formation exceeds that of regeneration of GSH. By these considerations, the pathophysiological manifestation of disorders in GSH metabolism and pentose-phosphate shunt may be explained.

With regard to its low specificity for hydroperoxides, GSH peroxidase could be involved in various metabolic events such as H_2O_2 removal in compartments low in catalase, hydroperoxide-mediated mutagenesis, protection of unsaturated lipids in biomembranes, prostaglandin biosynthesis, and regulation of prostacyclin formation.

Discovered in 1957 (Mills 1957), glutathione peroxidase (glutathione: H_2O_2 oxido-reductase; EC 1.11.1.9) has now come of age. In the meantime, progress in this field has become slow, and a time of re-evaluation, reinterpretation and review has begun (Ganther *et al.* 1976; Flohé *et al.* 1976; Flohé 1976). In this paper I shall briefly summarize the essential characteristics of the enzyme which have been established during the last two decades and touch on some obviously unresolved problems which are presently under investigation or have not yet been approached. Finally, I shall examine various possible roles of GSH peroxidase in biological systems to avoid the impression that an enzyme no longer deserves our interest once the data required for an enzymological handbook have been compiled. These considerations will reveal that the function of GSH peroxidase might be relevant to both acute and chronic alterations of mammalian tissue.

THE ENZYMOLOGICAL DATA

Type of reaction and substrate specificity

Glutathione (GSH) peroxidase catalyses the reduction of various hydroperoxides to alcohols. The only efficient physiological reductant appears to be GSH. The hydroperoxides accepted as substrates include H_2O_2, ethyl hydroperoxide, t-butyl hydroperoxide, cumene hydroperoxide, thymine hydroperoxide, hydroperoxides of unsaturated fatty acids and the corresponding esters, hydroperoxides of steroids and nucleic acids, and prostaglandin G_2, the primary intermediate of prostaglandin biosynthesis (for review see Flohé & Günzler 1974; Flohé *et al.* 1976). It is thus tempting to conclude that the enzyme reacts unspecifically with every hydroperoxide unless the hydroperoxy group is not sterically accessible, as in 25-hydroperoxycholesterol. Dialkyl peroxides are obviously not metabolized by GSH peroxidase.

Specificity studies with various SH compounds indicate that both carboxylic groups of the GSH molecule contribute to substrate binding, since the activity decreases considerably if the γ-glutamyl residue of GSH is substituted by a β-aspartyl or *N*-acetyl residue or if the glycine residue is replaced by a methoxy or amide group (Table 1; Flohé *et al.* 1971*b*).

TABLE 1

Some characteristic examples of RSH oxidation by H_2O_2 catalysed by GSH peroxidase. The results were obtained at H_2O_2 concentrations (1 mmol/l) yielding apparent maximum velocity. Data are taken from Flohé *et al.* (1971*b*)

Glutathione analogue	*Catalytic activity (%)*
(*a*) Variations of the γ-glutamyl residue	
γ-Glu-Cys-Gly (GSH)	100.0
β-Asp-Cys-Gly	7.6
Cys-Gly	6.8
N-Ac-Cys-Gly	2.7
(*b*) Variations of the glycine residue	
γ-Glu-Cys-Gly	100.0
γ-Glu-Cys-OMe	26.0
γ-Glu-Cys-NH_2	1.4

Molecular weights and subunits

The molecular weights of GSH peroxidases derived from different sources average around 85 000. The enzyme consists of four subunits which are not covalently linked and appear to be identical or very similar according to the following criteria: SDS electrophoresis (Flohé *et al.* 1971*a*), number of tryptic peptides, number of active sites (Flohé *et al.* 1976) and preliminary X-ray crystallography (Ladenstein & Epp 1977).

Chemical composition

Several investigators have reported on the amino acid composition of GSH peroxidase. Owing to special circumstances (see below), the content of cysteine residues is uncertain. The remaining data do not show any characteristic deviations from the amino acid pattern of other proteins (Günzler 1974; Nakamura *et al.* 1974). The amino acid sequence is still unknown. The presence of coenzymes typical for other peroxidases such as haems or possibly flavins can be excluded (Flohé *et al.* 1971*c*). Instead, 4 g atom of selenium/mol were found in GSH peroxidase isolated from bovine (Flohé *et al.* 1973), ovine (Oh *et al.* 1974) and human blood (Awasthi *et al.* 1975) and rat liver (Nakamura *et al.* 1974). Wendel *et al.* (1975) were able by means of ESCA spectroscopy to demonstrate that enzyme-bound selenium can change its redox state when the natural substrates are added. The ESCA spectra support the assumption that the selenium of the GSH-reduced enzyme is present in oxidation state 0 or -2 and shifts to a higher state of oxidation on addition of H_2O_2. This finding is in accordance with the recent report by Forstrom *et al.* (1978) suggesting selenocysteine as the active site of GSH peroxidase. The ratio of cysteine to selenocysteine has to be re-examined.

Kinetics

The kinetics of GSH peroxidase can be described by an initial rate equation (1) analogous to that developed for other peroxidases:

$$\frac{d[\mathrm{A}]}{\mathrm{d}t} = v = [E_0]\left(\frac{1}{k_{+1}[\mathrm{A}]} + \frac{1}{k_{+2}[\mathrm{B}]} + \frac{1}{k_{+3}[\mathrm{B}]}\right)^{-1} \tag{1}$$

where [A] is the concentration of hydroperoxide; [B] is the concentration of reductant, in this case GSH; k_{+1} is the constant for the reaction of reduced enzyme with the hydroperoxide; k_{+2} and k_{+3} are the rate constants of the oxidized enzyme species with GSH; and $[E_0]$ is the total enzyme concentration. By means of the definitions (2) and (3) a transformation into equation (4), the Dalziel equation (1957), is possible. This equation, which is satisfied by the experimental

$$\phi_1 = \frac{1}{k_{+1}} \tag{2}$$

$$\phi_2 = \frac{1}{k_{+2}} + \frac{1}{k_{+3}} \tag{3}$$

$$\frac{[E_0]}{v} = \frac{\phi_1}{[A]} + \frac{\phi_2}{[B]} \tag{4}$$

data for GSH peroxidase from both bovine blood (Flohé *et al.* 1972; Günzler *et al.* 1972) and rat liver (Chiu *et al.* 1975), describes a ping-pong mechanism without kinetically-relevant central complexes and is best interpreted by assuming that the enzyme goes through consecutive steps of oxidation and reduction during the catalytic cycle. The kinetic parameters obtained with the bovine enzyme are summarized in Table 2, p. 100.

HYPOTHETICAL MECHANISM OF ACTION

The present view of the mechanism of GSH peroxidase is based on the following facts:

(*a*) lack of specificity with respect to the hydroperoxide (Flohé *et al.* 1976);

(*b*) high specificity for GSH (Flohé *et al.* 1971*b*);

(*c*) selective inhibition by iodoacetate of the substrate-reduced enzyme only (Flohé & Günzler 1974);

(*d*) increased binding of *p*-chloromercuribenzoate by the enzyme on reduction by GSH (Flohé *et al.* 1971*c*);

(*e*) the ping-pong kinetics (Flohé *et al.* 1972; Günzler *et al.* 1972; Chiu *et al.* 1975);

(*f*) the identification of a selenol as functional group (Forstrom *et al.* 1978);

(*g*) the reactivity of the enzyme-bound selenium with the physiological substrates (Wendel *et al.* 1975).

The catalytic cycle can be formulated as follows (Fig. 1): a reduced form of the enzyme (E) reacts with a hydroperoxide in an uncomplicated bimolecular reaction. Neither the kinetics nor substrate-specificity studies indicate that an enzyme–substrate complex is involved in this catalytic step. The oxidized enzyme (F) then forms a complex with GSH. This complex, however, is rapidly transformed into a new intermediate (G) in an intramolecular reaction. Complex formation in the second step is supported by the high specificity of the enzyme for GSH, but not by the kinetic analysis. The intriguing observation that limiting Michaelis constants or limiting maximum velocities for GSH cannot be achieved despite high specificity, however, is compatible with the assumption

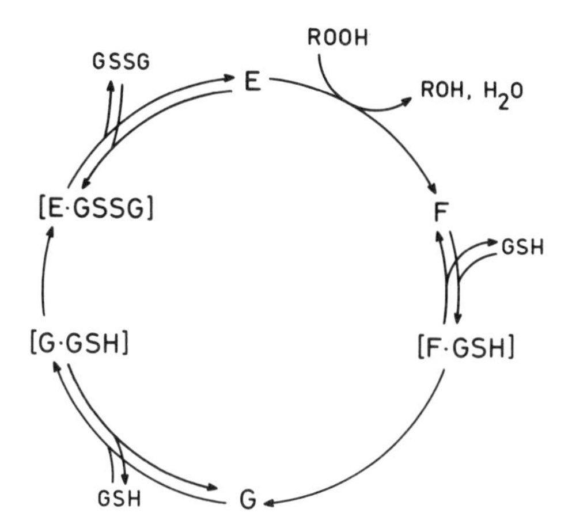

FIG. 1. Schematic representation of the GSH peroxidase reaction: E, reduced enzyme; F, oxidized enzyme; G, intermediate (see text).

that the formation of the complex (F·GSH) is much slower than the intramolecular transformation into the intermediate G. In the last steps the second molecule of GSH regenerates E from the intermediate G, whereby a process analogous to step 2 has to be assumed.

As to the chemical nature of the three enzyme forms, E, F and G, the following ideas can be proposed: E most likely represents the enzyme containing a largely dissociated selenol function of the selenocysteine residue (E—Se^-). The selective inhibition of E by iodoacetate, which does not show the pH dependence typical for the reaction with SH groups (Günzler 1974), strongly supports this assumption. The results of the ESCA studies (Wendel *et al.* 1975) and the binding of an additional 4 mol of *p*-chloromercuribenzoate to the GSH-reduced enzyme (Flohé *et al.* 1971*c*) also point to a selenol function. Finally, preliminary X-ray crystallography has revealed that the enzyme-bound selenium is exposed at the surface of the molecule (R. Ladenstein, personal communication) and is thus in an excellent position to react with various hydroperoxides.

In F the selenium may be oxidized to a selenenic acid derivative (E—SeOH) (Ganther *et al.* 1976). However, a mixed selenosulphide (RSe·SR′) cannot be excluded at the moment as an alternative. Both forms would readily react with thiols to generate finally disulphides and selenols. The most likely intermediate in both types of reactions would be a mixed selenosulphide consisting of the enzyme-bound selenium and GSH (E—Se·SG). Such a compound may represent the kinetic entity G. The present X-ray crystallographic studies (at the Max-Planck-Institut für Biochemie in Martinsried) will have to be completed before the reaction mechanism can be described more precisely.

TABLE 2

Kinetic constants of GSH peroxidase of bovine erythrocytes. For definition of ϕ_1, ϕ_2 and k_{+1}, see equations (1)–(4) and Fig. 1. Data are taken from Flohé *et al.* (1972) and Günzler *et al.* (1972)

Substrate	*Buffer system*	*pH*	$\phi_1/10^{-8} mol\ s\ l^{-1}$	$\phi_2/10^{-6}\ mol\ s\ l^{-1}$	$k_{+1}/10^7\ l\ mol^{-1}\ s^{-1}$
Hydrogen peroxide	0.05M-Potassium phosphate	7.0	0.56	1.27	17.86
Hydrogen peroxide	0.25M-MOPS	7.7	0.94	0.83	10.6
Hydrogen peroxide	0.25M-MOPS	6.7	1.70	2.19	5.88
Ethyl hydroperoxide	0.25M-MOPS	6.7	3.3	2.24	3.09
Cumene hydroperoxide	0.25M-MOPS	6.7	7.8	2.24	1.28
t-Butyl hydroperoxide	0.25M-MOPS	6.7	13.5	2.24	0.75

ESTABLISHED FUNCTIONS OF GSH PEROXIDASE

The definitely established functions of GSH peroxidase in living systems are few. From experiments of nature such as genetic disorders we know that human red blood cells deficient in GSH peroxidase are highly susceptible to pro-oxidative drug metabolites or xenobiotics. GSH peroxidase deficiency results in a clinical condition very similar to favism, i.e. G6PDH deficiency (Necheles 1974). Similarly, rat erythrocytes made deficient in GSH peroxidase by means of a diet low in selenium are prone to peroxide-induced haemolysis (Rotruck *et al.* 1972). These observations prove that GSH peroxidase contributes essentially to the integrity of the red cell membrane. Perfusion studies have demonstrated that exogenous hydroperoxides including H_2O_2 are metabolized by rat liver via GSH peroxidase (Sies *et al.* 1972, 1974). The hydroperoxides infused into the liver might to some degree mimic hydroperoxides originating endogenously outside the peroxisomal compartment. H_2O_2 generated within the peroxisomes by urate infusion usually results in compound I formation of catalase without the decrease of NADPH-dependent fluorescence and without the marked GSSG release typical for GSH peroxidase function. These experiments were interpreted as showing that GSH peroxidase is responsible for removing H_2O_2 (as well as other hydroperoxides) in cell compartments low in or free of catalase. In rat liver the compartments primarily protected by GSH peroxidase are the cytosol and the mitochondrial matrix space, and the enzyme is hardly detectable in microsomes, nuclei and the peroxisomes, which probably contain the entire catalase of the cell (Flohé & Schlegel 1971).

VARIABLES OF THE GSH PEROXIDASE SYSTEM

The removal of hydroperoxides is a reaction of obvious relevance and may, for convenience, be declared the biological sense of the GSH peroxidase system. The rate equation (5) for this crucial step is quite simple. The velocity of

$$v = \frac{\mathrm{d[ROOH]}}{\mathrm{d}t} = k_{+1}\mathrm{[ROOH][E]} \qquad (5)$$

hydroperoxide removal in biological systems is nevertheless not easily predicted, since usually none of the three factors determining the rate is experimentally available. [E] stands for the molarity of the reduced enzyme and can be calculated from the total enzyme concentration only if a complete kinetic analysis of the special GSH peroxidase under consideration has been performed. At present, unfortunately, [E] can only be estimated for bovine blood. [ROOH], the concentration of a biologically formed hydroperoxide, can be only roughly estimated, and frequently R in ROOH cannot be identified. A few values of k_{+1} have been determined, but again for bovine blood GSH peroxidase only.

It appears justified, however, to derive some basic rules relevant to the *in vivo* conditions from the kinetic analysis. These deductions may appear to be trivial, but should be repeated here, because unjustified extrapolations from early kinetic measurements have frequently led to erroneous conclusions. In the routine testing of GSH peroxidase, the turnover is largely independent of the concentration and nature of the hydroperoxide but directly proportional to the concentration of GSH in the test tube. In living systems just the opposite situation should prevail. To clarify this point, Figs. 2 and 3 show computer-simulated plots of velocity against substrate concentration based on the kinetic parameters of bovine blood GSH peroxidase (see Table 2). From Fig. 2 it is evident that, at a fixed concentration of 1mM-GSH, differences in turnover rates are hardly detectable at millimolar concentrations of different hydroperoxides (*in vitro* conditions), whereas at hydroperoxide concentrations below micromolar (*in vivo* conditions) the rate depends on [ROOH] and on R, as is described by equation (5). Fig. 3 also demonstrates that, at low concentrations of H_2O_2

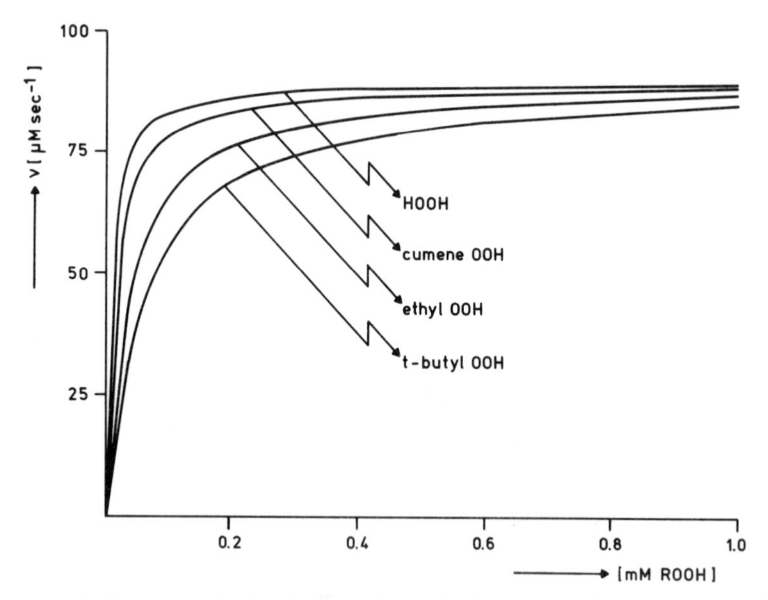

FIG. 2. Computer-simulated plots of velocity v against substrate concentration S for the GSH peroxidase reaction. The plots demonstrate the dependence of the reaction rate on the concentrations of four different hydroperoxides. The calculations are based on equation (4) and the coefficients given in Table 2 for pH 6.7. An enzyme concentration of 0.2 μmol/l, corresponding roughly to that of bovine red blood cell, and a GSH concentration of 1 mmol/l are assumed.

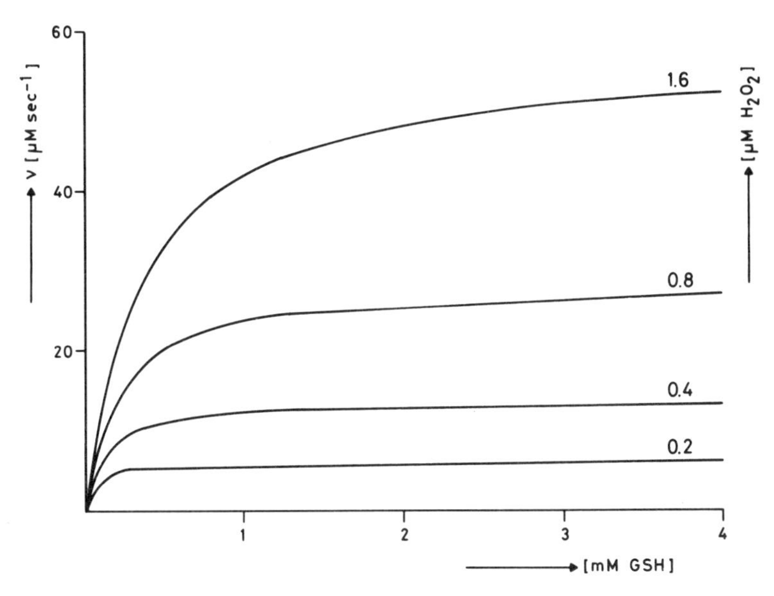

FIG. 3. Computer-simulated plots of velocity v against substrate concentration S for the GSH peroxidase reaction. The calculation is based on equation (4) and the coefficients $\phi_1 = 0.56 \times 10^{-8}$ mol s l^{-1} and $\phi_2 = 1.27 \times 10^{-6}$ mol s l^{-1} (Flohé *et al.* 1972). An enzyme concentration of 0.2 μmol/l is assumed. The curves demonstrate the dependence of the reaction rate on GSH concentration at different constant levels of H_2O_2 in μmol/l (numbers above the curves).

(which, however, may be considered maximum physiological limits) and physiological GSH concentrations (above 1 mmol/l), the rate increases more or less linearly with the hydroperoxide concentration. It is further evident from this plot that, with reasonable hydroperoxide levels, the velocity is largely independent of the concentration of GSH, unless it falls to less than one tenth of the usual levels. In other words, *in vitro* the regeneration rate of reduced enzyme by GSH is measured; *in vivo*, the available glutathione guarantees that the enzyme exists almost entirely in the reduced state, and the rate depends on the rate constant for a given hydroperoxide, its concentration and the total enzyme molarity.

From a pathophysiological point of view two conditions must primarily be considered: (*a*) a deficiency in GSH peroxidase itself due to either a genetic disorder or an inadequate selenium supply and (*b*) an insufficient regeneration rate of GSH from GSSG due to either genetic or alimentary deficiencies (e.g. deficiencies of the biosynthesis of GSH, GSSG reductase, glucose-6-phosphate dehydrogenase [G6PDH] and 6-phosphogluconate dehydrogenase [6PGDH]). As all these deficiencies in red blood cells result in similar clinical conditions, namely drug-induced non-spherocytic haemolytic disorders, it seems justified to interpret the functional impairment of the GSH peroxidase reaction as the final pathophysiological event in these conditions (Flohé 1971). The minor variations and variabilities of these disorders are easily explained when the above analysis is taken into account. Decreased GSH peroxidase activity will certainly lower the velocity of hydroperoxide removal according to equation (5), whereby hydroperoxides are allowed to react unspecifically with cell constituents, corresponding to the degree of insufficiency. The resulting cumulative defects will finally culminate in a shortened half-life for the red blood cell, but dramatic events like intravasal haemolysis cannot be expected. Glutathione reductase obviously has a large excess of reducing capacity, and Benöhr & Waller (1974) demonstrated that a substantial decrease does not make human erythrocytes more susceptible to oxidative stress. This observation agrees with our hypothesis that slight-to-moderate variations of the cellular GSH content do not at all affect the GSH-dependent hydroperoxide metabolism. A homozygous deficiency in GSH biosyntheses, however, may result in GSH levels below 10% of normal (Boivin *et al.* 1974). These low mean values probably do not reflect the GSH content of individual cells, and it seems reasonable to assume a nearly complete lack of GSH, especially in the older cell population. Correspondingly, this condition is consistently associated with haemolytic disorders, but heterozygous cases with a GSH level around 50% of normal remain completely asymptomatic (Boivin *et al.* 1974). G6PDH deficiency is the most puzzling condition in this context since it may be associated with a life-threatening haemolytic crisis in spite of a hardly-detectable or slight decrease in cellular GSH. From early investigations we know that patients with the Mediterranean type of defect may experience massive haemolysis after the first meal of Fava beans, whereas they tolerate a second a few days later without clinical manifestations. How do these observations fit into our concept? The slight decrease in mean cellular GSH can obviously not account for a collapse of the GSH peroxidase system.

Instead, the capacity of the individual cell to generate NADPH seems to limit GSSG reduction. Most types of G6PDH deficiency are characterized by the lability of the enzyme variant, resulting in a shortened half-life for G6PDH (Beutler *et al.* 1968). In older red blood cells with a low residual G6PDH activity exposed to a prooxidative xenobiotic, the rate of peroxide-dependent GSSG formation may exceed the rate of NADP reduction via the pentose phosphate shunt much more readily than in younger cells. Thus, at a critical rate of hydroperoxide production, the GSH level in the older cells will suddenly fall to zero and the excess of hydroperoxide, which can no longer be metabolized, will directly or indirectly destroy the cell (see below), whereas no alteration can be expected as long as the NADPH supply can compensate for the formation of GSSG. This interpretation accounts for the acute and severe manifestation as well as the variable and self-limiting character of the disease.

QUESTIONS, VIEWS AND SPECULATIONS

GSH peroxidase plays a key role in modulating the GSH/GSSG ratio and indirectly affects the NADP/NADPH quotient of the cell. The enzyme may thereby regulate multiple cellular functions such as cell division (Kosower & Kosower 1974*b*, 1976), pentose-phosphate shunt (Paniker *et al.* 1970; Egglestone & Krebs 1974; Flohé 1976), gluconeogenesis (Sies *et al.* 1974), mitochondrial oxidation of α-oxo-acids (Sies & Moss 1978) and others (Kosower & Kosower 1974*a*). A detailed discussion of these aspects is beyond the scope of this article. Some possible and/or probable roles of GSH peroxidase itself which might be relevant to most topical biochemical problems should be mentioned: (1) the protection of unsaturated lipids in biomembranes, (2) the prevention of chemical mutagenesis and (3) the interaction with the arachidonic acid cascade leading to the various prostaglandins.

Prevention of lipid peroxidation by GSH peroxidase

In vitro GSH peroxidase consistently prevents the oxidative break-down of unsaturated lipids of biomembranes, and I see no reason to rule out the possibility that this enzymic ability plays a part in the defence against oxidative damage of organisms living in aerobic conditions (Flohé *et al.* 1976). This view is predominantly based on the following observations from my former laboratory in Tübingen and from other investigators:

(*a*) GSH peroxidase can reduce esters of hydroperoxy-fatty acids (Little & O'Brien 1968);

(*b*) endogenous mitochondrial GSH peroxidase prevents lipid peroxidation and irreversible high-amplitude swelling of rat liver mitochondria (Flohé & Zimmermann 1970);

(*c*) in isolated inner membranes of rat liver mitochondria purified GSH peroxidase prevents the oxidative degradation of phospholipids and the concomitant formation of malonaldehyde (Flohé & Zimmermann 1974);

(*d*) bovine blood GSH peroxidase added to illuminated chloroplasts inhibits swelling and malonaldehyde formation (Flohé & Menzel 1971);

(*e*) *in vivo* inhibition of GSH peroxidase by repeated administration of cadmium salts results in an accumulation of degradation products of unsaturated lipids in rat testes (Omaye *et al.* 1975);

(*f*) conditions requiring a high rate of lipid peroxide removal, such as the ingestion of lipid peroxides (Reddy & Tappel 1974) or exposure to ozone (Chow & Tappel 1972), lead to increased GSH peroxidase activity;

(*g*) in rats deficient in selenium and consequently in GSH peroxidase lipid peroxidation can be detected *in vivo* by monitoring the evolution of ethane. The effect can be inhibited partially by selenium alone and more consistently by a combined treatment with selenium and tocopherol (Hafeman & Hoekstra 1977);

(*h*) low selenium and GSH peroxidase levels were detected in Finnish children suffering from neuronal ceroid lipofuscinosis (Westermarck 1977).

In spite of this overwhelming though admittedly indirect evidence, the hypothesis that GSH peroxidase acts directly on hydroperoxy groups of biomembrane lipids has been repeatedly questioned (McCay *et al.* 1976; Burk *et al.* 1978). It may be difficult to imagine that an enzyme which at least after conventional cell fractionation appears to be entirely soluble (Flohé & Schlegel 1971) may have access to the hydrophobic lipid bilayers. The same reasoning, however, applies to a second type of GSH peroxidase which has recently been considered as a possible substitute for the selenoenzyme in this special function (Burk *et al.* 1978). Besides, apart from being a poor GSH peroxidase, this selenium-independent enzyme is identical with ligandin (Arias *et al.* 1976) and glutathione transferase (EC 2.5.1.18; Prohaska & Ganther 1977), also catalyses the isomerization of PGH_2 to PGE_2 (Christ-Hazelhof *et al.* 1976) and thus does not necessarily need 'job enrichment'. McCay *et al.* (1976) claimed that GSH peroxidase cannot possibly reduce the hydroperoxides of membrane lipids directly, since they detected a lipid peroxide in a microsomal system and were unable to find a corresponding amount of hydroxy-lipid in the presence of GSH peroxidase, although this enzyme, in contrast to catalase, prevented the oxidative lipid degradation as measured by the thiobarbituric acid procedure. To explain these findings the authors had to exhaust first the much-quoted Haber–Weiss cycle (Haber & Weiss 1934) and secondly the argument that GSH peroxidase is superior to catalase in destroying low concentrations of H_2O_2. This interpretation cannot be accepted for several reasons: (*a*) malonaldehyde is not derived predominantly from lipid hydroperoxides but from cyclic dialkyl peroxides (Dahle *et al.* 1962) which are not substrates of GSH peroxidase. (*b*) The lipid peroxide intermediate detected by McCay *et al.* (1976) was not necessarily a hydroperoxide, since the method of identification does not distinguish clearly between different types of peroxides (Stahl 1958). The chromatographic behaviour of the lipid peroxide did not suggest the presence of a hydrophilic hydroperoxy group (Tam & McCay 1970). (*c*) The rate constants for H_2O_2 removal of GSH peroxidase (Table 2) and catalase (Chance *et al.* 1952) are rather close. Minor differences in the respective constants cannot account for an all-or-none difference in the biological effectiveness of the two enzymes under consideration.

I should emphasize that the observations of McCay *et al.* (1976) are in good agreement with those of other studies on oxidative membrane damage. In the experiments with mitochondrial membranes GSH peroxidase on a molar basis is also much more effective than catalase in preventing oxidative lipid degradation (Flohé & Zimmermann 1974). Similarly, oxidative haemolysis occurs in human red cells which are deficient in GSH peroxidase (see above) but rich in catalase, but acatalatic erythrocytes are not particularly prone to haemolysis (Aebi & Suter 1974). An explanation for the superiority of GSH peroxidase in these systems which is consistent with the established enzymological data must take into consideration the fact that the enzyme not only reduces H_2O_2 as fast as catalase but in addition other hydroperoxides. The nature of this hydroperoxide, which obviously initiates membrane destruction, still must be established, but a lipid hydroperoxide remains a likely candidate.

GSH peroxidase and mutagenesis

Peroxides may be considered as mutagens, and enzymes interacting with peroxides may, therefore, decrease the incidence of mutagenetic events. This could be true in particular for GSH peroxidase, since it reacts with hydroperoxides of nucleic acids and their precursors (Christophersen 1969*a*). Thymine hydroperoxide, which is also metabolized by GSH peroxidase, was identified as a potent mutagen (Thomas *et al.* 1976). A weak point of the above hypothesis is that GSH peroxidase activity could not be definitely detected at the site of mutagenesis, i.e. in the nucleus (Flohé & Schlegel 1971). However, mutagenetic mechanisms are not absolutely restricted to the nucleus. The interest in the theoretical ability of GSH peroxidase to scavenge mutagens is reinforced by a number of observations relating selenium supply to the incidence of cancer, which may be viewed as some kind of somatic mutation. Since we are not aware of any well-defined role of selenium in mammals apart from its being an integral part of GSH peroxidase, the beneficial effects of selenium are suggestive of an optimized removal of hydroperoxide (Schwarz 1976). In this context the following intriguing observation has been made: supplementation with subtoxic doses of selenium consistently decreases the incidence of tumours in several models of chemical cancerogenesis as well as in mice developing spontaneous mammary tumours (Schrauzer 1976 and references therein; Griffin & Jacobs 1977; Jacobs *et al.* 1977). Worldwide epidemiological studies have revealed an inverse relationship between dietary selenium intake in man and cancer mortality (Shamberger 1976; Schrauzer *et al.* 1977) that is statistically significant ($P < 0.01$

or better) for malignant growth of the colon, rectum, prostate, breast and white blood cells (Schrauzer *et al.* 1977).

Interestingly, in Schrauzer's studies with C3H mice, selenite supplementation neither had any beneficial effect once the tumour was developed nor did it inhibit the growth of transplanted tumours (Schrauzer & Ismael 1974). This clearly indicates that the anticancer activity of selenium is not cytostatic but that it rather prevents some early step during cancerogenesis which could well be a peroxide-mediated mutation. Clearly, more investigations are needed to link the theoretical basis (removal of mutagenic hydroperoxides) with the empirical data derived from feeding experiments and epidemiological surveys.

GSH peroxidase and prostaglandin biosynthesis

A specific involvement of GSH peroxidase in prostaglandin biosynthesis has not yet been established. Recent progress in this field, however, has resulted in the discovery of several peroxo intermediates or by-products of the arachidonic acid cascade, and it seems highly unlikely that GSH peroxidase ignores the opportunity to react with these biologically highly-significant compounds. Fig. 4 shows an up-to-date scheme of prostaglandin biosynthesis including possible sites of GSH peroxidase involvement (numbers 1–5).

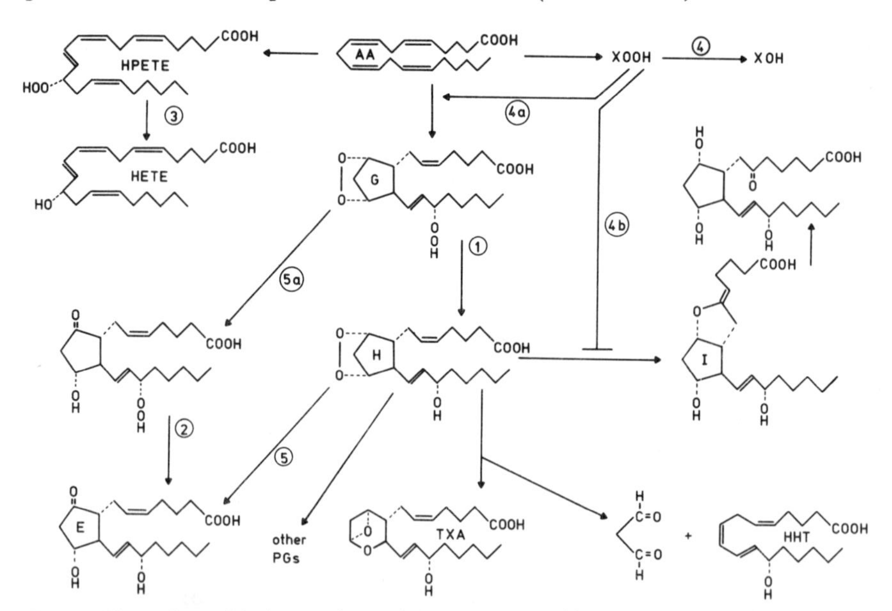

FIG. 4. Sites of possible interactions of the GSH peroxidase reaction (numbers in circles) with the arachidonic acid (AA) cascade: HPETE, 12-hydroperoxyeicosatetraenoic acid; HETE, 12-hydroxyeicosatetraenoic acid; G, prostaglandin G_2; E, prostaglandin E_2; I, prostacyclin; TXA, thromboxane A_2; H, prostaglandin H_2; numbers indicate possible sites at which GSH peroxidase could exert an effect.

The unstable sources of prostaglandins, arachidonic acid (AA) or other polyunsaturated fatty acids, must be protected against unspecific oxidative degradation, and it can be assumed that GSH peroxidase together with catalase, superoxide dismutase and probably α-tocopherol comprises an effective task force guaranteeing the availability of these indispensable substrates.

The first intermediate of the cascade, prostaglandin G_2 (G), can be metabolized to prostaglandin H_2 by GSH peroxidase (Nugteren & Hazelhof 1973; reaction 1). Reaction 2, the reduction of prostaglandin E_2 15-hydroperoxide to prostaglandin E_2 is analogous and thus should also be catalysed by GSH peroxidase. The biological significance of these reactions is still unclarified, since Christ-Hazelhof *et al.* (1976) reported that in biosynthetic assays prostaglandin H_2 is formed from arachidonic acid even in the absence of GSH peroxidase. This 'one-step' formation of 15-hydroxyprostaglandins, however, might also result from non-enzymic free-radical reactions analogous to those described by Haber & Weiss (1934) for H_2O_2 (e.g. 6).

$$ROOH + O_2^- + H^+ \rightarrow ROH + O_2 + OH^\bullet \qquad (6)$$

Such a reaction, of course, would not be suited to the role of keeping the reaction sequence in defined channels and, therefore, a specific catalysis of the reduction of the 15-hydroperoxy group by GSH peroxidase remains an attractive hypothesis.

The reduction of open-chain hydroperoxy-acids by GSH peroxidase as shown in reaction 3 in Fig. 4 has been definitely established for the analogous hydroperoxides of linoleic and linolenic acids by Christophersen (1968, 1969*b*). The product of reaction 3, 12-hydroxyeicosatetraenic acid (HETE), exhibits chemotactic activity and could contribute to inflammatory processes. In a further analogy, we may expect GSH peroxidase to reduce various hydroperoxides produced by the cyclooxygenase and lipoxygenase reactions (XOOH; reaction 4). Such hydroperoxides are believed to interfere significantly with the multiple steps of prostaglandin biosynthesis. Cook & Lands (1976) observed that arachidonic acid-dependent oxygen consumption in a sheep vesicular gland preparation can be largely suppressed by the addition of GSH and GSH peroxidase. The authors conclude from these experiments that cyclooxygenase is allosterically activated by some hydroperoxy product which can be scavenged by GSH peroxidase (reaction 4*a*). Thus, the GSH peroxidase system would at least take part in the regulation of prostaglandin biosynthesis. An even more attractive hypothesis may be derived from the discovery by Gryglewski *et al.* (1976) that prostacyclin (I) formation is inhibited by hydroperoxides of fatty acids, in particular by 15-hydroperoxyarachidonic acid. GSH peroxidase could keep the steady-state of such a hydroperoxide low enough to guarantee unimpaired prostacyclin formation. Since the prostacyclin of the arterial endothelium antagonizes the platelet adhesion and aggregation induced by thromboxane A_2 (TXA), a deficiency of GSH peroxidase in endothelial cells might favour a pathologically increased adhesion of platelets and finally the development of atheromatosis. In this context it seems revealing that, according to epidemiological studies made in the USA, the death rates from cardiovascular diseases correlate inversely with selenium supply (Shamberger 1976).

Reaction 5, the isomerization of the 9,11-endoperoxo bridge of prostaglandin G and H, is not catalysed by selenium-dependent GSH peroxidase but, according to Christ-Hazelhof *et al.* (1976), by glutathione transferase, which, however, has some GSH peroxidase activity (Prohaska & Ganther 1977). This reaction requires GSH and may thus be indirectly modulated by GSH peroxidase.

In conclusion I must point out that none of the observations and ideas discussed here has so far been checked for biological relevance. The number of possible interactions of GSH peroxidase and prostaglandin biosynthesis makes it highly unlikely that the arachidonic acid cascade is not in some way or another controlled by this enzyme. A thorough investigation of the hypotheses advanced may yield a better understanding of the more complex types of tissue injuries such as inflammation or atherosclerosis.

References

AEBI, H. & SUTER, H. (1974) Protective function of reduced glutathione (GSH) against the effect of prooxidative substances and of irradiation in the red cell, in *Glutathione* (Flohé, L., Benöhr, H. Ch., Sies H., Waller, H. D. & Wendel, A., eds.), pp. 192–201, Georg Thieme, Stuttgart

ARIAS, I. M., FLEISCHNER, G., KIRSCH, R., MISHKIN, S. & GATMAITAN, Z. (1976) On the structure, regulation, and function of ligandin, in *Glutathione* (Arias, I. M. & Jakoby, W. B., eds.), pp. 175–188, Raven Press, New York

AWASTHI, Y. C., BEUTLER, E. & SRIVASTAVA, S. K. (1975) Purification and properties of human erythrocyte glutathione peroxidase. *J. Biol. Chem. 250*, 5144–5149

BENÖHR, H. CH. & WALLER, H. D. (1974) Hematological manifestations in enzymatic deficiencies of glutathione reduction, in *Glutathione* (Flohé, L., Benöhr, H. Ch., Sies, H. Waller, H. D. & Wendel, A., eds.), pp. 184–191, Georg Thieme, Stuttgart

BEUTLER, E., MATHAI, C. K. & SMITH, J. E. (1968) Biochemical variants of glucose-6-phosphate dehydrogenase giving rise to congenital nonspherocytic hemolytic disease. *Blood 31*, 131–150

BOIVIN, P., GALAND, C. & BERNARD, J. F. (1974) Deficiencies in GSH biosynthesis, in *Glutathione* (Flohé, L., Benöhr, H. Ch., Sies, H., Waller, H. D. & Wendel, A., eds.), pp. 146–157, Georg Thieme, Stuttgart

BURK, R. F., NISHIKI, K., LAWRENCE, R. A. & CHANCE, B. (1978) Peroxide removal by selenium-dependent and selenium-independent glutathione peroxidases in hemoglobin-free perfused rat liver. *J. Biol. Chem. 253*, 43–46

CHANCE, B., GREENSTEIN, D. S. & ROUGHTON, F. J. W. (1952) The mechanism of catalase action. I. Steady-state analysis. *Arch. Biochem. Biophys. 37*, 301–321

CHIU, D., FLETCHER, B., STULTS, F., ZAKOWSKI, J. & TAPPEL, A. L. (1975) Properties of selenium-glutathione peroxidase. *Fed. Proc. 34*, 925 (abstr. 3996)

CHOW, C. K. & TAPPEL, A. L. (1972) An enzymatic protective mechanism against lipid peroxidation damage to lungs of ozone-exposed rats. *Lipids 7*, 518–524

CHRIST-HAZELHOF, E., NUGTEREN, D. H. & VAN DORP, D. A. (1976) Conversions of prostaglandin endoperoxides by glutathione *S*-transferases and serum albumins. *Biochim. Biophys. Acta 450*, 450–461

CHRISTOPHERSEN, B. O. (1968) Formation of monohydroxypolyenic fatty acids from lipid peroxides by a glutathione peroxidase. *Biochim. Biophys. Acta 164*, 35–46

CHRISTOPHERSEN, B. O. (1969*a*) Reduction of X-ray-induced DNA and thymine hydroperoxides by rat liver glutathione peroxidase. *Biochim. Biophys. Acta 186*, 387–389

CHRISTOPHERSEN, B. O. (1969*b*) Reduction of linolenic acid hydroperoxide by a glutathione peroxidase. *Biochim. Biophys. Acta 176*, 463–470

COOK, H. W. & LANDS, W. E. M. (1976) Mechanism for suppression of cellular biosynthesis of prostaglandins. *Nature (Lond.) 260*, 630–632

DAHLE, L. K., HILL, E. G. & HOLLMANN, R. T. (1962) The thiobarbituric acid reaction and the autoxidation of polyunsaturated fatty acid methyl esters. *Arch. Biochem. Biophys. 98*, 253–261

DALZIEL, K. (1957) Initial steady state velocities in the evaluation of enzyme–coenzyme–substrate reaction mechanisms. *Acta Chem. Scand. 11*, 1706–1723

EGGLESTONE, L. V. & KREBS, H. A. (1974) Regulation of the pentose phosphate cycle. *Biochem. J. 183*, 425–435

FLOHÉ, L. (1971) Die Glutathionperoxidase: Enzymologie und biologische Aspekte. *Klin. Wochenschr. 49*, 669–683

FLOHÉ, L. (1976) Role of selenium in hydroperoxide metabolism, in *Proceedings of the Symposium on Selenium-Tellurium in the Environment* (Industrial Health Foundation, Inc., ed.), pp. 138–157, Pittsburgh

FLOHÉ, L. & ZIMMERMANN, R. (1970) The role of GSH peroxidase in protecting the membrane of rat liver mitochondria. *Biochim. Biophys. Acta 223*, 210–213

FLOHÉ, L. & MENZEL, H. (1971) The influence of glutathione upon light-induced high-amplitude swelling and lipid peroxide formation of spinach chloroplasts. *Plant Cell Physiol. 12*, 325–333

FLOHÉ, L. & SCHLEGEL, W. (1971) Glutathion-Peroxidase, IV. Intrazelluläre Verteilung des Glutathion-Peroxidase-Systems in der Rattenleber. *Hoppe Seyler's Z. Physiol. Chem. 352*, 1401–1410

FLOHÉ, L. & GÜNZLER, W. A. (1974) Glutathione peroxidase, in *Glutathione* (Flohé, L., Benöhr, H. Ch., Sies, H., Waller, H. D. & Wendel, A., eds.), pp. 132–145, Georg Thieme, Stuttgart

FLOHÉ, L. & ZIMMERMANN, R. (1974) GSH-induced high-amplitude swelling of mitochondria, in *Glutathione* (Flohé, L., Benöhr, H. Ch., Sies, H., Waller, H. D. & Wendel, A., eds.), pp. 245–260, Georg Thieme, Stuttgart

FLOHÉ, L., EISELE, B. & WENDEL, A. (1971*a*) Glutathion-Peroxidase, I. Reindarstellung und Molekulargewichtsbestimmungen. *Hoppe Seyler's Z. Physiol. Chem. 352*, 151–158

FLOHÉ, L. GÜNZLER, W., JUNG, G., SCHAICH, E. & SCHNEIDER, F. (1971*b*) Glutathion-Peroxidase, II. Substratspezifität und Hemmbarkeit durch Substratanaloge. *Hoppe Seyler's Z. Physiol. Chem. 352*, 159–169

FLOHÉ, L., SCHAICH, E., VOELTER, W. & WENDEL, A. (1971*c*) Glutathion-Peroxidase, III. Spektrale Charakteristika und Versuche zum Reaktionsmechanismus. *Hoppe Seyler's Z. Physiol. Chem. 352*, 170–180

FLOHÉ, L., LOSCHEN, G., GÜNZLER, W. A. & EICHELE, E. (1972) Glutathione peroxidase, V. The kinetic mechanism. *Hoppe Seyler's Z. Physiol. Chem. 353*, 987–999

FLOHÉ, L., GÜNZLER, W. A. & SCHOCK, H. H. (1973) Glutathione peroxidase: a selenoenzyme. *FEBS (Fed. Eur. Biochem. Soc.) Lett. 32*, 132–134

FLOHÉ, L., GÜNZLER, W. A. & LADENSTEIN, R. (1976) Glutathione peroxidase, in *Glutathione* (Arias, I. M. & Jakoby, W. B., eds.), pp. 115–138, Raven Press, New York

FORSTROM, J. W., ZAKOWSKI, J. J. & TAPPEL, A. L. (1978) Identification of the catalytic site of rat liver glutathione peroxidase as selenocysteine. *Biochemistry 17*, 2639–2644

GANTHER, H. E., HAFEMAN, D. G., LAWRENCE, R. A., SERFASS, R. E. & HOEKSTRA, W. G. (1976) Selenium and glutathione peroxidase in health and disease: a review, in *Trace Elements in Human Health and Disease*, vol. II (Prasad, A., ed.), Academic Press, New York

GRIFFIN, A. C. & JACOBS, M. M. (1977) Effects of selenium on azo dye hepatocarcinogenesis. *Cancer Lett. 3*, 177–181

GRYGLEWSKI, R. J., BUNTING, S., MONCADA, S., FLOWER, R. J. & VANE, J. R. (1976) Arterial walls are protected against deposition of platelet thrombi by a substance (Prostaglandin X) which they make from prostaglandin endoperoxides. *Prostaglandins 12*, 685–713

GÜNZLER, W. A. (1974) *Glutathionperoxidase, Kristallisation, Selengehalt, Aminosäurezusammensetzung und Modellvorstellungen zum Reaktionsmechanismus*, Dissertation, Tübingen

GÜNZLER, W. A., VERGIN, H., MÜLLER, I. & FLOHÉ, L. (1972) Glutathion-Peroxidase, VI. Die Reaktion der Glutathion–Peroxidase mit verschiedenen Hydroperoxiden. *Hoppe Seyler's Z. Physiol. Chem. 353*, 1001–1004

HABER, F. & WEISS, J. (1934) The catalytic decomposition of hydrogen peroxide by iron salts. *Proc. R. Soc. Lond. 147*, 332–351

HAFEMAN, D. G. & HOEKSTRA, W. G. (1977) Lipid peroxidation *in vivo* during vitamin E and selenium deficiency in the rat as monitored by ethane evolution. *J. Nutr. 107*, 666–672

JACOBS, M. M., JANSSON, B. & GRIFFIN, A. C. (1977) Inhibitory effect of selenium on 1,2-dimethylhydrazine and methylazoxymethanol acetate induction of colon tumors. *Cancer Lett. 2*, 133–137

KOSOWER, E. M. & KOSOWER, N. S. (1974*a*) Manifestations of changes in the GSH–GSSG status of biological systems, in *Glutathione* (Flohé, L., Benöhr, H. Ch., Sies, H., Waller, H. D. & Wendel, A., eds.), pp. 287–297, Georg Thieme, Stuttgart

KOSOWER, N. S. & KOSOWER, E. M. (1974*b*) Effect of GSSG on protein synthesis, in *Glutathione* (Flohé, L., Benöhr, H. Ch., Sies, H., Waller, H. D. & Wendel, A., eds.), pp. 276–287, Georg Thieme, Stuttgart

KOSOWER, N. S. & KOSOWER, E. M. (1976) Functional aspects of glutathione disulfide and hidden forms of glutathione, in *Glutathione* (Arias, I. M. & Jakoby, W. B., eds.), pp. 159–172, Raven Press, New York

LADENSTEIN, R. & EPP, O. (1977) X-ray diffraction studies on the selenoenzyme glutathione peroxidase. *Hoppe Seyler's Z. Physiol. Chem. 358*, 1237–1238

LITTLE, C. & O'BRIEN, P. J. (1968) An intracellular GSH-peroxidase with a lipid peroxide substrate. *Biochem. Biophys. Res. Commun. 31*, 145–150

MCCAY, P. B., GIBSON, D. D., FONG, K. L. & HORNBROOK, K. R. (1976) Effect of glutathione peroxidase activity on lipid peroxidation in biological membranes. *Biochim. Biophys. Acta 431*, 459–468

MILLS, G. C. (1957) Hemoglobin catabolism. I. Glutathione peroxidase, an erythrocyte enzyme which protects hemoglobin from oxidative breakdown. *J. Biol. Chem. 229*, 189–197

NAKAMURA, W., HOSODA, S. & HAYASHI, K. (1974) Purification and properties of rat liver glutathione peroxidase. *Biochim. Biophys. Acta 358*, 251–261

NECHELES, T. F. (1974) The clinical spectrum of glutathione-peroxidase deficiency, in *Glutathione* (Flohé, L., Benöhr, H. Ch., Sies, H., Waller, H. D. & Wendel, A., eds.), pp. 173–180, Georg Thieme, Stuttgart

NUGTEREN, D. H. & HAZELHOF, E. (1973) Isolation and properties of intermediates in prostaglandin biosynthesis. *Biochim. Biophys. Acta 326*, 448–461

OH, S. H., GANTHER, H. E. & HOEKSTRA, W. G. (1974) Selenium as a component of glutathione peroxidase isolated from ovine erythrocytes. *Biochemistry 13*, 1825–1829

OMAYE, S. T., TAYLOR, S. L., FORSTROM, J. W. & TAPPEL, A. L. (1975) Lipid peroxidation and reactions of glutathione peroxidase. *Fed. Proc. 34*, 538 (abstr. 1797)

PANIKER, N. V., SRIVASTAVA, S. K. & BEUTLER, E. (1970) Glutathione metabolism of the red cells. Effect of glutathione reductase deficiency on the stimulation of hexose monophosphate shunt under oxidative stress. *Biochim. Biophys. Acta 215*, 456–460

PROHASKA, J. R. & GANTHER, H. E. (1977) Glutathione peroxidase activity of glutathione-*S*-transferases purified from rat liver. *Biochem. Biophys. Res. Commun. 76*, 437–445

REDDY, K. & TAPPEL, A. L. (1974) Effect of dietary selenium and autoxidized lipids on the glutathione peroxidase system of gastrointestinal tract and other tissues in the rat. *J. Nutr. 104*, 1069–1078

ROTRUCK, J. T., HOEKSTRA, W. G., POPE, A. L., GANTHER, H., SWANSON, A. & HAFEMAN, D. (1972) Relationship of selenium to GSH peroxidase. *Fed. Proc. 31*, 691

SCHRAUZER, G. N. (1976) Selenium: anticarcinogenic action of an essential trace element, in *Proceedings of the Symposium on Selenium-Tellurium in the Environment* (Industrial Health Foundation, Inc., ed.), pp. 293–299, Pittsburgh

SCHRAUZER, G. N. & ISMAEL, D. (1974) Effects of selenium and of arsenic on the genesis of spontaneous mammary tumors in inbred C_3H mice. *Ann. Clin. Lab. Sci. 4*, 441–447

SCHRAUZER, G. N., WHITE, D. A. & SCHNEIDER, C. J. (1977) Cancer mortality correlation studies — III. Statistical associations with dietary selenium intakes. *Bioinorg. Chem. 7*, 23–34

SCHWARZ, K. (1976) The discovery of the essentiality of selenium, and related topics, in *Proceedings of the Symposium on Selenium-Tellurium in the Environment* (Industrial Health Foundation, Inc., ed.), pp. 349–376, Pittsburgh

SHAMBERGER, R. J. (1976) Selenium in health and disease, in *Proceedings of the Symposium on Selenium-Tellurium in the Environment* (Industrial Health Foundation, Inc., ed.), pp. 253–267, Pittsburgh

SIES, H. & MOSS, K. M. (1978) A role of mitochondrial glutathione peroxidase in modulating mitochondrial oxidations in liver. *Eur. J. Biochem. 84*, 377–383

SIES, H., GERSTENECKER, C., MENZEL, H. & FLOHÉ, L. (1972) Oxidation in the NADP system and release of GSSG from hemoglobin-free perfused rat liver during peroxidatic oxidation of glutathione by hydroperoxides. *FEBS (Fed. Eur. Biochem. Soc.) Lett. 27*, 171–175

SIES, H., GERSTENECKER, C., SUMMER, K. H., MENZEL, H. & FLOHÉ, L. (1974) Glutathione-dependent hydroperoxide metabolism and associated metabolic transitions in hemoglobin-free perfused rat liver, in *Glutathione* (Flohé, L., Benöhr, H. Ch., Sies, H., Waller, H. D. & Wendel, A., eds.), pp. 261–276, Georg Thieme, Stuttgart

STAHL, E. (1958) Dünnschicht-Chromatographie. II. Standardisierung, Sichtbarmachung, Dokumentation und Anwendung. *Chem. Ztg. 82*, 323–329

TAM, B. K. & MCCAY, P. B. (1970) Reduced triphosphopyridine nucleotide oxidase-catalyzed alterations of membrane phospholipids. *J. Biol. Chem. 245*, 2295–2300

THOMAS, H. F., HERRIOTT, R. M., HAHN, B. S. & WANG, S. Y. (1976) Thymine hydroperoxide as a mediator in ionising radiation mutagenesis. *Nature (Lond.) 259*, 341–343

WENDEL, A., PILZ, W., LADENSTEIN, R., SAWATZKI, G. & WESER, U. (1975) Substrate-induced redox change of selenium in glutathione peroxidase studied by X-ray photoelectron spectroscopy. *Biochim. Biophys. Acta 377*, 211–215

WESTERMARCK, T. (1977) Selenium content of tissues in Finnish infants and adults with various diseases, and studies on the effects of selenium supplementation in neuronal ceroid lipofuscinosis patients. *Acta Pharmacol. Toxicol. 41*, 121–128

Discussion

Dormandy: For many years before you showed that selenium formed an essential part of glutathione peroxidase selenium had been recognized as an important antioxidant; but nobody knew how it acted. Can we now assume that the antioxidant importance of selenium resides in glutathione peroxidase or is that just one of several mechanisms?

Flohé: Yes, but selenium may have other functions in mammalian systems. All symptoms described in selenium deficiency, for example, are not necessarily due to lack of the glutathione peroxidase. Diplock *et al.* (1971, 1973) describe the distribution of selenium in liver cells; there is an acid-volatile selenium, the function of which is unknown. It is not distributed in the same way as glutathione peroxidase.

Cohen: Glutathione peroxidase removes peroxide in compartments that are poor in catalase; but does it not lower the level of peroxide in compartments that are rich in catalase?

Flohé: Not necessarily. You are probably reminding me of your work on the human red blood cells (Cohen & Hochstein 1963), which, incidentally, persuaded me to work on GSH peroxidase. The problem is that we still cannot explain your results theoretically. In principle, the fate of H_2O_2 depends on the ratio of the molar concentrations of catalase and GSH peroxidase, on the relevant rate constants of these enzymes and also on the functional state of GSH peroxidase in the tissue under consideration. However, we should always bear in mind the fact that only one type of enzyme has been investigated in detail. The *bovine* enzyme is roughly as effective as catalase in removing H_2O_2, but nobody has yet been able to tell me the molar concentration of GSH peroxidase in *human* blood, although it is generally agreed to be very low. Without molecular kinetic data the molar concentration cannot be deduced from activity measurements. Calculating the functional state of the enzyme is also not justified, since it is hazardous not to consider large species-related differences in rate constants of GSH peroxidase. So you still might be right: human blood cells might contain a reasonable amount of GSH peroxidase in spite of the low activity reflected by GSH removal. It could also be very effective in removing H_2O_2.

Cohen: There is a way to measure the steady-state concentration of H_2O_2 within cells in order to estimate the contribution of GSH peroxidase. The complex of peroxide with catalase (complex I) reacts with 3-amino-1,2,4-triazole in such a way that catalase is inactivated. The rate of inhibition presents an estimate of the steady-state concentration of the catalase–H_2O_2 complex and, hence, indirectly of the steady-state concentration of H_2O_2 within the cell. From such studies (e.g., Liebowitz & Cohen 1968) it is clear that the activity of glutathione peroxidase regulates the H_2O_2 concentration within a catalase-rich compartment, that is, in the human red blood cells.

Flohé: Is this test specific for hydrogen peroxide?

Fridovich: I am fairly sure it is.

Flohé: Do organic peroxides, for instance, interfere with it?

Cohen: I am not certain if that has been studied.

Flohé: What I was discussing in the context of lipid peroxidation may apply to red blood cells, but direct evidence for increased lipid peroxidation in GSH-peroxidase-deficient erythrocytes has still not been found. In *in vitro* studies with isolated or subcellular organelles, however, it has been clearly demonstrated that GSH peroxidase is superior to catalase in preventing lipid peroxidation (Neubert *et al.* 1962; Flohé & Zimmermann 1974; McCay *et al.* 1976). Since in at least some of the experiments enzymes with well defined kinetics were used (Flohé & Zimmermann 1974), we may conclude that in these experiments prevention of lipid peroxidation is not due to removal of H_2O_2 by GSH peroxidase and that another hydroperoxide substrate of GSH peroxidase thus obviously triggers lipid peroxidation in these systems.

Cohen: In some experiments we exposed normal human red blood cells to H_2O_2 by means of a vapour-state diffusion technique. In these conditions the red blood cells showed no evidence of lipid peroxides, at least as judged by the malonaldehyde reaction (unpublished results). When GSH peroxidase activity was impaired, either by removal of GSH (with *N*-ethylmaleimide) or by preventing the reduction of GSSG (by omitting glucose from the medium), the steady-state concentration of H_2O_2 rose (Liebowitz & Cohen 1968). So, in these experiments GSH peroxidase regulated the steady-state level of added reagent H_2O_2.

Winterhalter: Why do rats have much more glutathione (by about a factor of four) in their red cells than humans? The kinetics show that the velocity of reaction is not influenced by more glutathione at these concentrations.

Flohé: I don't know.

Roos: Glutathione does more than just act as a substrate for glutathione peroxidase: it can directly react, for instance, with free radicals; it can reduce disulphide bridges; it is involved in amino acid transport via the γ-glutamyl cycle, and in the protection against electrophilic compounds via *S*-transferases; and it can act as a cofactor in several enzymic reactions.

Flohé: In red blood cells, too?

Roos: Yes, it certainly acts as a cofactor for glyoxalase in erythrocytes (Prins *et al.* 1966) and possibly also it reacts directly with free radicals and disulphides in these cells.

Goldstein: You raised the important question of whether glutathione peroxidase reacts *in vivo* with lipid peroxides. Without doubt, once an oxygen free-radical reaction starts within a cell, glutathione peroxidase protects the membrane against lipid peroxidation. But once lipid peroxidation is initiated in a membrane, does glutathione peroxidase prevent the further effects of lipid peroxides on either the membrane or the interior of the cell? There is the problem of compartmentalization. How does glutathione peroxidase get to the membrane and, conversely, how do the lipid peroxides get out? McCay *et al.* (1976), working with peroxidizing liver microsomes, could not find the fatty acid alcohol derivatives that Christopherson (1968) suggested should be formed. McCay *et al.* further showed that if, instead of hydroperoxylinoleic acid (which does form the alcohol), one used a fatty acid hydroperoxide attached to a phospholipid (which is the form of fatty acid in the membrane), the latter did not serve as a substrate for glutathione peroxidase. Consequently, without a mechanism for removing the fatty acid, it would be extremely difficult for the fatty acid hydroperoxide and the glutathione peroxidase to come together.

Flohé: I agree that kinetic data are lacking on the interaction of glutathione peroxidase with the complex lipid hydroperoxides and that there is the problem of compartmentalization. The enzyme we use is not associated with membranes, but Tappel's group now claims that the rat liver enzyme may be associated to some degree with membranes (A. L. Tappel, personal communication). The proof that it cannot use other peroxides as substrates is not convincing, and we still have to assume that it uses other substances in addition to H_2O_2.

Hayaishi: What is the electron-donor specificity of glutathione peroxidase? Do other reducing agents, such as adrenalin and guaiacol, work?

Fridovich: Not at all.

Flohé: Glutathione is a good donor. Removal of the glycine residue reduces the reactivity; removal or substitution of the γ-glutamyl residue also reduces the activity (see p. 96). The enzyme is highly specific. Neither cysteine nor homocysteine is a good substrate (below 5%). Typical donor substrates of haem-containing peroxidases are not metabolized by GSH peroxidase (Mills 1959).

Hayaishi: We have some data on the specificity of prostaglandin hydroperoxidase (see Fig. 1), which was purified about 700-fold from bovine vesicular glands. A plot of rate (as measured by the difference in optical density) against concentration of peroxide shows that prostaglandin G and its analogues are by far the best substrates in terms of both K_m and V_{max}, when guaiacol (2-methoxyphenol) is used as a hydrogen donor. Table 1 shows that, for glutathione peroxidase with glutathione as the hydrogen donor, cumene hydroperoxide is by far the best substrate whereas prostaglandin G_1 is a poor substrate. With our purified prostaglandin hydroperoxidase glutathione is a poor hydrogen donor but with other hydrogen donors, such as adrenalin, hydroquinone, guaiacol or even tryptophan, prostaglandin G_1 is converted into prostaglandin H_1 very efficiently. So, tentatively, we are forced to conclude that prostaglandin hydroperoxidase is a specific enzyme which works poorly with glutathione as a hydrogen donor and is highly specific for prostaglandin-like structures as a hydrogen donor.

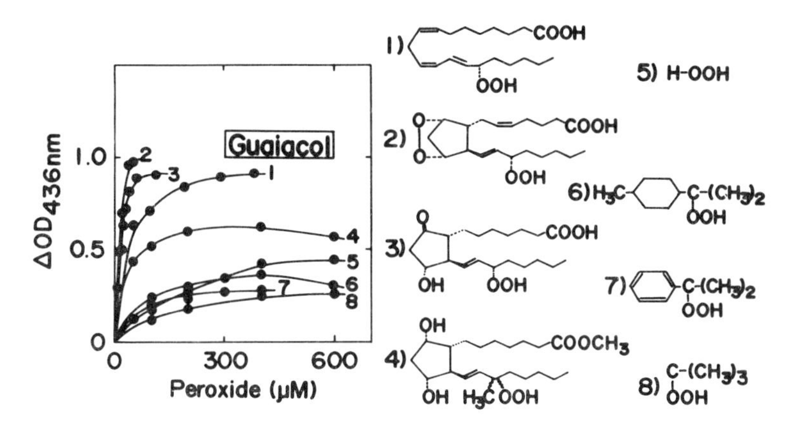

FIG. 1 (Hayaishi). Specificity of prostaglandin hydroperoxidase for the hydroperoxides 1–8. The activity was measured as the difference in the optical density at 436 nm, with guaiacol as hydrogen donor in the system.

TABLE 1 (Hayaishi)
Activity of glutathione (GSH) peroxidase and prostaglandin (PG) hydroperoxidase with different hydrogen donors and acceptors

Enzyme	*H donor*	*H acceptor*	*Activity (nmol/min)*	
GSH peroxidase	GSH	Cumene-OOH	NADPH oxidized	59.5
GSH peroxidase	GSH	PGG_1	PGH_1 formed	2.4
PG hydroperoxidase	GSH	Cumene-OOH	NADPH oxidized	0.0
PG hydroperoxidase	GSH	PGG_1	PGH_1 formed	0.3
PG hydroperoxidase	(Trp)	PGG_1	PGH_1 formed	2.3

Flohé: These data favour the conclusion that different enzymes are at work.

Willson: You quoted the epidemiological evidence of Schrauzer *et al.* and Shamberger for the protective effect of selenium. These reports have aroused considerable publicity. They should be qualified, however. Although the results of reduced tumour incidence in mice fed selenite are of interest, the subsequent epidemiological extrapolations and statistical analysis are questionable. The inverse linear correlation between incidence of breast tumour and selenium intake is based in one study on assessments of average diets in various regions and countries throughout the world and on estimates of the selenium content of the dietary components. In another study (Schrauzer *et al.* 1977) measurements of selenium concentrations in blood samples have been made. However, the statement 'it is postulated that the cancer mortalities in the USA and other western industrialized nations would decline significantly if the dietary selenium intakes were increased to approximately twice the current average amount supplied by the US diet' unfortunately is groundless, being based on a misuse of statistical analysis.

Flohé: It is hard to make any judgement on retrospective statistics. Such investigations can never reach the reliability of well controlled prospective studies, as some environmental factors might have either been neglected or over-emphasized. The statistical approach you mentioned is not the only one (see the references given in my paper and further references therein). My main point was only that it is plausible that GSH peroxidase scavenges mutagenic hydroperoxides. This enzymic potential may be responsible for the observation that selenium inhibits the development of cancer in some animal models. Different epidemiological approaches suggest that the results obtained with the pure enzyme and with animals might be relevant to corresponding human health problems. For that reason, the idea is worth considering, even though the pertinent human data may not be as reliable as we should like.

Smith: Gibson & Cagen (1977) demonstrated that the organ-selective effects of paraquat could be altered from the lung to the liver by administering non-lethal doses of paraquat to selenium-deficient rats. They suggested a relationship between paraquat toxicity to the liver and glutathione peroxidase activity.

Flohé: That implies that the liver damage observed in selenium-deficient animals is induced by peroxide.

Crichton: The role of glutathione peroxidase inside the cell resembles that of ceruloplasmin outside the cell. Ceruloplasmin has for years been a protein in search of a function. One knows that in copper deficiency iron metabolism goes haywire. Also, ceruloplasmin is remarkably effective at mobilizing iron out of the liver. But its real role in plasma is a question mark. Glutathione peroxidase is found in the cell sap. By implication, it is connected with lipid peroxidation of membranes and with the protection of nucleic acids in the nucleus against peroxidation. Obviously, these two proteins are not in the cell for no good purpose; glutathione peroxidase has to be involved in some way in regulation of the aberrations of oxygen activation within the cell but we seem to be a long way from pinning down its exact point of attack. Might it just have the same kind of role as catalase?

Flohé: No. In all highly organized cells glutathione peroxidase is present in different compartments from catalase; catalase is restricted to the peroxisomes. The cytosol and the matrix space of the mitochondria are the site of action of glutathione peroxidase, and so there is no overlap in function. Several tissues are rich in catalase and poor in glutathione peroxidase and *vice versa.* No catalase can be found in the crystalline lens, for instance, whereas glutathione peroxidase is abundant. In this case, its primary role is to scavenge H_2O_2, but from *in vitro* experiments on specificity, biomembrane protection and removal of exogenous organic hydroperoxides (see p. 105) we may deduce that it could do much more than that. However, since we do not have clear data for the occurrence of well defined hydroperoxides *in vivo*, we cannot say which peroxides are the primary targets of the enzyme.

Dormandy: I cannot agree with Professor Crichton that ceruloplasmin is a protein without a function. It is the most important antioxidant in serum (Al-Timimi & Dormandy 1977; Dormandy 1978).

Crichton: Several functions have been attributed to it, including that of a ferrooxidase.

Michelson: It is not quite true that thymine 6-hydroperoxide is a mutagen (p. 107). Rather, the formation of the hydroperoxide is a mutagenic event because insertion of the 6-hydroperoxide group changes the conformation from *anti* to *syn*. If glutathione peroxidase uses this hydroperoxide as a substrate, it is correcting a mutation that has already happened.

Flohé: Thomas *et al.* (1976) added thymine hydroperoxide to transforming DNA to produce mutations and concluded that such modified bases cause mispairing of bases during replication. A contribution to the repair mechanisms of GSH peroxidase such as you propose could also be visualized.

Fridovich: One should keep in mind that the glutathione peroxidase reduces a hydroperoxide only to an alcohol and no further.

Cohen: You referred to the similarity between glutathione peroxidase deficiency and favism (p. 104), in which, although the amount of glutathione did not fall, a haemolytic episode ensued. However, a dramatic fall in reduced glutathione (GSH) ensues when primaquine is administered to a subject deficient in glucose-6-phosphate dehydrogenase (cf. Flanagan *et al.* 1958). Could the difference reflect the fact that fava-bean haemolysis is not necessarily oxidant haemolysis? After all, an individual subject to favism will undergo a haemolytic episode merely on walking through a field of fava beans without ingesting the material.

Winterhalter: Only when the plants are in bloom!

Cohen: Others have invoked immune phenomena as the cause of the disorder (inhalation of fava-bean pollen). This need not be inconsistent with mechanisms invoking peroxides and glutathione. The complex relationship between an immune factor in red cells and glucose-6-phosphate dehydrogenase deficiency remains obscure.

Flohé: Immunological complications certainly add to the variability of this clinical condition but, in principle, the active compounds from fava beans are prooxidants and cause lipid peroxidation and GSH depletion (Flohé *et al.* 1971).

Cohen: In primaquine haemolysis not only glutathione but haemoglobin is oxidized. In favism is haemoglobin oxidized also?

Winterhalter: Yes.

Fridovich: The fava bean compounds, vicine and convicine, are of the kind that could undergo a redox cycle.

Flohé: I did not mean to imply that glutathione is stable in glucose-6-phosphate dehydrogenase-deficient red blood cells. Its instability is routinely demonstrated by incubation with acetylphenylhydrazine *in vitro*. However, the GSH depletion appears to depend on the residual activity of glucose-6-phosphate dehydrogenase in the individual cell and would be most pronounced in those cells which cannot be detected any more after a haemolytic episode. The decrease of the *mean* concentration of glutathione in glucose-6-phosphate dehydrogenase-deficient erythrocytes is too small to account for the total collapse of the peroxidase system.

Roos: Cells in general probably have a critical concentration of glutathione. For instance, in glutathione synthetase-deficient leucocytes with 25% of the normal glutathione concentration, leucocyte function is severely impaired (Spielberg *et al.* 1978). However, at about 50%, as is found in phagocytosing leucocytes of heterozygotes for glutathione reductase-deficiency (D. Roos *et al.* unpublished work, 1978), the cells still function normally.

With regard to the supposed overlap between catalase and the glutathione system, I shall present some data (pp. 225–226) that these two systems function beside each other. When leucocytes that are homozygously deficient in glutathione reductase are stressed, either by phagocytosis (whereupon these cells produce hydrogen peroxide) or by incubation with exogenous hydrogen peroxide, they are rapidly inactivated but the concentration of reduced glutathione does not drop much below 40–50% of normal.

Flohé: What does that mean? A normal concentration in perhaps 50% of the cells? Leucocytes are not homogeneous.

Roos: But all the cells lose their functional activity. I do not know whether glutathione reductase-deficient cells with a normal amount of reduced glutathione are still present. Addition of azide (to inhibit catalase) rapidly lowers the amount of reduced glutathione almost to zero. On the other hand, when catalase-deficient leucocytes are stressed by large amounts of hydrogen peroxide, they are also inactivated, in contrast to normal leucocytes. It seems that hydrogen peroxide is detoxified by the catalase and glutathione systems, which operate next to each other.

Flohé: I agree. According to our kinetic model the critical GSH level should be reached earlier in white blood cells producing H_2O_2 at a high rate than in erythrocytes.

Flower: Does the amount of glutathione peroxidase in, say, red blood cells change much with age?

Flohé: In rats it increases with the age of the animal (Pinto & Bartley 1969; Flohé & Zimmermann 1970).

Fridovich: The amount of most enzymes in red cells declines with the age of the red cell.

Flohé: The enzyme inside the red cell seems to be more stable than, for instance, glucose-6-phosphate dehydrogenase, the decline of which appears to limit the lifespan of normal erythrocytes.

Fridovich: With regard to mechanism it is generally accepted that the oxidation of horseradish peroxidase by hydrogen peroxide is a bivalent event and that the reduction is a sequence of two monovalent events proceeding through substrate radicals as intermediates. Do you consider that glutathione peroxidase makes the $GS^{\bullet}$ radical as an intermediate, by analogy?

Flohé: No, but we did not really consider this possibility, since the analogy of these two reactions is restricted to the kinetics. The molecular reaction mechanism of horseradish peroxidase must be entirely different. Horseradish peroxidase yields at least 20 different identifiable products due to the monovalent reduction of the donor substrate, whereas GSH peroxidase produces exclusively GSSG.

Fridovich: If $GS^{\bullet}$ radicals were made and came together, there would be only one product.

Flohé: Your suggestion would be plausible if the enzyme prevented $GS^{\bullet}$ from diffusing from the active site.

Fridovich: Why is that necessary?

Flohé: If the $GS^{\bullet}$ could diffuse, it would be able to react with many cellular constituents.

Fridovich: But what about *in vitro* in a system containing pure enzyme, pure hydroperoxide and GSH (which I presume, is the system you use to study the kinetics)?

Flohé: In purified systems and with unpurified enzyme probes, which contain many other materials, the only product is GSSG. That does not eliminate the possibility of a free-radical mechanism, but it makes it highly unlikely.

Willson: Is this necessarily so? Even if $GS^{\bullet}$ does diffuse a little way its environment will still be rich in GSH, some of which (no matter how little) will be in the form of GS^{-}. $GS^{\bullet}$ will react with GS^{-} extremely rapidly to give $GSSG^{-}$ which in turn will react rapidly with oxygen to give GSSG. So does $GS^{\bullet}$ necessarily diffuse a long way and enter into other reactions or does it not subsequently lead to GSSG?

Flohé: Such a scheme does not seem likely to me. Besides, selenium in the active site (now identified as a selenol) is not prone to monovalent oxidation-reduction.

Fridovich: Until somebody studies this by a method such as e.s.r. spectroscopy we shall not know; certainly, a substrate radical is possible.

References

AL-TIMIMI, D. J. & DORMANDY, T. L. (1977) Inhibition of lipid autoxidation by human caeruloplasmin. *Biochem. J. 168*, 283–288

COHEN, G. & HOCHSTEIN, P. (1963) Glutathione peroxidase: the primary agent for the elimination of hydrogen peroxide in erythrocytes. *Biochemistry 2*, 1420–1428

CHRISTOPHERSEN, B. D. (1968) Formation of monohydroxy polyenic fatty acids from lipid peroxides by a glutathione peroxidase. *Biochim. Biophys. Acta 164*, 35–36

DIPLOCK, A. T., BAUM, H. & LUCY, J. A. (1971) The effect of Vitamin E on the oxidation state of selenium in rat liver. *Biochem. J. 123*, 721–729

DIPLOCK, A. T., CAYGILL, C. P. J., JEFFERY, E. H. & THOMAS, C. (1973) The nature of acid-volatile selenium in the liver of the male rat. *Biochem. J. 134*, 283–293

DORMANDY, T. L. (1978) Free-radical oxidation and antioxidants. *Lancet 1*, 647–650

FLANAGAN, C. L., SCHREIR, S. L., CARSON, P. E. & ALVING, A. S. (1958) The hemolytic effect of primaquine VIII. The effect of drug administration on parameters of primaquine sensitivity. *J. Lab. Clin. Med. 51*, 600

FLOHÉ, L. & ZIMMERMANN, R. (1970) The role of GSH peroxidase in protecting the membrane of rat liver mitochondria. *Biochim. Biophys. Acta 223*, 210–213

FLOHÉ, L. & ZIMMERMANN, R. (1974) GSH-induced high-amplitude swelling of mitochondria, in *Glutathione* (Flohé, L., Benöhr, H. Ch., Sies, H., Waller, H. D. & Wendel, A., eds.), pp. 245–260, Georg Thieme, Stuttgart

FLOHÉ, L., NIEBCH, G. & REIBER, H. (1971) Zur Wirkung von Divicin in menschlichen Erythrocyten. *Z. Klin. Chem. Klin. Biochem. 9*, 431–437

GIBSON, J. E. & CAGEN, S. Z. (1977) Paraquat induced functional changes in kidney and liver, in *Biochemical Mechanisms of Paraquat Toxicity* (Autor, A. P., ed.), pp. 117–136, Academic Press, New York

LIEBOWITZ, J. & COHEN, G. (1968) Increased hydrogen peroxide levels in glucose 6-phosphate dehydrogenase deficient erythrocytes exposed to acetylphenylhydrazine. *Biochem. Pharmacol. 17*, 983–988

MCCAY, P. B., GIBSON, D. D., FONG, K. L. & HORNBROOK, K. R. (1976) Effect of glutathione peroxidase activity on lipid peroxidation in biological membranes. *Biochim. Biophys. Acta 431*, 459–468

MILLS, G. C. (1959) The purification and properties of glutathione peroxidase of erythrocytes. *J. Biol. Chem. 234*, 502–506

NEUBERT, D., WOJTCZAK, A. B. & LEHNINGER, A. L. (1962) Purification and enzymatic identity of mitochondrial contraction-factors I and II. *Proc. Natl. Acad. Sci. U.S.A. 48*, 1651–1658

PINTO, R. E. & BARTLEY, W. (1969) The effect of age and sex on glutathione reductase and glutathione peroxidase activities and on aerobic glutathione oxidation in rat liver homogenates. *Biochem. J. 112*, 109–115

PRINS, H. K., OORT, M., LOOS, J. A., ZÜRCHER, C. & BECKERS, T. A. (1966) Congenital non-spherocytic hemolytic anemia associated with glutathione deficiency of the erythrocytes. Hematologic, biochemical and genetic studies. *Blood 27*, 145–166

SCHRAUZER, G. N., WHITE, D. A. & SCHNEIDER, C. J. (1977) Cancer mortality correlation studies III. Statistical associations with dietary selenium intakes. *Bioinorg. Chem. 7*, 23–34

SPIELBERG, S. P., BOXER, L. A., OLIVER, J. M., ALLEN, J. M. & SCHULMAN, J. D. (1978) Oxidative damage to neutrophils in glutathione synthetase deficiency. *Br. J. Haematol.*, in press

THOMAS, H. F., HERRIOTT, R. M., HAHN, B. S. & WANG, S. Y. (1976) Thymine hydroperoxide as a mediator in ionising radiation mutagenesis. *Nature (Lond.) 259*, 341–342

Biosynthesis of prostaglandins

RODERICK J. FLOWER

Wellcome Research Laboratories, Beckenham, Kent

Abstract The generation of prostaglandins is catalysed by a membrane-bound multienzyme complex. The first reaction of the biosynthetic sequence is the generation (by the enzyme 'fatty acid cyclooxygenase') of prostaglandin endoperoxides, a reaction which involves the incorporation of two moles of oxygen: this reaction probably proceeds by an *ene* reaction rather than a free-radical mechanism. After biosynthesis the endoperoxides can be metabolized in various ways depending on the cell-type. For example, in platelets they may be transformed into non-prostanoid compounds called thromboxanes, whereas vascular endothelium and many other tissues generate another derivative, prostacyclin. In other tissues, the 'classical' prostaglandins E, F or D may be generated.

Some products of the cyclooxygenase (e.g. hydroperoxides, malonaldehyde) may have a direct toxic action on cells: prostaglandins themselves do not, but some types (especially those of the E series) are probably responsible for many of the clinical signs and symptoms of inflammation.

Amongst the most exciting events in the pharmacological world during the past few decades has been the discovery of the prostaglandins and their close relatives the thromboxanes and prostacyclins. Prostaglandins are generated by the oxidation of polyunsaturated fatty acids such as arachidonic acid. They are biosynthesized whenever cells are injured, whether this injury is mechanical, chemical, immunological or bacteriological. The appearance of prostaglandins seems to be a most sensitive index of cell damage and the prostaglandins so generated are almost certainly important causative agents in the clinical signs and symptoms of inflammation.

In this paper I shall describe the biosynthesis of the various prostaglandins, thromboxanes and prostacyclins, and conclude by outlining the inflammogenic and toxic effects of the various end products of the oxygenation.

PROSTAGLANDIN NOMENCLATURE

The prostaglandins may be considered derivatives of prostanoic acid (see Fig. 1) and the nomenclature is based on that carbon skeleton. There are several groups of prostaglandins distinguished by the nature and geometry of the substituent groups in the ring (for example E, F, D) and the number of double bonds in the side-chains (for example E_1, E_2, E_3; F_1, F_2, F_3 etc.). Fig. 1 shows the structures of these primary prostaglandins as well as the structures of two non-prostanoid transformation products of the endoperoxide intermediate: the thromboxanes (A and B) and prostacyclin. Also shown are the side-chain modifications of the principal metabolites.

NATURE, LOCATION AND DISTRIBUTION OF THE PROSTAGLANDIN BIOSYNTHETIC ENZYMES

The enzyme system which catalyses the oxidation of polyunsaturated fatty acids to endoperoxides and their subsequent transformation to prostaglandins

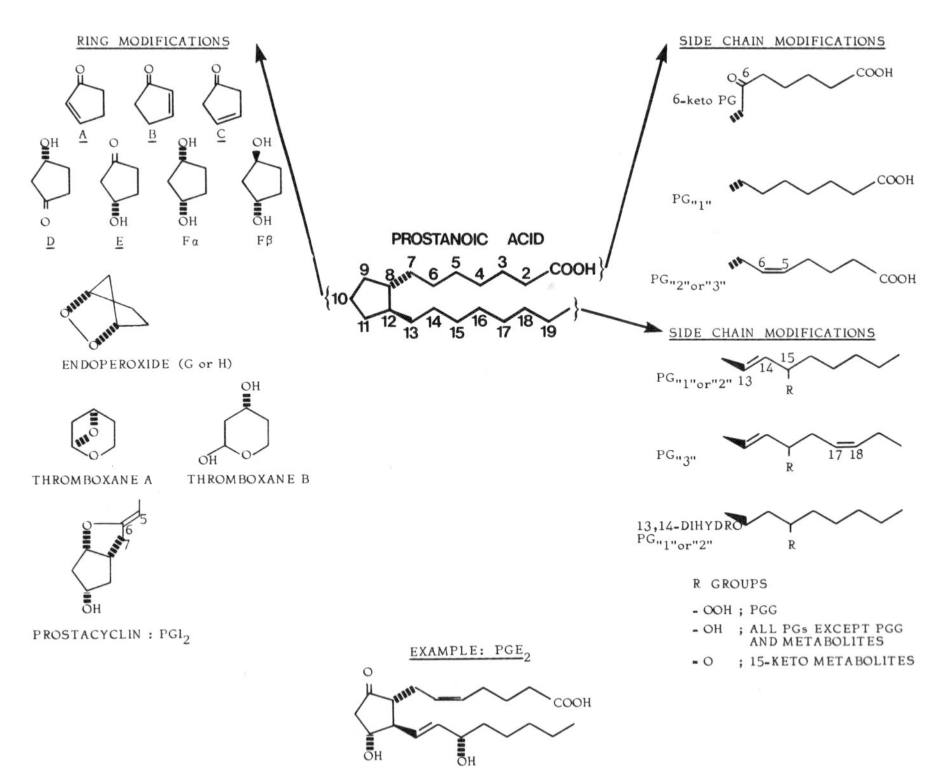

FIG. 1. Prostanoic acid and its derivatives the prostaglandin thromboxanes and prostacyclins.

and other products is a multienzyme complex located in the high-speed particulate fraction of cells known as the *fatty acid cyclooxygenase* (old nomenclature; prostaglandin synthase, EC 1.14.99.1). It is not clear to which membrane fraction the enzyme is bound although in a series of careful experiments Bohman & Larsson (1975) found that most of the cyclooxygenase activity in rabbit kidney was associated with the endoplasmic reticulum membranes.

The membrane-bound nature of the enzyme complex has frustrated most efforts to characterize the individual components. Recently, however, an important advance has been made: Professor Hayaishi and his colleagues (Miyamoto *et al.* 1974; Yoshimoto *et al.* 1977) have succeeded in solubilizing the cyclooxygenase system from bovine seminal vesicles and from platelets using Tween 20 as a detergent. In the former case they found that the soluble fraction could be further separated by DEAE-cellulose chromatography into two fractions: 'fraction 1', a dioxygenase enzyme which catalysed the generation from dihomo-γ-linolenic acid of the prostaglandin endoperoxide PGH_1 (see later), and 'fraction 2', which could be eluted from the column with increased (200mM-)phosphate buffer concentrations and catalysed the isomerization of the endoperoxide to PGE_1.

Cyclooxygenase activity appears to be present in every mammalian tissue so far investigated, as well as several non-mammalian tissues also (Horton 1969; Ramwell & Shaw 1970; Christ & Van Dorp 1972, 1973; Light & Samuelsson 1972; Miyares-Cao & Menendez-Cepero 1976).

NATURE AND ORIGIN OF FATTY ACID SUBSTRATES FOR THE CYCLOOXYGENASE

Substrates for the cyclooxygenase belong to the group of so-called essential fatty acids. The substrate requirements for the sheep vesicular gland cyclooxygenase were tested by Van Dorp (1967) who surmised that the structural requirements for conversion of fatty acids into prostaglandins is the presence of *cis*-double bonds at least at positions 8, 11, 14 and — since the methyl esters are not substrates — a free carboxy group also. For optimal activity, a chain length of 20 carbon atoms is required with *cis*-double bonds in positions 8, 11, 14 or 5, 8, 11, 14.

Prostaglandins, and other cyclooxygenase products, are not stored within cells (Piper & Vane 1971) and so biosynthesis must immediately precede release. For synthesis the substrate must be in a non-esterified form (Lands & Samuelsson 1968; Vonkeman & Van Dorp 1968) and probably originates in the phospholipid fraction of the cell, under the influence of the hydrolytic enzyme phospholipase

A_2 (EC 3.1.1.4: cf. Lands & Samuelsson 1968; Vonkeman & Van Dorp 1968; Kunze & Vogt 1971; Flower & Blackwell 1976; Blackwell *et al.* 1977).

CHEMICAL TRANSFORMATIONS CATALYSED BY THE CYCLOOXYGENASE

The enzymic conversion of essential fatty acids into prostaglandins was demonstrated in 1964 by two groups: Van Dorp *et al.* (1964*a,b*) in Holland and Samuelsson's (Bergström *et al.* 1964*a,b*) in Sweden. Most of the subsequent biochemistry of prostaglandin synthesis has also been pioneered by these two groups. Although much of the original experimental work on the reaction mechanism used the cyclooxygenase from sheep seminal vesicles (because of the high conversion of substrate), there is no reason to suspect that the generation of endoperoxides catalysed by enzymes from other tissues proceeds by a radically-different mechanism. As I have pointed out above, fatty acid cyclooxygenase is a multienzyme complex which catalyses several reactions and, furthermore, the end products of the reaction vary from tissue to tissue. For this reason the description of the reaction mechanism which follows is broken down into different sections.

Generation of the cyclic endoperoxides

Evidence for the formation of an endoperoxide structure as an intermediate in the cyclooxygenase reaction first came from experiments with $^{18}O_2$–$^{16}O_2$ by Samuelsson (1965) who observed that the prostaglandin products E and F contained either $^{16}O_2$ or $^{18}O_2$ in the cyclopentane ring but never a mixture. In subsequent experiments with guinea-pig lung Änggård & Samuelsson (1965) observed that PGE and PGF could not be interconverted, a fact that indicates their formation from a common intermediate, and this was confirmed later by the studies of Wlodawer & Samuelsson (1973).

The initial step of this dioxygenase reaction (Hamberg & Samuelsson 1967*a*; Nugteren *et al.* 1966) is initiated by a stereospecific (L) removal of the $\omega-8$ proton and the conversion of the substrate into an $\omega-10$ hydroperoxide (see Fig. 2). This step is somewhat reminiscent of the reaction catalysed by the plant enzyme soyabean lipoxidase which also removes this hydrogen stereospecifically although in this case an $\omega-6$ hydroperoxide is formed (Hamberg & Samuelsson 1967*b*). The next stage is a concerted reaction; the addition of oxygen at C-15 is followed by an isomerization of the C-12 double bond, ring closure between C-8 and C-12 and an attack by an oxygen radical (of the C-11 hydroperoxide) at C-9 thus forming a cyclic endoperoxide. The resulting intermediate is referred to as PGG by Samuelsson's group (see Hamberg & Samuelsson 1973; Hamberg

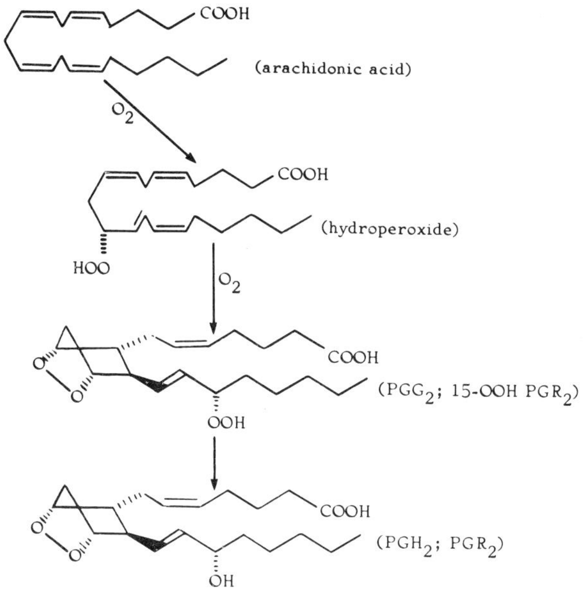

FIG. 2. Transformation (catalysed by the dioxygenase component of the cyclooxygenase) of arachidonic acid to the unstable endoperoxides G_2 and H_2.

et al. 1974*b*) and as 15-hydroperoxy PGR by Nugteren & Hazelhof (1973). The consensus thus favours a synchronous *ene* (rather than a free-radical) mechanism.

The next stage involves reduction of the 15-hydroperoxy group to the corresponding hydroxy group, giving the compound known as PGH by Samuelsson's group or as PGR by Nugteren & Hazelhof. Both endoperoxide intermediates are unstable, spontaneously decomposing (in a matter of minutes) in aqueous solutions to mixtures of prostaglandins. Temperature and pH are important in determining their half-life in aqueous solutions (Nugteren & Hazelhof 1973). Notwithstanding their instability, the endoperoxides can be isolated in organic solvents and are stable in dry acetone at -20 °C for some weeks.

There is evidence that the cyclooxygenase (*in vitro* at least) 'self-destructs' during this initial conversion, perhaps because of free radicals released during reduction of prostaglandin G_2 to H_2 (Egan *et al.* 1976; Kuehl *et al.* 1977).

Transformation of the endoperoxide to HHT

The 17-monohydroxy fatty acid HHT (L-12-hydroxy-5,8,10-heptadecatrienoic acid, see Fig. 3) had been isolated from incubation mixtures of arachidonic acid and sheep seminal vesicle homogenates by Wlodawer & Samuelsson (1973)

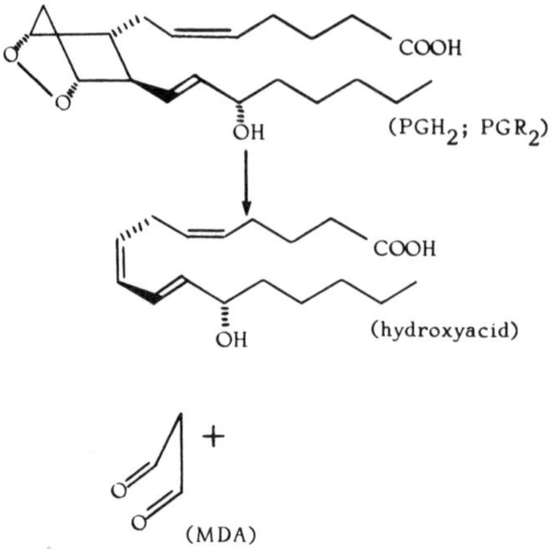

FIG. 3. Transformation of the endoperoxide to a C-17 fatty acid and malonaldehyde (MDA). Both enzymic and non-enzymic mechanisms probably exist to generate these products.

and was also found to be a by-product of chemical reduction of the endoperoxides PGG_2 and PGH_2 by $SnCl_2$ (Hamberg *et al.* 1974*a,b*). Formation by sheep seminal vesicle homogenates of the analogous hydroxy fatty acid from dihomo-γ-linolenic acid had previously been demonstrated by Nugteren *et al.* (1966), and by Hamberg & Samuelsson (1966, 1967*a,b*), who postulated that it was derived from the endoperoxide intermediate after the expulsion of malonaldehyde from PGG_2. The formation of HHT in platelets was observed in 1974 by Hamberg & Samuelsson, and later the same year Hamberg and his co-workers measured the release of HHT from aggregating human platelets and demonstrated that it represented a considerable proportion (about 30%) of total endoperoxide metabolism. In a further paper Hamberg & Samuelsson demonstrated that the transformation of PGG_2 to HHT also occurs in guinea-pig lungs as well and that here also it represents a major metabolic pathway.

The formation of HHT is accompanied by the generation of malonaldehyde. As this compound may be readily measured by a simple colorimetric method using thiobarbituric acid (TBA), this assay has been used (successfully) as an index of cyclooxygenase activity (cf. Flower *et al.* 1973). The TBA reaction has also been used for many years as an index of 'lipid peroxidation' (see for example Dahle *et al.* 1962; Schultz 1962 and references therein), and an interesting question is whether much of the malonaldehyde measured by early workers originated from the cyclooxygenase pathway. According to Niehaus & Samuelsson (1968) the formation of malonaldehyde which occurs during lipid peroxidation, although derived mainly from phospholipid arachidonate, proceeds by a different mechanism.

Transformation of the endoperoxide to prostaglandins

Opinion seems to be divided as to which of the endoperoxide intermediates is the precursor of PGE (see Fig. 4). Nugteren & Hazelhof (1973) propose that PGH is the precursor of PGE (as well as PGF and PGD), whereas Hamberg & Samuelsson (1974) believe that the reaction proceeds *via* the intermediate 15-hydroperoxy-PGE. However, as Nugteren & Hazelhof point out, when 15-hydroperoxy-PGE decomposes in the absence of a reducing agent it gives rise to significant quantities of 15-oxo-PGE, a fact which perhaps explains why this compound is sometimes found as a product of prostaglandin biosynthesis.

Using labelled PGH_2 as a precursor Nugteren & Hazelhof (1973) investigated the formation of labelled prostaglandins in different tissues. The enzyme responsible for the formation of PGE ('endoperoxide–PGE isomerase'; prostaglandin R_2E-isomerase, EC 5.3.99.3) was found by these authors to be particulate and to require GSH for maximal activity; it could perhaps be identical to glutathione peroxidase. The enzyme catalysing the formation of PGD ('endoperoxide–PGD isomerase') was apparently a soluble protein of molecular weight 36 000–42 000, which also required GSH for activity. Nugteren & Hazelhof concluded that the formation by PGF (by an 'endoperoxide reductase') could simply be a non-enzymic phenomenon since endoperoxides are reduced to PGF whenever tissue homogenates contain reducing agents (i.e. SH compounds, ferrihaem compounds, etc.) and an appreciable conversion of endoperoxide into PGH could be seen even when such homogenates were boiled, thus ruling out an enzymic conversion. Possibly, there is no 'endoperoxide reductase' and PGF occurs in tissues having a relative deficiency of 'E' or 'D' isomerase enzymes and a sufficient level of reducing factors. This

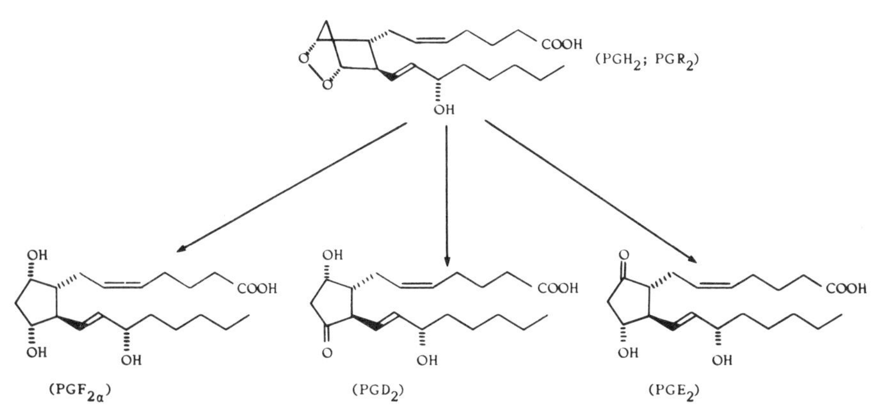

FIG. 4. Transformation by 'terminal isomerases' of the unstable endoperoxide H_2 to prostaglandins D and E. It is not clear whether the reduction to prostaglandin F is enzymic.

conclusion is also supported by the experiments of Chan *et al.* (1975) who demonstrated that chemical reducing agents promote PGF biosynthesis by bovine seminal vesicle microsomes.

Transformation of the endoperoxide to thromboxane A and B (PHD)

The first demonstration that PGG_2 could be transformed to 'thromboxanes' (see Fig. 5) was reported in 1974 by Hamberg & Samuelsson (1974*a*), although at that time the name was not yet coined. They described the transformation of PGG_2 in human platelets; the two major compounds formed from the endoperoxide were a C_{17} hydroxy fatty acid (HHT; see above) and a polar compound not previously described: the hemi-acetal of 9,12L-dihydroxy-8-(1-hydroxy-3-oxopropyl)-5,10-heptadecadienoic acid. Experiments with $^{18}O_2$ demonstrated that both the endoperoxide oxygen atoms were incorporated into the six-membered ring of this polar compound, although the exact mechanism of its formation was not elucidated at that time. Later the same year a further paper by Samuelsson's group appeared (Hamberg *et al.* 1974*a*) in which quantitative measurements of the formation of this polar product (provisionally called PHD), the hydroxy fatty acid HHT, and PGs were reported, as a result of multiple-ion analysis with deuteriated internal standards for the HHT and PHD estimations. When washed human platelets were aggregated by thrombin, large quantities of arachidonate oxidation products were released, about one third of which was PHD and the remainder comprised equal quantities of HHT and a lipoxygenase product. The amounts of prostaglandins (E_2 and $F_{2\alpha}$) formed were some two orders of magnitude less than the total amounts of

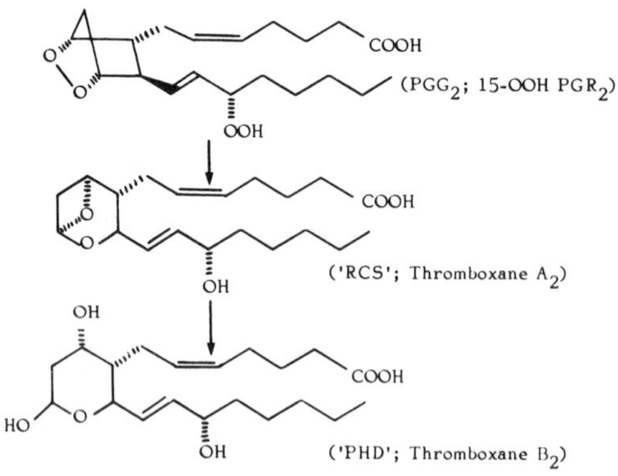

FIG. 5. Transformation of the endoperoxide G_2 to, first, the highly labile thromboxane A_2 and then the stable but biologically-inactive thromboxane B_2. The former but not the latter pathway is enzymically catalysed.

PGG_2 biosynthesized, thus indicating that prostaglandins as such were formed in insignificant amounts compared with other endoperoxide products.

The mechanism of the transformation of PGG_2 to this metabolite in platelets (and lungs) was not understood until Hamberg *et al.* (1975) gave an account of the enzymic reaction by which PGG_2 is converted into PHD *via* a highly unstable intermediate. Various nucleophilic reagents such as ethanol, methanol and sodium azide when added to the platelet incubation medium resulted in the formation of derivatives from which the structure of the intermediate was deduced. Because this intermediate had an oxane ring, was a potent platelet aggregating substance, and in this case was derived from arachidonic acid (and therefore had two double bonds) Samuelsson's group named it thromboxane A_2. The metabolite of TXA_2 which had been provisionally named PHD was then renamed thromboxane B_2. As with the initial work on transformations of PGG_2, this first demonstration of the generation of TXA_2 was made using human platelets.

Transformation of the endoperoxides to prostacyclin

The last major transformation product of the endoperoxides that I shall discuss is also the last addition to the prostaglandin family: prostacyclin (see

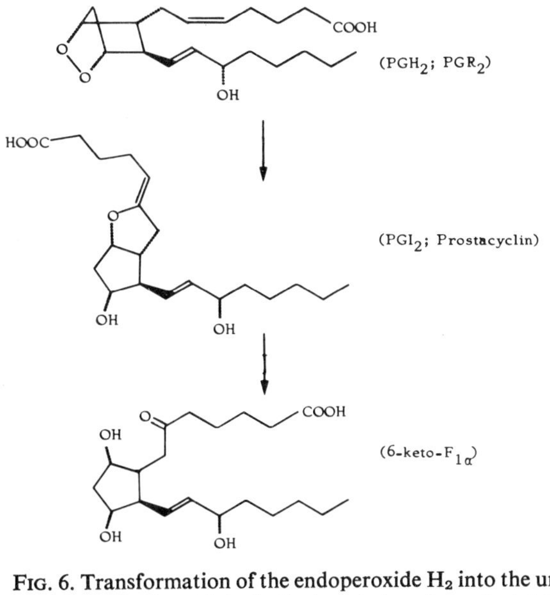

COOH
O
O
OH
(PGH_2; PGR_2)
HOOC
O
(PGI_2; Prostacyclin)
OH
OH
O
COOH
OH
(6-keto-$F_{1\alpha}$)
OH
OH

FIG. 6. Transformation of the endoperoxide H_2 into the unstable prostacyclin, and ultimately to the stable (but biologically-inactive) end product 6-keto-$PGF_{1\alpha}$. The latter reaction is probably not enzymic.

Fig. 6). This compound (also called prostaglandin I_2) was discovered in 1976 by a team working at the Wellcome Foundation and later identified by the Upjohn Company in the USA. It was observed that the incubation of prostaglandin endoperoxides with the 'microsomal' fraction of homogenized arteries led to the formation of a product which had potent anti-platelet-aggregating activity (Moncada *et al.* 1976*a*). It was demonstrated that this substance was different from other known products of prostaglandin metabolism, that it was unstable and that it was less spasmogenic than the parent endoperoxides (Gryglewski *et al.* 1976; Bunting *et al.* 1976) and that its formation was inhibited by lipid peroxides (Moncada *et al.* 1976*b*). The product was subsequently identified and given the name prostacyclin (Johnson *et al.* 1976) although this was later changed to PGI_2; the instability of prostacyclin was confirmed and its degradation product was identified as 6-keto-$PGF_{1\alpha}$.

TISSUE DAMAGE

It is the intention of this symposium to relate the formation of oxygenated derivatives to tissue damage and so I shall, in conclusion, discuss the ability of certain products of the cyclooxygenase to cause tissue damage and inflammation.

Occurrence of different lipid mediators in inflammation and tissue damage

Prostaglandins E or F can be recovered in increased quantities from many types of damaged or inflamed tissue including several forms of skin damage such as contact dermatitis (Greaves *et al.* 1971), u.v.-induced inflammation (Søndergaard & Greaves 1970) and burn injury (Änggård & Jonsson 1971; Hamberg & Jonsson 1973). They have also been recovered in experimental uveitis (Eakins *et al.* 1972) and other laboratory models of inflammation such as carrageenan-induced inflammation (Willis 1969; Di Rosa *et al.* 1971), monarticular arthritis (Blackham *et al.* 1974), as well as 'clinical' rheumatoid arthritis (Higgs *et al.* 1974). Immunological injury, too, releases prostaglandins (Piper & Vane 1969). Homogenates of inflamed tissue can generate large quantities of prostacyclin (Chang *et al.* 1976). As in peripheral inflammation, prostaglandin concentrations found in the cerebrospinal fluid of cats during fever may be increased several-fold over control levels (Feldberg & Gupta 1973).

Because they so readily decompose to prostaglandins it is not known whether the endoperoxides G_2 and H_2 themselves are found in inflammatory exudates. Malonaldehyde has been reported to occur in damaged tissues (Plaa & Witschi 1976).

Activity of lipid mediators in inflammation

Prostaglandins have several important inflammogenic properties; when injected subdermally in man or animals E prostaglandins cause vasodilatation (Solomon *et al.* 1968) and erythema (Ferreira 1972; Crunkhorn & Willis 1971; Juhlin & Michaelsson 1969) as well as an increase in vascular permeability and oedema (Crunkhorn & Willis 1971; Glenn *et al.* 1972), effects which strongly resemble the inflammatory process itself. When given concomitantly with carrageenan in the rat-foot oedema assay, E_1 and E_2 greatly potentiate the oedema due to this phlogistic agent (Moncada *et al.* 1973; Lewis *et al.* 1974). This experiment highlights an important property of prostaglandins, namely their ability to potentiate the inflammatory effects of other mediators. Another interesting and important property of prostaglandins is their long-lasting effect; the erythema caused by subdermal infusions of prostaglandin E_1 may last for as long as 10 h (Ferreira 1972). Nanogram amounts of E prostaglandins are effective in producing such effects.

Accompanying the inflammatory response is often pain and here, too, E prostaglandins can reproduce the signs and symptoms of inflammation. Large amounts of prostaglandins can cause pain (Ferreira 1972) but it is unlikely that such concentrations are ever likely to be found *in vivo*. It is observed, however, that very low (nanogram) amounts of prostaglandins can sensitize pain endings in the skin (Ferreira 1972) and elsewhere to the algesic effects of other stimuli, whether they are chemical or mechanical (Ferreira *et al.* 1973; Moncada *et al.* 1975; Willis & Cornelsen 1973). The hyperalgesia produced by subdermal infusions is extremely long lasting (Ferreira 1972). This property of E prostaglandins — their action as 'biochemical amplifiers' thus exacerbating the inflammogenic or algesic effects of other mediators — is undoubtedly one of the most important features of their action.

Prostaglandin F shares with prostaglandins of the E series the ability to cause erythema (Solomon *et al.* 1968; Crunkhorn & Willis 1971; Juhlin & Michaelsson 1969), but the doses required are some 2–3 orders of magnitude greater than those of E prostaglandins. In contradistinction to E prostaglandins, however, $F_{2\alpha}$ is a venoconstrictor (Sweet *et al.* 1971) and may antagonize the increase in vascular permeability caused by 5-hydroxytryptamine, histamine and E prostaglandins (Crunkhorn & Willis 1971). Because of this action, it has been suggested that prostaglandin $F_{2\alpha}$ is involved in termination of the inflammatory response (Velo *et al.* 1973).

Prostaglandin D_2 has never been found (or looked for!) in inflammation. In experimental situations it shares with E prostaglandins the ability to cause a long-lasting erythema and causes an increase in vascular permeability when injected subdermally in rats. Quantitatively, however, prostaglandin D_2 was much less active (by one order of magnitude) than E_1 or E_2 (Flower *et al.* 1976).

Because of the difficulty in handling the very labile prostaglandin endoperoxides and the rapidity with which they decompose, they have proved extremely difficult to test as inflammatory mediators. At least one study demonstrates that G_2 or H_2 possesses inflammogenic properties in the rat although it is unfortunately impossible to exclude the possibility that this effect was due to the prior conversion of these compounds into prostaglandins (S. H. Ferreira, M. F. Parsons & J. R. Vane, unpublished results, 1976). Kuehl and his coworkers have provided interesting evidence that prostaglandin G_2 is important in inflammation: investigating a novel anti-inflammatory compound designated MK-447, they found that it accelerated the reduction of prostaglandin G_2 to H_2 without decreasing the overall conversion of substrate. They speculated that MK-447 acted as a scavenger for proinflammatory free radicals released during the conversion.

The possibility that HHT contributes to the inflammatory response has not been investigated. There is ample evidence from the literature that malonaldehyde is cytotoxic (cf. Flower *et al.* 1973).

Amongst the undesirable sequelae of inflammation may be platelet aggregation, and there is now good evidence that thromboxane A_2 is an important effector of this process (Hamberg *et al.* 1975). Nanogram quantities of thromboxane A_2 cause irreversible aggregation of platelets accompanied by extrusion of granular 5-hydroxytryptamine (Hamberg *et al.* 1975).

PHARMACOLOGICAL CONTROL OF LIPID MEDIATOR PRODUCTION

Antiphospholipase drugs

By preventing the initial release of fatty acid substrates from their store (we are assuming this is generally phosphatides) the entire cascade of lipid mediator production may be prevented. Mepacrine inhibits phospholipase A_2 activity *in vitro* but, unfortunately, at the high doses required for its activity it also inhibits cyclooxygenase; this makes critical examination of its pharmacology difficult (cf. Flower & Blackwell 1976). There is now, however, increasing evidence that anti-inflammatory steroids can interfere with the release of fatty acids from phospholipids — perhaps by antagonizing in some way the action of cell-membrane phospholipase A_2 (cf. Nijkamp *et al.* 1976). These compounds are exceedingly potent anti-inflammatory drugs. Although a convincing mode of action of steroids has not yet been suggested, perhaps one facet of their action is the suppression of lipid-mediator formation. Since the formation of all the lipid mediators would be suppressed by antiphospholipase drugs one might expect them to have a rather more complete anti-inflammatory effect than the following group of drugs.

Cyclooxygenase inhibitors

In 1971 Vane, Ferreira *et al.* and Smith & Willis made the fundamental discovery that the so-called 'aspirin-like' drugs were powerful inhibitors of the cyclooxygenase. The chief pharmacological actions of this group of compounds are anti-inflammatory, antipyretic and analgesic. Since the proinflammatory pyretic and algesic actions of E prostaglandins have already been discussed it can be seen how neatly the theory proposed by Vane to account for the pharmacological activities of this group of drugs fits the observed facts. The observation that aspirin-like drugs reduce inflammation is excellent support for the role of prostaglandins and other lipid mediators of the cyclooxygenase pathway in inflammation.

SUMMARY AND CONCLUSIONS

I have briefly reviewed our current knowledge of the biosynthesis of that group of polyoxygenated arachidonic acid derivatives known as prostaglandins and thromboxanes. This field is undoubtedly one of the fast growth points in the biomedical world and it is evident that the cyclooxygenase is a ubiquitous and important enzyme system, whose products may be important modulators of tissue damage, and which is, therefore, important target enzyme for anti-inflammatory drugs.

References

ÄNGGÅRD, E. & JONSSON, C. E. (1971) Efflux of prostaglandins in lymph from scalded tissue. *Acta Physiol. Scand. 81*, 440–447

ÄNGGÅRD, E. & SAMUELSSON, B. (1965) Biosynthesis of prostaglandins from arachidonic acid in guinea pig lung. Prostaglandins and related factors 38. *J. Biol. Chem. 240*, 3518–3521

BERGSTRÖM. S., DANIELSSON, H., KLENBERG, D. & SAMUELSSON, B. (1964*a*) The enzymatic conversion of essential fatty acids into prostaglandins. *J. Biol. Chem. 239*, 4006

BERGSTRÖM, S., DANIELSSON, H. & SAMUELSSON, B. (1964*b*) The enzymatic formation of prostaglandin E_2 from arachidonic acid. Prostaglandins and related factors. 32. *Biochim. Biophys. Acta 90*, 207–210

BLACKHAM, A., FARMER, J. B., RADZIOWONIK, H. & WESTWICK, J. (1974) The role of prostaglandins in rabbit monoarticular arthritis. *Br. J. Pharmacol. 51*, 35–44

BLACKWELL, G. J., DUNCOMBE, W. G., FLOWER, R. J., PARSONS, M. F. & VANE, J. R. (1977) The distribution and metabolism of arachidonic acid in rabbit platelets during aggregation and its modification by drugs. *Br. J. Pharmacol. 59*, 353–366

BOHMAN, S. O. & LARSSON, C. A. (1975) Prostaglandin synthesis in membrane fractions from the rabbit renal medulla. *Acta Physiol. Scand. 94*, 244–258

BUNTING, S., GRYGLEWSKI, R. J., MONCADA, S. & VANE, J. R. (1976) Arterial walls generate from prostaglandin endoperoxides a substance (Prostaglandin X) which relaxes strips of mesenteric and coeliac arteries and inhibits platelet aggregation. *Prostaglandins 12*, 897–913

CHAN, J. A., NAGASAWA, N., TAKEGUCHI, C. & SIH, C. J. (1975) On agents favouring prostaglandin F formation during biosynthesis. *Biochemistry 14*, 2987–2991

CHANG, W. C., MUROTA, S. I., MATSUO, M. & TSURUFUJI, S. (1976) A new prostaglandin transformed from arachidonic acid in carrageenin-induced granuloma. *Biochem. Biophys. Res. Commun. 72*, 1259–1264

CHRIST, E. J. & VAN DORP, D. A. (1972) Comparative aspects of prostaglandin biosynthesis in animal tissues. *Biochim. Biophys. Acta 270*, 537–545

CHRIST, E. J. & VAN DORP, D. A. (1973) Comparative aspects of prostaglandin biosynthesis in animal tissues, in *Supplementum to Advances in the Biosciences* (*Int. Conf. on Prostaglandins*, Vienna), Viewig, Braunschweig, Pergamon Press

CRUNKHORN, P. & WILLIS, A. L. (1971) Interaction between prostaglandins E and F given intradermally in the rat. *Br. J. Pharmacol. 41*, 507–512

DAHLE, I. K., HILL, E. G. & HOLMAN, R. T. (1962) The thiobarbituric acid reaction and the auto-oxidation of polyunsaturated fatty acid methyl esters. *Arch. Biochem. Biophys. 98*, 253–261

DI ROSA, M., GIROUD, J. P. & WILLOUGHBY, D. A. (1971) Studies of the mediators of acute inflammatory response induced in cats in different sites by carrageenan and turpentine. *J. Pathol. 104*, 15–29

EGAN, R. W., PAXTON, J. & KUEHL, F. A. (1976) Mechanism for irreversible self deactivation of prostaglandin synthetase. *J. Biol. Chem. 257*, 7329–7339

EAKINS, K. E., WHITELOCK, R. A. F., PERKINS, E. S., BENNETT, A. & UNGAR, W. G. (1972) Release of prostaglandins in ocular inflammation of the rabbit. *Nat. New Biol. 239*, 248–249

FELDBERG, W. & GUPTA, K. P. (1973) Pyrogen fever and prostaglandin like activity in cerebrospinal fluid. *J. Physiol. 228*, 41–48

FERREIRA, S. H. (1972) Prostaglandins, aspirin-like drugs and analgesia. *Nat. New Biol. 240*, 200–203

FERREIRA, S. H., MONCADA, S. & VANE, J. R. (1971) Indomethacin and aspirin abolish prostaglandin release from the spleen. *Nat. New Biol. 231*, 237–239

FERREIRA, S. H., MONCADA, S. & VANE, J. R. (1973) Prostaglandins and the mechanism of analgesia produced by aspirin-like drugs. *Br. J. Pharmacol. 49*, 86–97

FLOWER, R. J. & BLACKWELL, G. J. (1976) The importance of phospholipase A2 in prostaglandin biosynthesis. *Biochem. Pharmacol. 25*, 285–291

FLOWER, R. J., CHEUNG, H. S. & CUSHMAN, D. W. (1973) Quantitative determination of prostaglandins and malondialdehyde formed by the arachidonate oxygenase system of bovine seminal vesicles. *Prostaglandins 4*, 325–341

FLOWER, R. J., HARVEY, E. A. & KINGSTON, W. P. (1976) Inflammatory effects of prostaglandin D2 in rat and human skin. *Br. J. Pharmacol. 56*, 229–233

GLENN, E. M., BOWMAN, B. J. & ROHLOFF, N. A. (1972) Proinflammatory effects of certain prostaglandins, in *Prostaglandins in Cellular Biology*, Vol. 1 (Ramwell, P. W. & Pharris, B. B., eds.), pp. 329–373, Plenum Press, New York & London

GREAVES, M. W., SØNDERGAARD, J. & MCDONALD-GIBSON, W. (1971) Recovery of prostaglandins in human cutaneous inflammation. *Br. Med. J. 2*, 258–260

GRYGLEWSKI, R. J., BUNTING, S., MONCADA, S., FLOWER, R. J. & VANE, J. R. (1976) Arterial walls are protected against deposition of platelet thrombi by a substance (Prostaglandin X) which they make from prostaglandin endoperoxides. *Prostaglandins 12*, 658–714

HAMBERG, M. & JONSSON, C. E. (1973) Increased synthesis of prostaglandins in the guinea pig following scalding injury. *Acta Physiol. Scand. 87*, 240–245

HAMBERG, M. & SAMUELSSON, B. (1966) Novel biological transformations of 8,11,14-eicosatrienoic acid. *J. Am. Chem. Soc. 88*, 2349–2350

HAMBERG, M. & SAMUELSSON, B. (1967*a*) On the mechanism of the biosynthesis of prostaglandins E_1 and $F_{1\alpha}$. *J. Biol. Chem. 242*, 5336–5343

HAMBERG, M. & SAMUELSSON, B. (1967*b*) On the specificity of the oxygenation of unsaturated fatty acids catalysed by the soybean lipoxidase. *J. Biol. Chem. 242*, 5329–5335

HAMBERG, M. & SAMUELSSON, B. (1976*c*) Oxygenation of unsaturated fatty acids by vesicular gland of sheep. *J. Biol. Chem. 242*, 5344–5354

HAMBERG, M. & SAMUELSSON, B. (1973) Detection and isolation of an endoperoxide intermediate in prostaglandin biosynthesis. *Proc. Natl. Acad. Sci. U.S.A. 70*, 899–903

HAMBERG, M. & SAMUELSSON, B. (1974*a*) Prostaglandin endoperoxides. Novel transformations of arachidonic acid in human platelets. *Proc. Natl. Acad. Sci. U.S.A. 71*, 3400–3404

HAMBERG, M. & SAMUELSSON, B. (1974*b*) Prostaglandin endoperoxides VII. Novel transformations of arachidonic acid in guinea pig lung. *Biochem. Biophys. Res. Commun. 61*, 942–949

HAMBERG, M., SVENSSON, J. & SAMUELSSON, B. (1974*a*) Prostaglandin endoperoxides. A new concept concerning the mode of action and release of prostaglandins. *Proc. Natl. Acad. Sci. U.S.A. 71*, 3824–3828

HAMBERG, M., SVENSSON, J., WAKABAYASHI, T. & SAMUELSSON, B. (1974*b*) Isolation and structure of two prostaglandin endoperoxides that cause platelet aggregation. *Proc. Natl. Acad. Sci. U.S.A. 71*, 345–349

HAMBERG, M., SVENSSON, J. & SAMUELSSON, B. (1975) Thromboxanes: a new group of biologically active compounds derived from prostaglandin endoperoxides. *Proc. Natl. Acad. Sci. U.S.A. 72*, 2994–2998

HIGGS, G. A., VANE, J. R., HART, F. D. & WOJTULEWSKI, J. A. (1974) Effects of anti-inflammatory drugs on prostaglandins in rheumatoid arthritis, in *Prostaglandin Synthetase Inhibitors* (Robinson, H. J. & Vane, J. R., eds.), pp. 165–174, Raven Press, New York

HORTON, E. W. (1969) Hypothesis on physiological roles of prostaglandins. *Physiol. Rev. 49*, 122–161

JOHNSON, R. A., MORTON, D. R., KINNER, J. H., GORMAN, R. R., MCGUIRE, J. R., SUN. F. F., WHITTAKER, N., BUNTING, S., SALMON, J. A., MONCADA, S. & VANE, J. R. (1976) The chemical structure of prostaglandin X (prostacyclin). *Prostaglandins 12*, 915–928

JUHLIN, S. & MICHAELSSON, G. (1969) Cutaneous vascular reactions to prostaglandins in healthy subjects and in patients with urticaria and atopic dermatitis. *Act. Derm-Venereol. 49*, 251–261

KUEHL, F. A., HUMES, J. L., EGAN, R. W., HAM, E. A., BEVERIDGE, G. C. & VAN ARMAN, G. (1977) Role of prostaglandin endoperoxide PGG_2 in inflammatory processes. *Nature (Lond.) 265*, 170–173

KUNZE, H. & VOGT, W. (1971) Significance of phospholipase A for prostaglandin formation. *Ann. N.Y. Acad. Sci. 180*, 123–125

LANDS, W. E. M. & SAMUELSSON, B. (1968) Phospholipid precursors of prostaglandins. *Biochim. Biophys. Acta 164*, 426–429

LEWIS, A. J., NELSON, D. J. & SUGRUE, M. F. (1974) Potentiation by prostaglandin E1 and arachidonic acid of oedema in the rat paw induced by various phlogogenic agents. *Br. J. Pharmacol. 50*, 468P-469P

LIGHT, R. J. & SAMUELSSON, B. (1972) Identification of prostaglandins in the Gorgonian *plexaura homomalla. Eur. J. Biochem. 28*, 232–240

MIYAMOTO, T., YAMAMOTO, S. & HAYAISHI, O. (1974) Prostaglandin synthetase system: resolution into oxygenase and isomerase components. *Proc. Natl. Acad. Sci. U.S.A. 71*, 3645–3648

MIYARES-CAO, C. M. & MENENDEZ-CEPERO, E. (1976) Identification of prostaglandin-like substances in plants, in *Advances in Prostaglandin and Thromboxane Research*, New York, Raven Press

MONCADA, S., FERREIRA, S. H. & VANE, J. R. (1973) Prostaglandins, aspirin-like drugs and the oedema of inflammation. *Nature (Lond.) 246*, 217–218

MONCADA, S., FERREIRA, S. H. & VANE, J. R. (1975) Inhibition of prostaglandin biosynthesis as the mechanism of analgesia of aspirin-like drugs in the dog knee joint. *Eur. J. Pharmacol. 31*, 250–260

MONCADA, S., GRYGLEWSKI, R., BUNTING, S. & VANE, J. R. (1976*a*) An enzyme isolated from arteries transforms prostaglandin endoperoxides to an unstable substance that inhibits platelet aggregation. *Nature (Lond.) 263*, 663–665

MONCADA, S., GRYGLEWSKI, R. J., BUNTING, S. & VANE, J. R. (1976*b*) A lipid peroxide inhibits the enzyme in blood vessel microsomes that generates from prostaglandin endoperoxides the substance (Prostaglandin X) which prevents platelet aggregation. *Prostaglandins 12*, 715–733

NIEHAUS, W. G. & SAMUELSSON, B. (1968) Formation of malondialdehyde from phospholipid arachidonate during microsomal lipid peroxidation. *Eur. J. Biochem. 6*, 126–130

NIJKAMP, F. J., FLOWER, R. J., MONCADA, S. & VANE, J. R. (1976) Partial purification of RCS-RF (rabbit aorta contracting substance releasing factor) and inhibition of its activity by anti-inflammatory steroids. *Nature (Lond.) 263*, 479–482

NUGTEREN, D. H. & HAZELHOF, E. (1973) Isolation and properties of intermediates in prostaglandin biosynthesis. *Biochim. Biophys. Acta 326*, 448–461

NUGTEREN, D. H., BEERTHUIS, R. K. & VAN DORP, D. A. (1966) The enzymic conversion of all-*cis* 8,11,14 eicosatrienoic acid into prostaglandin E_1. *Rec. Trav. Chim. Pays-Bas 85*, 405–419

PIPER, P. J. & VANE, J. R. (1969) Release of additional factors in anaphylaxis and its antagonism by anti-inflammatory drugs. *Nature (Lond.) 223*, 29–35

PIPER, P. J. & VANE, J. R. (1971) The release of prostaglandins from lung and other tissues. *Ann. N.Y. Acad. Sci. 180*, 363–385

PLAA, G. L. & WITSCHI, H. (1976) Chemicals, drugs and lipid peroxidation. *Annu. Rev. Pharm. Toxicol. 16*, 125–141

RAMWELL, P. W. & SHAW, J. E. (1970) Biological significance of the prostaglandins. *Rec. Prog. Horm. Res. 26*, 139–187

SAMUELSSON, B. (1965) On the incorporation of oxygen in the conversion of 8,11,14-eicosatrienoic acid to prostaglandin E_1. *J. Am. Chem. Soc. 87*, 3011–3013

SAMUELSSON, B. & HAMBERG, M. (1976) Role of endoperoxides in the biosynthesis and action of prostaglandins, in *Prostaglandin Synthetase Inhibitors*, New York, Raven Press, pp. 107–119

SCHULTZ, H. W. (1962) *Symposium on Foods: Lipids and their Oxidation*, The AVI Publishing Co., Westport, Connecticut

SMITH, J. B. & WILLIS, A. L. (1971) Aspirin selectively inhibits prostaglandin production in human platelets. *Nat. New Biol. 231*, 235–237

SOLOMON, L. M., JUHLIN, L. & KIRSCHBAUM, M. M. (1968) Prostaglandins in cutaneous vasculature. *J. Invest. Dermatol. 51*, 280–282

SØNDERGAARD, J. & GREAVES, M. W. (1970) Pharmacological studies in inflammation due to exposure to ultraviolet radiation. *J. Pathol. 101*, 93–97

SWEET, C. S., KADOWITZ, P. J. & BRODY, M. J. (1971) A hypertensive response to infusion of prostaglandin $F_{2\alpha}$ into the vertebral artery of the conscious dog. *Eur. J. Pharmacol. 16*, 229–232

VAN DORP, D. A. (1967) Aspects of the biosynthesis of prostaglandins. *Prog. Biochem. Pharmacol. 3*, 71–82

VAN DORP, D. A., BEERTHUIS, R. K., NUGTEREN, D. H. & VONKEMAN, H. (1964*a*) Enzymatic conversion of all-*cis*-polyunsaturated fatty acids into prostaglandins. *Nature (Lond.) 203*, 839–843

VAN DORP, D. A., BEERTHUIS, R. K., NUGTEREN, D. H. & VONKEMAN, H. (1964*b*) The biosynthesis of prostaglandins. *Biochim. Biophys. Acta 90*, 204–207

VANE, J. R. (1971) Inhibition of prostaglandin synthesis as a mechanism of action for aspirin-like drugs. *Nat. New Biol. 231*, 232–235

VELO, G. P., DUNN, C. J., GIROUD, J. P., TIMSIT, J. & WILLOUGHBY, D. A. (1973) Distribution of prostaglandins in inflammatory exudate. *J. Pathol. 111*, 149–158

VONKEMAN, H. & VAN DORP, D. A. (1968) The action of prostaglandin synthetase on 2-arachidonyl-lecithin. *Biochim. Biophys. Acta 164*, 430–432

WILLIS, A. L. (1969) A release of histamine, kinin and prostaglandins during carrageenin-induced inflammation in the rat, in *Prostaglandins, Peptides and Amines* (Mantegazza, G. & Horton, E. W., eds.), pp. 31–38, Academic Press, London

WILLIS, A. L. & CORNELSEN, M. (1973) Repeated injection of prostaglandin E_2 in rat paws induces chronic swelling and marked decrease in pain threshold. *Prostaglandins 3*, 353–357

WLODAWER, P. & SAMUELSSON, B. (1973) On the organization and mechanism of prostaglandin synthetase. *J. Biol. Chem. 248*, 5673–5678

YOSHIMOTO, T., YAMAMOTO, S., OKUMA, M. & HAYAISHI, O. (1977) Stabilisation and resolution of thromboxane synthetising system from microsomes of bovine blood platelets. *J. Biol. Chem. 25*, 5871–5874

Discussion

Allison: Kuehl *et al.* (1977) have shown that the synthetase system gives a good free-radical signal in the e.s.r. spectrum, which has not yet been identified. Sulindac, a fluoro(methylsulphinylphenylmethylene)indeneacetic acid, a potent anti-inflammatory agent, abolishes this signal and apparently converts G_2 into H_2. Is that pathway important? Is G_2 an important mediator of pharmacological activity, as they claim?

Flower: According to the prevailing dogma, PGG_2 is reduced almost immediately to PGH_2, which is in turn transformed into the other prostaglandins (see Figs. 2 and 4) which may be important in the inflammatory response. On the basis of the results you mentioned, Kuehl *et al.* suggested that the liberation of cytotoxic free-radical oxygen during the conversion of PGG_2 into PGH_2 was important for the development of inflammation and that drugs such as MK-447 which accelerated this conversion and trapped the free-radicals constituted an entirely new class of anti-inflammatory agent (Kuehl *et al.* 1977). An obvious experiment would be to compare the inflammatory actions of PGG_2 and PGH_2. However, according to Vane (1976) there is little difference between the two.

Willson: Does MK-447 complex metals? If so, iron, for instance, might be involved in this reaction.

Flower: I don't know.

Allison: Promethazine and metazianic acid are also efficient anti-inflammatory agents and have the same effects as MK-447. They interact efficiently with radicals.

Willson: Anti-inflammatory agents of the phenothiazine group avidly mop up hydroxyl and related electrophilic radicals. But how are the radicals initially generated from the hydroperoxide? How is the reaction catalysed: enzymically or non-enzymically (by iron, for instance)?

Allison: The same phenothiazines that mop up radicals convert G_2 into H_2 very efficiently.

Flohé: Diczfalusy *et al.* (1977) recently claimed that the malonaldehyde is not a non-enzymic metabolite but the product of the thromboxane synthetase. They reported a stoichiometric correlation of products (HHT, malonaldehyde and thromboxane B_2) from PGH_2.

Flower: Malonaldehyde may arise from cyclooxygenase products in two ways: by a simple non-enzymic degradation of the endoperoxides (as shown in Fig. 3) and also as a by-product of thromboxane biosynthesis, since the partially purified enzyme which converts PGH_2 into TXA_2 also generates the 17-hydroxy acid (HHT), presumably with simultaneous formation of malonaldehyde (Yoshimoto *et al.* 1977).

Flohé: Do cells that do not form the thromboxane produce malonaldehyde *in vivo*?

Flower: Theoretically they could, because malonaldehyde may also arise from non-enzymic peroxidation of lipid arachidonate (Niehaus & Samuelsson 1968).

Flohé: If malonaldehyde formation is not due to the thromboxane synthetase reaction, is it not inhibited by some mechanism *in vivo*?

Flower: Cyclooxygenase preparations such as ram seminal vesicle microsomes do not make thromboxane, but nevertheless generate a considerable amount of malonaldehyde; its formation, in fact, is an excellent index of enzyme activity (Flower *et al.* 1973).

Flohé: But usually these systems contain many artificial activators which might trigger some free-radical chain reactions. These incubation systems may not reflect the *in vivo* situation.

Keberle: Anti-inflammatory agents share the same properties only to some extent: many, like the salicylates, can chelate metallic ions; others, e.g. pyrazolones, are prone to auto-oxidation; there is a third group, the arylacetic acids, made up of compounds having an acid character in common. One property shared by almost all known antirheumatic agents is a special affinity for proteins in the plasma and the liver. As a result, their distribution volume in the body is relatively small. Could it be, perhaps, that a further common feature of these substances is a propensity for scavenging radicals?

Allison: I don't believe so, but our information is at present limited.

Goldstein: The pharmacological effects of the various prostaglandins you described are extensive but I guess that at least half a dozen people in this symposium have shut off these pathways just by ingesting aspirin in the last 24 h. Yet, with the possible exception of a slightly greater skin bruisability aspirin does not really have drastic effects.

Flower: Impaired platelet aggregation would be one of the most noticeable effects if one had taken aspirin within the last three or four days. However, platelet aggregation can proceed by other routes as well as the cyclooxygenase-dependent mechanism.

Goldstein: I still wonder how important these prostaglandin compounds are. Granted, there is a slight increase in bruisability but beside that what physiological significance for humans *in vivo* do these compounds have? There is no question that they are extremely active in the test tube.

Flower: Their importance in platelet aggregation has already been touched upon, but the more one studies habitual aspirin takers the more odd effects one uncovers which previously went unnoticed. For example, labour seems to be delayed in women who are chronic aspirin takers (Lewis & Schulman 1973).

Certainly, persistent taking of aspirin has no apparent physiological effect in *healthy* individuals, unless some of the side-effects of anti-inflammatory drug abuse (gastric ulceration, papillary necrosis) are associated in some way with chronic prostaglandin deficiency.

Chvapil: Do aspirin-like drugs inhibit synthesis of thromboxane A_2 as well as that of prostacyclin?

Flower: Yes.

Chvapil: In that case would you not expect that the fine balance between that prostaglandin aggregation and the other preventing adhesion to be destroyed?

Flower: Yes.

Flohé: The balance between thromboxanes and prostacyclins is shifted by aspirin. When thromboxane synthetase is knocked out, the whole prostaglandin biosynthesis in platelets is irreversibly shut down for the remaining lifetime of the platelet. In contrast, the endothelial wall can rebuild the enzyme (protein synthesis is still operating). The effect of aspirin is thus longer lasting in platelets than in endothelial cells (Livio *et al.* 1978).

Williams: How much is known about the enzymes in this pathway? Are they membrane-bound? What size are they?

Flower: They are membrane-bound, but Professor Hayaishi is best qualified to answer this question, since he has purified almost all of them.

Hayaishi: They have been solubilized and purified but the enzyme which catalyses the first step, the cyclooxygenase, and that which catalyses the second step, the so-called prostaglandin hydroperoxidase, cannot be separated — they exist in a 'multienzyme complex'. Its molecular weight has not yet been estimated.

Williams: Does it contain any metals?

Hayaishi: It contains no metal, as far as we can tell, but both reactions require haem for activity. The haem does not seem to bind tightly to these proteins, so it is a strange haemoprotein or maybe a unique haem-requiring enzyme.

Williams: Is this free haem?

Hayaishi: Free or bound haem; haematin and bound haem such as haemoglobin or myoglobin are just as effective.

Williams: Is it a totally different sort of protein from the lipoxygenases?

Hayaishi: Yes; lipoxygenases contain non-haem iron and do not require haem for activity.

Winterhalter: Perhaps you can advise me on the outcome of a simple-minded experiment Dr C. Richter and I did (unpublished results). One can hammer rats' paws in some graded way and measure the degree of inflammation by determining the increase in volume through an appropriate device. When we pre-treated rats with diethyldithiocarbamate (an inhibitor of superoxide dismutase), we expected the inflammation to become greater but we found the converse: the inflammation (as measured by oedema) was considerably weaker. This curious effect is repeatable and clearly demonstrable.

Flower: I cannot explain your data.

Hill: Inhibition of superoxide dismutase is not the only possible reaction of dithiocarbamates *in vivo*.

Michelson: They cause megamitochondria, for example, and many other things.

References

DICZFALUSY, U., FALARDEAU, P. & HAMMARSTRÖM, S. (1977) Conversion of prostaglandin endoperoxides to C_{17}-hydroxy acids catalyzed by human platelet thromboxane synthetase. *FEBS (Fed. Eur. Biochem. Soc.) Lett. 84*, 271–274

FLOWER, R. J., CHEUNG, H. S. & CUSHMAN, D. W. (1973) Quantitative determination of prostaglandins and malondialdehyde formed by the arachidonate oxygenase (prostaglandin synthetase) system of bovine seminal vesicle. *Prostaglandins 4*, 325–241

KUEHL, F. A., HUMES, J. L., EGAN, R. W., HAM, E. A., BEVERIDGE, G. C. & VAN ARMAN, C. G. (1977) Role of prostaglandin endoperoxide PGG_2 in inflammatory processes. *Nature (Lond.) 265*, 170–174

LEWIS, R. B. & SCHULMAN, J. D. (1973) Influence of acetylsalicylic acid, an inhibitor of prostaglandin synthesis, on the duration of human gestation and labour. *Lancet 2*, 1159–1161

LIVIO, M., VILLA, S. & DE GAETANO, G. (1978) Aspirin, thromboxane, and prostacyclin in rats: a dilemma resolved? *Lancet 1*, 1307

NIEHAUS, W. G. & SAMUELSSON, B. (1968) Formation of malondialdehyde from phospholipid arachidonate during microsomal lipid peroxidation. *Eur. J. Biochem. 6*, 126–130

VANE, J. R. (1976) Prostaglandins as mediators of inflammation, in *Advances in Prostaglandin and Thromboxane Research*, vol. 2 (Samuelsson, B. & Paoletti, R., eds.), pp. 791–801, Raven Press, New York

YOSHIMOTO, T., YAMAMOTO, S., OKUMA, M. & HAYAISHI, O. (1977) Solubilization and resolution of thromboxane synthesising system from microsomes of bovine blood platelets. *J. Biol. Chem. 252*, 5871–5874

Mechanisms of protection against the damage produced in biological systems by oxygen-derived radicals

T. F. SLATER

Biochemistry Department, Brunel University, Uxbridge, Middlesex

Abstract This paper will concentrate on the damage to liver endoplasmic reticulum and plasma membrane fractions that results from exposure to O_2-derived radicals and lipid peroxidation.

Lipid peroxidation in rat liver endoplasmic reticulum can be produced in various ways involving electron flow out of the NADPH-cytochrome P450 electron-transport chain; analogous reactions occur also in liver plasma membrane suspensions. The subsequent damaging reactions of oxygen-derived radicals and of lipid peroxidation on biological components can be attenuated by various free-radical scavengers and a survey of more than 50 such scavengers in four different systems involving lipid peroxidation has been made.

The conditions required for such scavenging reactions to be effective will be outlined, and the difficulties inherent in using such scavengers *in vivo* will be discussed.

One major feature of oxygen metabolism is its interaction (either as molecular oxygen or as a derived free radical) with polyunsaturated fatty acids (PUFAH) or derived free radicals ($PUFA^{\cdot}$) in the process described as lipid peroxidation. When lipid peroxidation occurs in biological membranes (for example, endoplasmic reticulum, the mitochondrial membranes, or the plasma membrane) there may be gross disturbances in structural organization (for detailed references see Slater 1972) and in associated enzymic function (Lewis & Wills 1962). Numerous examples are now known where lipid peroxidation *in vivo* or *in vitro* produces irreversible damage to membrane systems which often results in the death of the affected cells: for example, high-energy irradiation (Desai *et al.* 1964), photosensitization (Slater & Riley 1966), exposure to ozone (Goldstein *et al.* 1969), and administration of CCl_4 (see Slater 1968*b*, 1972; Recknagel 1967) or of paraquat (Bus *et al.* 1975). In these examples other damaging reactions do also occur but lipid peroxidation is, without doubt, a major feature of the overall injury.

Lipid peroxidation involves an initiation step in which a polyunsaturated fatty acid interacts with a reactive oxidizing radical ($R^•$) which abstracts a proton to form the fatty acid radical $PUFA^•$ (reaction 1). This initiation step is followed by chain-propagation steps, of which the first is the formation of the

$$R^• + PUFAH \rightarrow RH + PUFA^• \quad (1)$$

fatty acid peroxy-radical (reaction 2). The reactive initiating radical $R^•$ may be

$$PUFA^• + O_2 \rightarrow PUFAO_2^• \quad (2)$$

$OH^•$ or $CCl_3O_2^•$ (singlet oxygen can also initiate peroxidation; Baird *et al.* 1977) so that both initiation and propagation stages of lipid peroxidation involve oxygen or oxygen-derived radicals of various types: $^1O_2$, $OH^•$, $CCl_3O_2^•$, $PUFAO_2^•$. In view of the reactive oxygen species involved as reactants in peroxidation and the damaging nature of lipid peroxidation on cellular behaviour in general, lipid peroxidation is directly relevant to this symposium.

Reactions (1)–(5) represent the general peroxidative breakdown of a polyunsaturated fatty acid; the later stages of degradation of $PUFAO_2^•$ and

$$PUFAO_2^• \rightarrow \text{Bond rearrangement, diene formation} \quad (3)$$

$$PUFAO_2^• + PUFAH \rightarrow PUFAO_2H + PUFA^• \quad (4)$$

$$PUFAO_2^•, PUFAO_2H \rightarrow \text{Degradation products: malonaldehyde, ethane, etc.} \quad (5)$$

$PUFAO_2H$ involve the formation of a rich variety of low-molecular-weight, water-soluble products (Schauenstein 1967) including aldehydes (such as malonaldehyde), and of diene conjugates, fluorescent products and gases like ethane (Riely *et al.* 1974). Several reviews on lipid peroxidation have been published (Slater 1975; Plaa & Witschi 1976; Mead 1976; Dormandy 1978; Dianzani & Ugazio 1978).

The most generally used methods that are used to follow the progress of lipid peroxidation in tissue samples or in whole animals are: diene conjugation measurements on extracted lipids (Recknagel & Ghoshal 1966*b*); the thiobarbituric acid reaction with malonaldehyde (see Slater 1972); and the appearance of fluorescent high-molecular-weight products (Fletcher *et al.* 1973). *In vivo*, determinations of lipid hydroperoxides in blood have been proposed but the most suitable method appears to be the measurement of ethane in expired air (Lindstrom & Anders 1978; Cohen, this volume). Other methods that may be useful in certain conditions include: iodometry and other titrimetric methods (see Johnson & Siddiqi 1970); coupling of lipid dienes with tetracyanoethylene to give a fluorescent product (Waller & Recknagel 1977); and

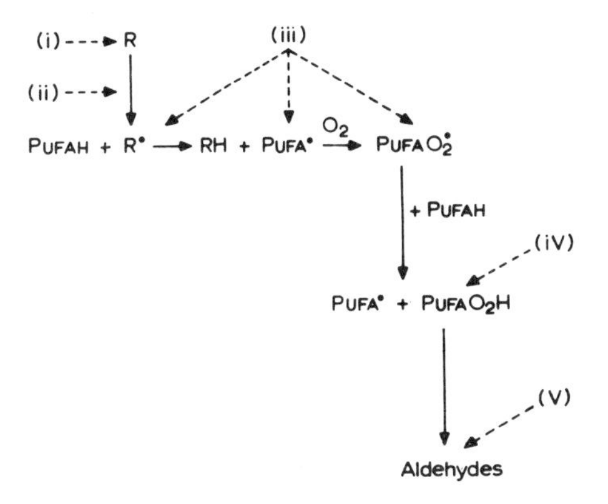

FIG. 1. Main ways of attenuating lipid peroxidation and its toxic consequences for a polyunsaturated fatty acid (PUFAH); peroxidation is initiated by the reactive free radical R˙. Sites at which attenuation can be attempted are indicated by broken arrows. In (i), the attenuation would be to reduce exposure to R or to divert its metabolism into another pathway; in (ii) the attenuating procedure would be to decrease activation of R to R˙ (e.g. by decreasing body temperature, by inhibiting any necessary enzyme step, etc.). In (iii), R˙ and secondary radicals are scavenged; in (iv) and (v) the toxic products of lipid peroxidation are metabolized by appropriate enzymes (GSH peroxidase, aldehyde dehydrogenases, etc.).

an enzymic assay of lipid hydroperoxides based on glutathione peroxidase (Heath & Tappel 1976).

As lipid peroxidation is a major feature of many types of cell injury, studies on the mechanisms that result in an attenuation of peroxidative reactions are of some considerable practical importance as well as of theoretical interest. Fig. 1 summarizes the main ways in which lipid peroxidation can be attenuated and some possible ways in which the metabolic consequences of peroxidation on the intact cell can be decreased. These processes fall into three main groups: (i) decreases in the supply of the primary reacting species (substrate, and initiating radicals); (ii) scavenging of the initiating radical and propagating radicals; (iii) metabolic removal of toxic products of lipid peroxidation such as peroxides and aldehydes.

(i) *Decreases in the supply of primary reacting species*

Under this heading come such procedures as decreasing the proportion of polyunsaturated fatty acids present in the material under study (for example, see Burdino *et al.* 1976); decreasing the exposure of the toxic agent (for example, CCl_4 or ozone) that gives rise to the initiating radical R˙; and decreasing the rate of any metabolic step(s) that activate the incoming toxic agent to the interactive species R˙ (for example, the metabolism of CCl_4 to CCl_3˙ by the microsomal NADPH-cytochrome P450 electron-transport chain). Experimental manipulations that lead to decreased lipid peroxidation through these mechanisms have been reviewed (Slater 1978).

(ii) *Scavenging of free-radical intermediates*

Precise solubility and intracellular distribution properties of the potential scavenger may become dominant here. The initiating radical $R^{\bullet}$ may be formed, for example, in the aqueous phase surrounding a membrane, whereas the propagating radicals $PUFA^{\bullet}$ and $PUFAO_2^{\bullet}$ will be confined to the hydrophobic membrane interior (and, perhaps, to one particular region of the membrane). Effective scavenging of such radicals requires the scavenger to penetrate to the immediate vicinity of formation of the damaging radical species, as well as having requisite chemical scavenging reactivity. In the latter respect, the one-electron redox potential is an important parameter (see Willson 1978).

(iii) *Metabolism of toxic products of lipid peroxidation*

Major attention has concentrated in this respect on (*a*) GSH peroxidase and (*b*) aldehyde dehydrogenases. GSH peroxidase, a selenoenzyme (Rotruck *et al.* 1973), metabolizes lipid peroxides to hydroxy-fatty acids (Christopherson 1968). It depends on a suitable supply of reduced GSH; this supply, in turn, depends on GSH reductase and NADPH. Oxidation of GSH and NADPH by experimental manipulations should result in an accumulation of lipid peroxide in conditions known to be conducive to peroxidation. When such changes in GSH and NADPH were produced in isolated rat liver hepatocytes exposed to CCl_4, however, no increase in lipid peroxide as measured iodometrically or enzymically was detected (Table 1). Other routes of metabolism of lipid peroxides seem to be operative, therefore, in these conditions; indeed, some evidence is available for metal-catalysed decomposition of lipid peroxides in liver microsomes (T. F. Slater, unpublished data).

The metabolism of aldehydic products produced by lipid peroxidation involves several enzymes. Malonaldehyde, for example, is metabolized *in vivo* (Placer *et al.* 1965); an enzyme active in this respect is found in rat liver mitochondria (Recknagel & Ghoshal 1966*a*) so that malonaldehyde levels in liver homogenates reflect not only production rate (e.g. in microsomes) but the removal rate by mitochondrial catabolism. Keto-aldehydes can be metabolized by glyoxalase and by a dehydrogenase that is widely distributed (Monder 1967). In addition, reactive aldehydes can have their pharmacological properties modified by adduct formation with thiols such as cysteine or GSH (see Schauenstein *et al.* 1977).

In this paper, I shall concentrate on only one of the three main ways of

TABLE 1

Effects of metabolic changes on lipid peroxidation in isolated rat hepatocytes. The hepatocytes (7×10^6/ml) were incubated in buffered medium for 60 min at 37 °C with and without CCl_4 (final concentration 125 μmol/l). Incubations were also carried out with diamide (Sigma Chemical, London, UK) (0.2 mmol/l) to oxidize glutathione, and D-xylose (2 mmol/l) to oxidize NADPH. At the end of the incubation, lipid peroxide was estimated iodometrically, and malonaldehyde was measured by the thiobarbituric acid reaction. Results are unpublished data of P. H. Beswick, G. Poli & T. F. Slater.

Incubation mixture	*Additions*	*Lipid peroxide (iodometric)*	*Malonaldehyde production (nmol/10⁶ cells)*
Control		ldl[a]	ldl
Control + CCl_4		ldl	4.04
Control + CCl_4	+ diamide	ldl	4.16
Control + CCl_4	+ xylose	ldl	3.98
Control + CCl_4	+ diamide + xylose	ldl	4.10

[a] ldl, lower than detection limit.

attenuating lipid peroxidation outlined above, emphasizing free-radical scavengers that interact with the initiating and chain-propagating radicals. The data, which illustrate several basic concepts, were obtained by studies of three model systems involving different primary radical species: (*a*) the NADH–phenazine methosulphate (PMS)-mediated reduction of nitroblue tetrazolium chloride (NBT) (the 'NADH/PMS/NBT system'). There is evidence (Nishikimi *et al.* 1972) that in aerobic conditions the reduction of NBT proceeds by interaction with O_2^- formed by the autoxidation of reduced PMS (reactions 6–8). The

$$NADH + H^+ + PMS \rightarrow NAD^+ + PMSH_2 \quad (6)$$

$$PMSH_2 + 2O_2 \rightarrow PMS + 2H^+ + 2O_2^- \quad (7)$$

$$4O_2^- + NBT^{2+}2Cl^- \rightarrow \text{diformazan} + 4O_2 + 2HCl \quad (8)$$

experimental conditions (Ponti *et al.* 1978) were as follows: 0.016M-Tris–HCl buffer, pH 8.0; 73μM-NADH, pH 8.0 in Tris buffer; 5.2μM-PMS in water; 80μM-NBT in water; final volume containing the above final concentrations 3.1 ml; temperature, ambient; initial rate of reaction was obtained at 560 nm.

Model system (*c*) was a liver microsomal system in which lipid peroxidation was stimulated by the addition of NADPH and a mixture of ADP and iron(II) sulphate (Hochstein & Ernster 1963) (the 'NADPH–ADP/Fe^{2+}-dependent lipid peroxidation system'). According to Fong *et al.* (1973) this system involves the hydroxyl radical ($OH^\cdot$) as a major reactive intermediate. The conditions (Slater 1968*a*) were: 33mM-Tris buffer, pH 7.4; 100mM-KCl; 9μM-$FeSO_4$; 0.83mM-ADP; 118μM-NADPH; liver microsomes, about 1 mg protein; temperature,

TABLE 2

Effects of free-radical scavengers and radical-trapping agents on four model systems involving free-radical intermediates (for details, see text); the results are shown as percentage inhibitions relative to controls (stimulations are shown marked with an asterisk). Each agent was studied over a range of concentrations; the results represent work by Barbara Sawyer, K. Cheeseman, V. Ponti & T. F. Slater.

Agent	Model system studied							
	NADH/PMS/NBT		NADPH–lipid peroxidation		NADPH–ADP/Fe^{2+} lipid peroxidation		CCl_4-stimulated lipid peroxidation	
	Effect (%)	Concentration (μmol/l)	Effect (%)	Concentration (μmol/l)	Effect (%)	Concentration (μmol/l)	Effect (%)	Concentration (μmol/l)
Promethazine[a]	15*	100	53	1.66	50	25	79	1.66
Propyl gallate[b]	50	5	63	20	50	4	27 100	1 20
(+)-Catechin[c]	65	117	30	100	50	7	82	40
Vitamin E	26*	160					60* 90	(5×10^{-3}) 5.5
Menadione (K_3)			67	1	65	28	71	1
1,4-Benzoquinone	100	59	no effect	30			53	30
Diphenylfuran[d]	14	50	no effect	100	27	350	no effect	100
β-Carotene[b]	43	50	no effect	2.3×10^3			20	2.3×10^3
Superoxide dismutase[e]	50	1 IU	no effect	100 IU	26	200 IU	10	100 IU

Suppliers: [a] May & Baker (Dagenham, Essex, UK); [b] Sigma Chemical (London, UK); [c] Zyma (Nyon, Switzerland); [d] Eastman Kodak (Liverpool, UK); [e] Fisons (Loughborough, UK).

25 °C; oxygen uptake was measured using a Gilson oxygraph model K−1C.

Model systems (*b*) and (*d*) also used liver microsomes but the microsomes were incubated at 37 °C in a complex stock-medium either with or without the addition of CCl_4 — model systems (*d*) and (*b*), respectively. The experimental details were the same as those given in Slater & Sawyer (1971*a*,*b*). At the end of a 15 min incubation, the reaction was stopped and malonaldehyde production was measured by the thiobarbituric acid reaction. These systems are abbreviated to 'NADPH-dependent lipid peroxidation system' (in the absence of CCl_4),and the 'CCl_4-stimulated lipid peroxidation system' (in the presence of CCl_4). There is evidence that lipid peroxidation is initiated in the presence of CCl_4 by $CCl_3^{\bullet}$ (see Slater 1972), or, more probably, by $CCl_3O_2^{\bullet}$ (see Packer *et al.* 1978).

These model systems provide opportunity, therefore, for the study of the effects of scavengers on systems involving $O_2^{-\bullet}$, $OH^{\bullet}$, and $CCl_3^{\bullet}$ (or $CCl_3O_2^{\bullet}$). The free-radical scavengers and other substances discussed here fall into two main divisions: (i) 'classical' free-radical scavengers and free-radical-trapping substances; (ii) metal ions and metal chelators.

Table 2 lists the effects of various free-radical scavengers and trapping agents (for some structures, see Fig. 2) on the model systems studied. Promethazine, propyl gallate, catechin, and vitamin E (above 1 μmol/l) are effective inhibitors of model systems (*b*), (*c*) and (*d*); with model system (*a*), however, promethazine and vitamin E stimulated rather than inhibited. Moreover, vitamin E provided a small stimulation of the CCl_4-stimulated peroxidation system when present in low concentration (about 5 nmol/l); in this system, therefore, vitamin E exhibits a 'cross-over' behaviour.

The results with promethazine show that this phenothiazine interacts only a little with the presumed O_2^--dependent system (system *a*) but strongly inhibits

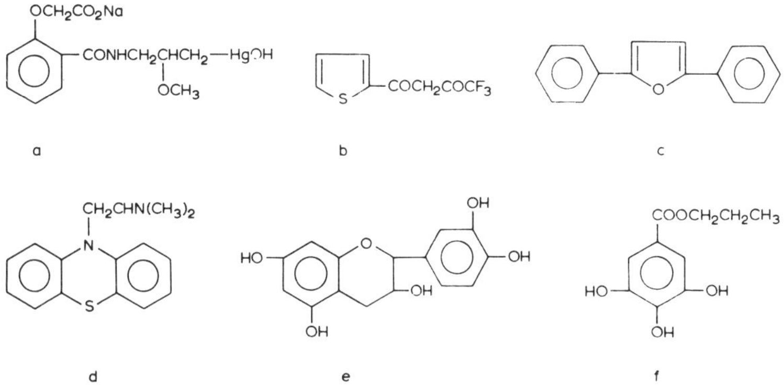

FIG. 2. Structure of (*a*) mersalyl; (*b*) thenoyltrifluoroacetone (Eastman Kodak, Liverpool, UK); (*c*) 2,5-diphenylfuran; (*d*) promethazine; (*e*) (+)-catechin; (*f*) propyl gallate.

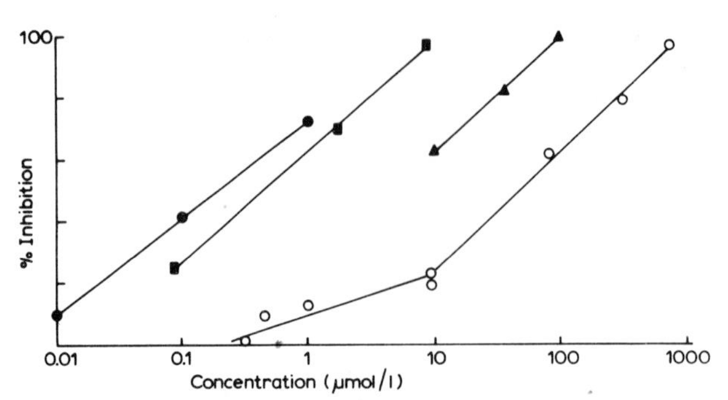

FIG. 3. Effects of promethazine (●), (+)-catechin (▲), and propyl gallate (■) on the CCl_4-stimulated lipid peroxidation system in rat liver microsomes (model system *d*). Results are shown as percentage inhibitions relative to controls; temperature, 37 °C. For experimental details of the method see Slater & Sawyer (1971*a*). Also shown is the effect of promethazine (○) on aminopyrene metabolism in rat liver microsomes to indicate the different order of magnitude for the effects of promethazine on lipid peroxidation compared to drug metabolism.

the $CCl_3^{\bullet}$-dependent system (system *d*). The latter effect is shown in more detail in Fig. 3 in which the dose–response behaviour of promethazine in model system (*d*) is illustrated and compared with related data for propyl gallate and catechin. The high reactivity of promethazine in attenuating the rate of peroxidation in model system (*d*) has been ascribed (Willson & Slater 1975) to a fast scavenging of $CCl_3^{\bullet}$ by promethazine (Pr) according to reaction (9). The rate constant

$$CCl_3^{\bullet} + Pr \rightarrow CCl_3^{-} + Pr^{+\bullet} \qquad (9)$$

measured for this reaction in aerobic conditions by pulse radiolysis was $1.1 \times 10^9\ l\ mol^{-1}\ s^{-1}$. However, when the reaction is done in strict anaerobic conditions, the rate constant is much reduced (Packer *et al.* 1978) and it seems possible that $CCl_3^{\bullet}$ reacts primarily with O_2 to form the peroxy-radical ($CCl_3O_2^{\bullet}$) before reacting with promethazine (reaction 10). The high reactivity of $CCl_3^{\bullet}$ with O_2

$$CCl_3^{\bullet} + O_2 \rightarrow CCl_3O_2^{\bullet} \xrightarrow{Pr} CCl_3O_2^{-} + Pr^{+\bullet} \qquad (10)$$

is probably a major reason for the very low steady-state concentration of $CCl_3^{\bullet}$ in liver after dosing with CCl_4 and which has frustrated attempts to demonstrate its presence directly by electron spin resonance spectroscopy (Keller *et al.* 1971; Slater 1972; Burdino *et al.* 1973). Although addition of the spin trap 2-methyl-2-nitrosopropane (Aldrich Chemical, Gillingham, Kent, UK) (see Perkins *et al.* 1970) to isolated hepatocytes or liver microsomes in the presence of CCl_4 leads to a triplet nitroxy-radical e.s.r. signal (Ingall *et al.* 1978; Fig. 4), the radical trapped is probably not $CCl_3^{\bullet}$ but a secondarily derived radical.

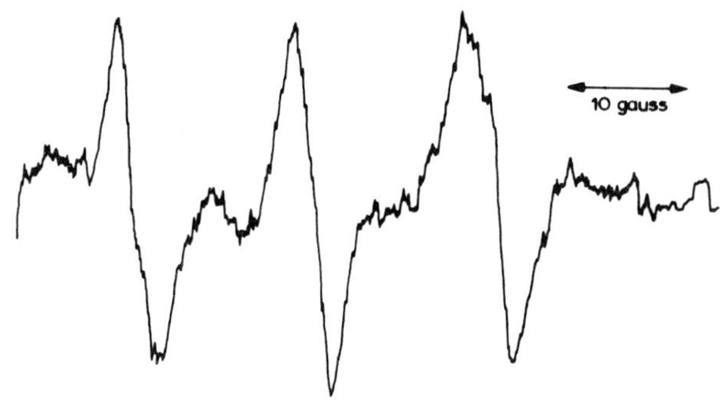

FIG. 4. Electron spin resonance spectrum of rat liver microsomes incubated at 37 °C with buffer and an NADPH-generating system for 5 min in the presence of 8mM-2-methyl-2-nitrosopropane as a spin-trap and CCl_4.

Model system (*c*), which is believed to involve the $OH^{\bullet}$ radical, is more strongly inhibited by catechin and propyl gallate than by promethazine (Table 2). Dose–response characteristics are illustrated in Fig. 5. Note that, although (+)-catechin is very active in inhibiting free-radical reactions in each of the systems studied here, it has little protective action on CCl_4-hepatotoxicity *in vivo*, except with high doses (Danni *et al.* 1977). The low activity of (+)-catechin *in vivo* is probably the result of its rapid excretion and its relatively high solubility in water; these results emphasize earlier remarks about the importance of scavenger penetration for effectiveness *in vivo*. An interesting study on a system

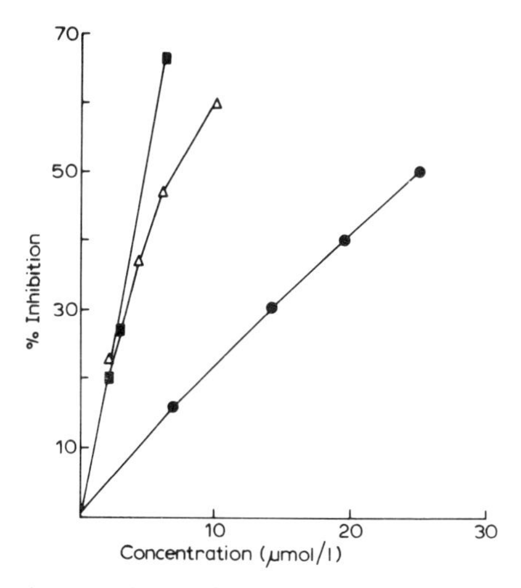

FIG. 5. Effects of promethazine (●), (+)-catechin (△) and propyl gallate (■) on the NADPH–ADP/Fe^{2+}-lipid peroxidation system (model system *c*) in rat liver microsomes. The effects are shown as percentage inhibitions relative to controls; temperature, 25 °C. For experimental details of the method see Slater (1968*a*).

TABLE 4

Effects of chelating agents and metal ions on four model systems involving free-radical intermediates (for details, see text); the results are shown as percentage inhibitions relative to controls (stimulations are shown marked with an asterisk). Each agent was studied over a range of concentrations; the results are largely from previously unpublished studies by K. Cheeseman, V. Ponti, B. Sawyer & T. F. Slater.

Agent	*Model system studied*							
	NADH/PMS/NBT		*NADPH–lipid peroxidation*		*NADPH–ADP/Fe^{2+} lipid peroxidation*		*CCl_4-stimulated lipid peroxidation*	
	Effect (%)	*Concentration (μmol/l)*	*Effect (%)*	*Concentration (μmol/l)*	*Effect (%)*	*Concentration (μmol/l)*	*Effect (%)*	*Concentration (μmol/l)*
EDTA	no effect	50	51	5			69	5
Desferrioxamine[a]			16	1			41	1
Thenoyltrifluoroacetone			{227* 358*	1×10^3 2×10^3			51	2×10^3
$CoCl_2$	50	35	51	35	43	25	64	35
Mn(II)	50	0.1	46	126	44	2	42	126
Zn^{2+}	nothing	1×10^3	{60* 58	180 1×10^3	61	100	{25 100	180 1×10^3
$CuSO_4$	50	2	{48* 23	10 20	62	16	51	20

[a] Supplier: CIBA-GEIGY (Basle, Switzerland)

TABLE 3

Effect of superoxide dismutase (SOD; 100 IU) on the NADH/PMS/NBT reaction in aerobic and anaerobic conditions (data from Ponti *et al.* 1978).

Reaction conditions	*Additions*	*Reaction rate* ($\Delta\epsilon_{560}$ nm/min)
Aerobic	none	0.42
Aerobic	SOD	0.00
Anaerobic	none	0.56
Anaerobic	SOD	0.41

closely related to model system (*c*) is that of Kappus *et al.* (1977) who investigated the effects of catechol and pyrogallol derivatives on an ADP–Fe^{2+}-stimulated lipid peroxidation in rat liver microsomes.

Table 2 also shows that only model system (*a*) is at all responsive to superoxide dismutase in aerobic conditions. Unexpectedly, Ponti *et al.* (1978) found that the NADH/PMS/NBT system also proceeds well in anaerobic conditions and is then largely insensitive to superoxide dismutase (Table 3). The lack of effect of superoxide dismutase on systems (*b*), (*c*) and (*d*) indicates that the radical species involved ($OH^{\cdot}$, $CCl_3^{\cdot}$, $CCl_3O_2^{\cdot}$) are either not scavenged by the dismutase or not accessible to it, or both.

Diphenylfuran and β-carotene, which are often used as scavengers for singlet oxygen (see Foote 1976), had only weak effects on model systems (*b*), (*c*) and (*d*) but β-carotene was moderately active on the NADH/PMS/NBT reaction. Doubts have recently been raised (Takayami *et al.* 1977) about the specificity of diphenylfuran for singlet oxygen in biological systems.

The effects of some metal chelating agents and of various metal salts on the model systems (*a*)–(*d*) are shown in Table 4. All chelating agents inhibited the CCl_4-stimulated peroxidation but, whereas EDTA and desferrioxamine inhibited the related NADPH-mediated peroxidation (model system *b*), thenoyltrifluoroacetone stimulated it. EDTA had no effect on the NADH/PMS/NBT reaction; the chelating agents were not tested on model system (*c*) as that system requires iron(II) ions for full activity.

The strongest inhibitions of the CCl_4-stimulated peroxidation by the metal ions tested were by Cu^{2+} and Co^{2+}; different relative reactivities were seen with model systems (*a*) and (*c*) in which Mn^{2+} ions were the most effective metal species. Dose–response behaviour for Cu^{2+}, Co^{2+} and Mn^{2+} in the NADH/PMS/NBT system are illustrated in Fig. 6. Of the metal ions tested, zinc was relatively inactive (in molar terms) in all models studied.

The data reported here show that free-radical reactions involving several different oxygen and oxygen-derived free radicals can be effectively attenuated

TABLE 5

Effects of thiol-binding agents on model systems (*b*), (*c*) and (*d*); results are given as percentage inhibitions relative to controls (stimulations are shown marked with an asterisk).

Agent	*Model system studied*					
	NADPH–lipid peroxidation		*NADPH–ADP/Fe^{2+}-lipid peroxidation*		*CCl_4-stimulated lipid peroxidation*	
	Effect (%)	*Concentration (mmol/l)*	*Effect (%)*	*Concentration (mmol/l)*	*Effect (%)*	*Concentration (mmol/l)*
p-Chloromercuribenzoate			43	0.100	21* 66	0.100 0.200
N-Ethylmaleimide	24	2	20	0.175	22*	2
Iodoacetate	no effect	1	17	5	no effect	1
Mersalyl	30*	0.2	46	0.025	46	0.200

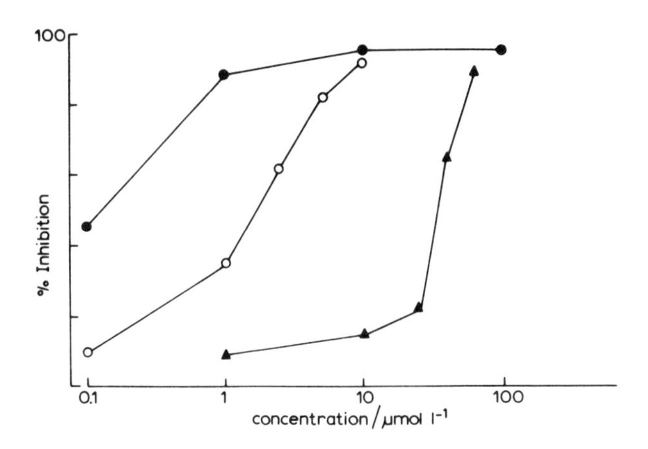

FIG. 6. Effects of manganese(II) ions (●), cupric ions (○) and cobalt(II) ions (▲) on the NADH/PMS/NBT system (model system *a*). Results are shown as percentage inhibitions relative to controls; temperature, ambient. For experimental details, see Ponti *et al.* (1978).

by a variety of agents in low concentration. Similar results have been found (Poli *et al.* 1978) with CCl_4 for promethazine, propyl gallate and (+)-catechin using isolated rat hepatocytes where the integrity of the cell is preserved so that it might be thought that the results *in vitro* can be extended directly to situations *in vivo*. However, whereas promethazine and propyl gallate are effective *in vivo* against CCl_4-hepatotoxicity, (+)-catechin and vitamin E are much less active (for reviews of protection produced by free-radical scavengers in relation to CCl_4-hepatotoxicity see Slater 1972, 1978). Such relative differences between protective effects seen *in vitro* and *in vivo* point to the importance of excretion rates, tissue distribution, and intracellular concentration of the scavenger *in vivo* in determining the efficiency of attenuation of relevant free-radical reactions known to be responsive to that scavenger *in vitro*. It is also evident from data presented here, and as is well known from classical free-radical chemistry studies, that scavengers can show marked specificity in their reactivity with different free-radical intermediates. An example of this specificity is promethazine (see Table 2), which effectively inhibits model systems (*b*), (*c*) and (*d*) but not (*a*).

If one bears in mind the high reactivity of several free-radical intermediates formed by metabolic reactions *in vivo* (e.g. $OH^\bullet$ and $CCl_3O_2^\bullet$), it seems likely that for effective protection the scavenger must gain access to the immediate sphere of influence of the primary free radical. For $CCl_3^\bullet$ it has been calculated, for example, that its half-life *in vivo* is less than 100 μs (Slater 1976) and that its radius of diffusion will be correspondingly small. The trichloromethyl radical is thereby trapped in its microenvironment, and effective scavenging requires the incoming scavenger to gain access to the locus of formation of $CCl_3^\bullet$ and to get there in sufficient concentration to compete for the $CCl_3^{\bullet}$ effectively against neighbouring groupings (unsaturated fatty acids, proteins, nucleotides). To be efficient *in vivo*, a scavenger that shows good activity *in vitro* has to get to the right intracellular site, at the right time, and in the right concentration.

The results of Tables 2 and 4 show that some scavengers have substantially different effects on model systems (*c*) and (*d*): for example, promethazine, menadione and manganese(II) ions. Although these differences may reflect the different reactivities of the $OH^{\bullet}$ and $CCl_3O_2^{\bullet}$ radicals involved in model systems (*c*) and (*d*), it is probable that other factors also contribute to the overall effect: for example, different loci of electron-flow out of the NADPH-cytochrome P450 chain in liver endoplasmic reticulum (see Slater 1975) with associated differences in accessibility for the scavenger to the primary site of free-radical formation. Evidence for the latter suggestion is provided by data in Table 5, where it can be seen that *p*-chloromercuribenzoate inhibits model system (*c*) at 100 μmol/l final concentration whereas it stimulates model (*d*) in similar conditions. The data given in Table 5 also indicate the weak activity of water-soluble thiol-binding reagents compared to the more lipophilic agents *p*-chloromercuribenzoate and mersalyl.

The results (for promethazine, vitamin E, *p*-chloromercuribenzoate) show that an agent may stimulate a free-radical-mediated reaction at low final concentrations, but may inhibit at higher concentrations. This 'cross-over' effect (see Slater 1969) shows the importance of studying scavengers and related substances over a range of concentrations in order to gain adequate information about their role as pro- or anti-oxidants in systems undergoing lipid peroxidation.

ACKNOWLEDGEMENTS

Work described here was generously supported over several years by CIBA-GEIGY Ltd., Cancer Research Campaign, Zyma (U.K.) Ltd., National Foundation for Cancer Research, and the Medical Research Council.

References

BAIRD, M. B., MASSIE, H. R. & PIEKIELNIAK, M. J. (1977) Formation of lipid peroxides in isolated rat liver microsomes by singlet molecular oxygen. *Chem. Biol. Interact. 16*, 145–153

BURDINO, E., GRAVELA, E., UGAZIO, G., VANNINI, V. & CALLIGARO, A. (1973) Initiation of free radical reactions and hepatotoxicity in rats poisoned with CCl_4 or $CBrCl_3$. *Agents Actions 4*, 244–253

BURDINO, E., DANNI, O., DIANZANI, M. U., MILILLO, P. A., POLI, G., SENA, L. M., TORRIELLI, M. V. & UGAZIO, G. (1976) Halogenoalkane toxicity in different experimental conditions. *Panminerva Med. 18*, 332–344

BUS, J. S., AUST, S. D. & GIBSON, J. E. (1975) Lipid peroxidation: a possible mechanism for paraquat toxicity. *Res. Commun. Chem. Pathol. Pharmacol. 11*, 31–38

CHRISTOPHERSON, B. O. (1968) Formation of monohydroxy-polyenoic fatty acids from lipid peroxides by a glutathione peroxidase. *Biochim. Biophys. Acta 164*, 35–46

COHEN, G. (1979) Lipid peroxidation: detection *in vivo* and *in vitro* through the formation of saturated hydrocarbon gases, in *This Volume*, pp. 177–182

DANNI, O., SAWYER, B. C. & SLATER, T. F. (1977) Effects of (+)-catechin *in vitro* and *in vivo* on disturbances produced in rat liver endoplasmic reticulum by CCl_4. *Biochem. Soc. Trans. 5*, 1029–1032

DESAI, I. D., SAWANT, P. L. & TAPPEL, A. L. (1964) Peroxidative and radiative damage to isolated lysosomes. *Biochim. Biophys. Acta 86*, 277–285

DIANZANI, M. U. & UGAZIO, G. (1978) Lipid peroxidation, in *Biochemical Mechanisms of Liver Injury* (Slater, T. F., ed.), pp. 669–707, Academic Press, London

DORMANDY, T. L. (1978) Free radical oxidation and antioxidants. *Lancet 1*, 647–650

FLETCHER, B. L., DILLARD, C. J. & TAPPEL, A. L. (1973) Measurement of fluorescent lipid peroxidation products in biological systems and tissues. *Anal. Biochem. 52*, 1–9

FONG, K.-L., MCCAY, P. B., POYER, J. L., KEELE, B. B. & MISRA, H. (1973) Evidence that peroxidation of lysosomal membranes is initiated by hydroxyl free radicals produced during flavin enzyme activity. *J. Biol. Chem. 248*, 7792–7797

FOOTE, C. S. (1976) Photosensitised oxidation and singlet oxygen: consequences in biological systems, in *Free Radicals in Biology*, vol. 2 (Pryor, W. A., ed.), pp. 85–133, Academic Press, New York

GOLDSTEIN, B. D., LODI, C., COLLINSON, C. & BALCHUM, O. J. (1969) Ozone and lipid peroxidation. *Arch. Environ. Health 18*, 631–635

HEATH, R. L. & TAPPEL, A. L. (1976) A new sensitive assay for the measurement of hydroperoxides. *Anal. Biochem. 76*, 184–191

HOCHSTEIN, P. & ERNSTER, L. (1963) ADP activated lipid peroxidation coupled to the TPNH system of microsomes. *Biochem. Biophys. Res. Commun. 12*, 388–394

INGALL, A., LOTT, K. A. K., SLATER, T. F., FINCH, S. & STIER, A. (1978) Metabolic activation of CCl_4 to a free radical product: studies using a spin trap. *Biochem. Soc. Trans. 6*, 962–964

JOHNSON, R. M. & SIDDIQI, I. W. (1970) *The Determination of Organic Peroxides*, Pergamon, Oxford

KAPPUS, H., KIECZKA, H., SCHEULEN, M. & REMMER, H. (1977) Molecular aspects of catechol and pyrogallol inhibition of liver microsomal lipid peroxidation stimulated by ferrous ion-ADP-complexes or by CCl_4. *Naunyn-Schmiedeberg's Arch. Pharmakol. 300*, 179–187

KELLER, F., SNYDER, A. B., PETRACEK, F. J. & SANCIER, K. M. (1971) Hepatic free radical levels in ethanol-treated and CCl_4-treated rats. *Biochem. Pharmacol. 20*, 2507–2511

LEWIS, S. E. & WILLS, E. D. (1962) The destruction of —SH groups of proteins and amino acids by peroxides of unsaturated fatty acids. *Biochem. Pharmacol. 11*, 901–912

LINDSTROM, T. D. & ANDERS, M. W. (1978) Effects of agents known to alter CCl_4 hepatotoxicity and cytochrome P450 levels on CCl_4-stimulated lipid peroxidation and ethane expiration in the intact rat. *Biochem. Pharmacol. 27*, 563–567

MEAD, J. F. (1976) Free radical mechanisms of lipid damage and consequences for cellular membranes, in *Free Radicals in Biology*, vol. 1 (Pryor, W. A., ed.), pp. 51–68, Academic Press, New York

MONDER, C. (1967) α-Keto aldehyde dehydrogenase, an enzyme that catalyses the enzymic oxidation of methyl glyoxal to pyruvate. *J. Biol. Chem. 242*, 4603–4609

NISHIKIMI, M., RAO, N. A. & YAGI, K. (1972) The occurrence of superoxide anion in the reaction of reduced phenazine methosulphate and molecular oxygen. *Biochem. Biophys. Res. Commun. 46*, 849–854

PACKER, J. E., SLATER, T. F. & WILLSON, R. L. (1978) Reactions of the CCl_4-related peroxy free radical ($CCl_3O_2^{\bullet}$) with amino acids: pulse radiolysis evidence. *Life Sci. 23*, 2617–2620

PERKINS, M. J., WARD, P. & HORSFIELD, A. (1970) A probe for homolytic reactions in solution. III. Radicals by hydrogen abstraction. *J. Chem. Soc. B*, 395–400
PLAA, G. L. & WITSCHI, H. (1976) Chemicals, drugs and lipid peroxidation. *Annu. Rev. Pharmacol. Toxicol. 16*, 125–141
PLACER, Z., VESELKOVA, A. & RATH, R. (1965) Kinetik des Malondialdehydes im Organismus. *Experientia 21*, 19–20
POLI, G., CHIONO, M. P., SLATER, T. F., DIANZANI, M. U. & GRAVELA, E. (1978) Effects of CCl_4 on isolated rat liver cells: stimulation of lipid peroxidation and inhibitory action of free radical scavengers. *Biochem. Soc. Trans. 6*, 589–591
PONTI, V., DIANZANI, M. U., CHEESEMAN, K. & SLATER, T. F. (1978) Studies on the reduction of nitroblue tetrazolium chloride mediated through the action of NADH and phenazine methosulphate. *Chem.-Biol. Interact. 23*, 281–291
RECKNAGEL, R. O. (1967) Carbon tetrachloride hepatotoxicity. *Pharmacol. Rev. 19*, 145–208
RECKNAGEL, R. O. & GHOSHAL, A. K. (1966*a*) New data on the question of lipoperoxidation in CCl_4 poisoning. *Exp. Mol. Pathol. 5*, 108–117
RECKNAGEL, R. O. & GHOSHAL, A. K. (1966*b*) Quantitative estimation of peroxidative degeneration of rat liver microsomal and mitochondrial lipids after CCl_4 poisoning. *Exp. Mol. Pathol. 5*, 413–426
RIELY, C. A., COHEN, G. & LIEBERMAN, H. (1974) Ethane evolution: a new index of lipid peroxidation. *Science (Wash. D.C.) 183*, 208–210
ROTRUCK, J. T., POPE, A. L., GANTHER, H. E., SWANSON, A. B., HAFEMAN, D. G. & HOEKSTRA, W. G. (1973) Selenium: biochemical role as a component of glutathione peroxidase. *Science (Wash. D.C.) 179*, 588–590
SCHAUENSTEIN, E. (1967) Autoxidation of polyunsaturated esters in water: chemical structure and biological activity of the products. *J. Lipid Res. 8*, 417–428
SCHAUENSTEIN, E., ESTERBAUER, H. & ZOLLNER, H. (1977) *Aldehydes in Biological Systems*, Pion, London
SLATER, T. F. (1968*a*) The inhibitory effects *in vitro* of phenothiazines and other drugs on lipid peroxidation systems in rat liver microsomes, and their relationship to the liver necrosis produced by CCl_4. *Biochem. J. 106*, 155–160
SLATER, T. F. (1968*b*) Aspects of cellular injury and recovery, in *The Biological Basis of Medicine*, vol. 1 (Bittar, E. E. & Bittar, N., eds.), pp. 369–414, Academic Press, New York
SLATER, T. F. (1969) Lysosomes and experimentally induced tissue injury, in *Lysosomes in Biology and Medicine*, vol. 1 (Dingle J. T. & Fell, H. B., eds.), pp. 469–492, North-Holland, Amsterdam
SLATER, T. F. (1972) in *Free Radical Mechanisms in Tissue Injury*, Pion, London
SLATER, T. F. (1975) The role of lipid peroxidation in liver injury, in *Pathogenesis and Mechanisms of Liver Cell Necrosis* (Keppler, D. ed.), 209–223, MTP, Lancaster
SLATER, T. F. (1976) Biochemical pathology in microtime, in *Recent Advances in Biochemical Pathology: toxic liver injury* (Dianzani, M. U., Ugazio, G. & Sena, L. M., eds.), pp. 99–108, Minerva Medica, Turin
SLATER, T. F. (1978) Mechanisms of protection, in *Biochemical Mechanisms of Liver Injury* (Slater, T. F., ed.), pp. 745–801, Academic Press, London
SLATER, T. F. & RILEY, P. A. (1966) Photosensitisation and lysosomal damage. *Nature (Lond.) 209*, 151–154
SLATER, T. F. & SAWYER, B. C. (1971*a*) The stimulatory effects of CCl_4 and other halogenoalkanes on peroxidative reactions in rat liver fractions *in vitro*: general features of the systems used. *Biochem. J. 123*, 805–814
SLATER, T. F. & SAWYER, B. C. (1971*b*) The stimulatory effects of CCl_4 on peroxidative reactions in rat liver fractions *in vitro:* inhibitory effects of free radical scavengers and other agents. *Biochem. J. 123*, 823–828
TAKAYAMA, K., NOGUCHI, T., NAKANO, M. & MIGITA, T. (1977) Reactivities of diphenylfuran (a singlet oxygen trap) with singlet oxygen and hydroxyl radical in aqueous systems. *Biochem. Biophys. Res. Commun. 75*, 1052–1058

WALLER, R. L. & RECKNAGEL, R. O. (1977) Determination of lipid conjugated dienes with tetracyano-ethylene-[14]C: significance for study of the pathology of lipid peroxidation. *Lipids 12*, 914–921

WILLSON, R. L. (1978) Free radicals and tissue damage: mechanistic evidence from radiation studies, in *Biochemical Mechanisms of Liver Injury* (Slater, T. F., ed.), pp. 123–224, Academic Press, London

WILLSON, R. L. & SLATER, T. F. (1975) Carbon tetrachloride and biological damage: pulse radiolysis studies of associated free radical reactions, in *Proceedings of the 5th L. H. Gray Memorial Conference: Fast Processes in Radiation Chemistry and Biology* (Adams, G. E., Fielden, E. M. & Michael, B. D., eds.), pp. 147–161, Wiley, Chichester

Discussion

Cohen: Did you measure the stimulatory effect of 5nM-vitamin E on lipid peroxidation by the production of malonaldehyde?

Slater: Yes. The results for the CCl_4-stimulated lipid peroxidation (model system *d*) were routinely obtained that way. The results for the NADPH–ADP/Fe^{2+}-stimulated lipid peroxidation (model system *c*) were measured by following oxygen uptake with a Clarke platinum electrode.

Cohen: α-Tocopherol has been reported to be protective *in vivo*, for example, against carbon tetrachloride hepatotoxicity.

Slater: In vivo administration of CCl_4 to rats results in a fatty degeneration and centrilobular necrosis in the liver (Cameron & Karunaratne 1936). Although these injurious processes are believed to be linked to the production of a $CCl_3^{\bullet}$ radical, the prior dosing of rats with vitamin E does not have much protective action *in vivo*, although there is some attenuation of the triglyceride accumulation (see Slater 1972). *In vitro*, however, vitamin E strongly protects against lipid peroxidation at micromolar concentrations (see Table 2). Presumably, when given parenterally or orally, vitamin E does not get to the precise site of formation of $CCl_3^{\bullet}$ *in vivo* (for discussion of this point see Slater 1976). In contrast, promethazine is a good protector *in vitro* and *in vivo* (see Slater 1972).

Cohen: You said that 5-hydroxypent-2-enal caused cross-linking of proteins?

Slater: Yes, but the cross-linking is almost certainly due largely to malonaldehyde and related dialdehydes (Chio & Tappel 1969*a,b*; Kwon *et al.* 1965). 4-Hydroxypent-2-enal significantly inhibits DNA synthesis in several cell systems (Conroy *et al.* 1977) and also reacts with glutathione to give the cyclic product (1). This reaction proceeds rapidly at neutral pH without any enzymic intervention. We are currently investigating whether the product (1) is of significance to the powerful actions *in vivo* of the unsaturated aldehyde.

Smith: Carbon tetrachloride *in vivo* affects the intracellular concentration of

$$CH_3-\underset{\displaystyle OH}{\underset{|}{CH}}-CH=CH-CHO + GSH \longrightarrow$$

(1)

NADPH. Might there not be some differences in the mechanism of toxicity between *in vivo* and *in vitro* studies where one uses malonaldehyde as the end-point for lipid peroxidation?

Slater: We have spent several years trying to elucidate the mechanism by which CCl_4 decreases the level of NADPH in rat liver. Within a short time of exposing rats to CCl_4 there is a substantial decrease in the liver's content of NADPH (Slater *et al.* 1964; Slater & Sawyer 1977). This is a net loss and not a simple oxidation; the sum of $NADP^+$ + NADPH also falls. We have shown (Slater & Jose 1969) that *in vitro* $CCl_3^{\bullet}$ free radicals can destroy NADH and NADPH possibly with the formation of the dimeric products $(NAD)_2$ and $(NADP)_2$ (see Land & Swallow 1968).

After administration of CCl_4 to rats only NADPH is destroyed; there is no decrease in NADH. This preferential destruction possibly reflects the formation of $CCl_3^{\bullet}$ near to the NADPH-binding site on the NADPH–flavoprotein (see Slater 1972).

In vivo the necrosis is localized to the central regions of the liver lobules and it is possible that NADPH is completely destroyed in those cells that later become necrotic; the loss of NADPH in the periportal zone may be much less so that, overall, the whole liver NADPH shows a decrease of about 30%.

We are now developing a microdensitometric method (similar to that described for cytochrome P450) to measure NADPH before and after exposure of rats to CCl_4.

Allison: Are you not confusing an observation with an interpretation? You observed a change in probe mobility after exposure of membranes to radical systems. You interpret this as being due to cross-linking of proteins. Do you have any evidence that this occurs?

Slater: We don't ourselves have any direct evidence but Tappel's group in particular have demonstrated interaction of aldehydes with proteins (Chio & Tappel 1969*a,b*); bifunctional aldehydes (e.g. malonaldehyde) can be expected *a priori* to produce some cross-linking, as indeed reported (Roubal & Tappel 1966*a,b*; Schaich & Karel 1976). We have found that lipid peroxidation, which is associated with the production of such aldehydes, causes a decrease in microsomal membrane fluidity (Ingall *et al.* 1979); we have speculated that this decreased fluidity is the result of aldehyde-induced cross-links between membrane components.

Stern: What is the evidence for the $CCl_3^{\cdot}$ radical reacting with oxygen?

Slater: As I have mentioned, promethazine reacts rapidly with $CCl_3^{\cdot}$ in aerobic conditions; the rate constant has been reported as $1.1 \times 10^9 \, l \, mol^{-1} \, s^{-1}$ (Willson & Slater 1975). However, quite unexpectedly, when the reaction is done in strictly anaerobic conditions, it is relatively slow. Dr John Packer (who is visiting my department on sabbatical leave from the University of Auckland) has evidence that the $CCl_3^{\cdot}$ radical reacts rapidly with O_2 possibly to yield the peroxy-radical $CCl_3O_2^{\cdot}$. (He is studying this reaction in detail by pulse radiolysis and electron spin resonance in collaboration with Dr Willson in Brunel and Professor Davies at University College, London.) We believe that in aerobic conditions $CCl_3^{\cdot}$ is rapidly converted into $CCl_3O_2^{\cdot}$.

Stern: Rather than a reduced product of O_2?

Slater: Yes; $CCl_3^{\cdot}$ is essentially an electrophilic reactant. I should not expect it to form superoxide on reaction with oxygen.

Fridovich: Results with the first model system (p. 147) call to mind the results that have been seen on NBT reduction (Beauchamp & Fridovich 1971): NBT can be reduced by xanthine oxidase anaerobically as well as aerobically. When one uses superoxide dismutase as a probe, one finds that aerobically it inhibits the reduction because O_2^- is then the major reductant but anaerobically it has no effect because NBT is then reduced directly at an enzyme active site.

Slater: The simple model system (*a*) had no enzymic component. I was surprised that the anaerobic reaction was a bit faster than the aerobic reduction of NBT; this suggests that reduced PMS directly reduces the tetrazolium salt without the intervention of superoxide radicals. However, the *aerobic* reaction is completely inhibited by superoxide dismutase. The following scheme explains the results:

$$\text{NADH} \xrightarrow{\text{PMS}} \text{PMSH}_2 \xrightarrow{O_2} O_2^- \xrightarrow{\text{NBT}} \text{Formazan}$$

$$\text{PMSH}_2 \xrightarrow{\text{NBT}} \text{Formazan}$$

The reaction of $PMSH_2$ with O_2 is not particularly rapid in the conditions used in the model (Ponti *et al.* 1978).

Fridovich: Oxygen, when present, should compete favourably against NBT and then O_2^- would reduce the NBT; but anaerobically the NBT should be reduced directly by the reduced PMS, as you show.

Michelson: FMN plus NADH will reduce NBT both aerobically and anaerobically but only the aerobic effect would be inhibited by superoxide dismutase. Nobody has ever taken reduction of NBT as an absolute criterion for superoxide ions; there is always a safeguard if the reduction that is seen is inhibited by SOD.

Keberle: You explain the centrilobular damage in the liver as being due to the higher concentration of metabolizing enzymes in this region. Drug-distribution studies, both *in vivo* and in the isolated perfused liver, have shown that many compounds — even some that are not metabolized — accumulate in the centrilobular tissue. Might it not be possible that, with many substances at least, this accumulation also contributes to the centrilobular damage?

Slater: Yes, but the centrilobular location of the injury in CCl_4 intoxication has several possible explanations, such as a preferential distribution of free-radical scavengers in the periportal zone or the centrilobular location of the P450-activating enzyme sequence. We know that cytochrome P450 is preferentially localized in the central region (Gooding *et al.* 1978); the gradient of distribution across the liver lobule is much more evident after phenobarbital induction, which increases CCl_4 toxicity (McLean & McLean 1966).

Smith: Does the distribution of the lesion in the liver correlate with the oxygen concentration in different parts of the liver? May this provide a teleological basis for protective enzymes or systems?

Slater: The idea that a differential oxygen gradient across the liver lobule is an important factor in determining the location of the injury was developed in particular by Himsworth (1950), although previous workers also had stressed the relevance of a central ischaemic hypoxia. It is probable that this factor is important in later stages of the liver injury produced by CCl_4 but not in the early events. In an important experiment Brauer (1965) reversed the direction of blood flow through an isolated perfused rat liver so that the perfusate entered the lobules through the central vein and left through the portal tract and still observed centrilobular damage after exposure to $CHCl_3$.

Williams: Vallee pointed out that the damage due to CCl_4 was strongly centralized in liver mitochondria which lost all their gradients of ions. Where does the CCl_4 act? With the P450 system is it more a microsomal effect than action on the mitochondrial membrane (because there is a P450 mitochondrial system as well)?

Slater: Vallee's group (Reynolds *et al.* 1962) studied the effects of CCl_4 on mitochondrial function and metal content but at late stages of the intoxication. Many secondary features appear to complicate interpretations after 3–6 h of CCl_4 intoxication. However, within 5–15 min of orally dosing rats with CCl_4 one observes marked disturbances in the endoplasmic reticulum, especially in cells of the centrilobular region. When one measures lipid peroxidation by diene conjugation, the change in liver endoplasmic reticulum is maximal by 15 min after dosing (Recknagel & Glende 1973); covalent binding of labelled

CCl_4 to microsomal components is maximal after an even shorter period of exposure. Mitochondrial disturbances appear much later, after several hours, and lysosomal changes after 12–18 h (Slater & Greenbaum 1965).

Williams: Is the mitochondrial injury due directly to CCl_4 or to different metabolites from some earlier injury?

Slater: Recent results (Roders *et al.* 1978; Benedetti *et al.* 1977*a*; Benedetti *et al.* 1977*b*) indicate that peroxidizing microsomes can exert damaging effects at a considerable distance on other intracellular and extracellular structures. Peroxidizing microsomes are placed inside a dialysis bag and target structures (e.g. red blood cells, microsomes, or mitochondria) are placed in the medium surrounding the dialysis sack. Using this technique, Comporti's group has shown that active materials produced during peroxidation of microsomes can diffuse across the dialysis membrane and cause haemolysis of red cells or enzyme inactivation in microsomes. These active substances appear to be moderately long-chain lipid factors, possibly aldehydes or peroxides or both. These substances can also cause systemic effects. Ugazio *et al.* (1976) have shown that peroxides and aldehydes can have significant effects on capillary permeability and on platelet aggregation.

Williams: In the reverse experiment in which the mitochondria are on the outside and the free-radical-generating system is inside the dialysis bag, is the microsomal system damaged?

Slater: Peroxidizing microsomes show a progressive decline in several enzymic activities (see Slater 1972; Recknagel & Glende 1973). Mitochondria (and microsomes) are not, however, significantly damaged by adding small quantities of CCl_4 in the absence of an activating system. Larger quantities of CCl_4 are inhibitory but this is probably by a simple lipid solvent-like effect.

ZINC AND NADPH-OXIDATION-DEPENDENT LIPID PEROXIDATION

Chvapil: Several pieces of evidence implicate singlet oxygen in the NADPH-dependent microsomal lipid peroxidation (King *et al.* 1975; Nakano *et al.* 1975). After acute as well as chronic administration of CCl_4 the lipid peroxidation in liver microsomes is enhanced parallel with other signs of hepatotoxicity, such as labilization of lysosomes, increased rate of synthesis and deposition of collagen in the liver tissue. Supplementation with zinc of rats treated with CCl_4 reduced the hepatotoxicity of the toxin. The formation of malonaldehyde by liver microsomes was also suppressed (Chvapil *et al.* 1973).

Studying the mechanism of the protective effect of zinc against CCl_4 hepatotoxicity we showed that the activity of mixed-function oxidases of smooth endoplasmic reticulum is inhibited by Zn^{2+}. As the activity of those enzymes

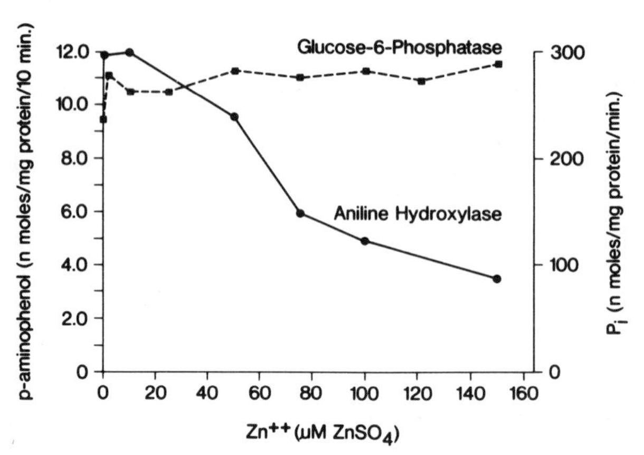

FIG. 1 (Chvapil). Effect of Zn^{2+} on the activity of aniline hydroxylase and glucose 6-phosphatase in liver microsomes. Aniline (1 mmol/l) was incubated for 10 min in a 3 ml incubation medium (pH 7.4) consisting of 5mM-$MgCl_2$, 12mM-glucose 6-phosphate, 1 unit glucose-6-phosphate dehydrogenase, 0.33mM-NADP, 6 mg of microsomal protein, 50mM-Tris and 154mM-KCl. Glucose 6-phosphate with microsomes was suspended in Tris-KCl and the P_i was determined by the colorimetric procedure of Fiske & Subbarow (1925). Each point represents the mean of two incubations.

independent of NADPH, such as glucose 6-phosphatase, was not affected by Zn^{2+} (Fig. 1), we postulated that zinc interacts with the enzymes of the mixed-function oxidases or with the substrate, NADPH, or both. My co-worker, Janet Ludwig, has shown that zinc ions inhibit the enzymic microsomal electron-transport chain essential to the drug-oxidizing system of liver cells *in vitro*. The initial step for mixed-function oxidases is the oxidation of NADPH. Fig. 2 shows that, in a microsomal preparation of rat liver, Zn^{2+} inhibits oxidation of NADPH. (The initial reaction was recorded so that Michaelis–Menton kinetics could be applied.) Lineweaver–Burk double-reciprocal plots of zinc concentration against velocity show a competitive mechanism for zinc inhibition (Fig. 3.) The K_i was calculated to be 7.22μM-Zn^{2+}. As Hochstein & Ernster (1963) showed the relationship between the oxidation of NADPH on the smooth endoplasmic reticulum and lipid peroxidation, this may be a mechanism by which zinc protects against CCl_4 hepatotoxicity.

The second possible mechanism by which zinc may inhibit the mixed-function oxidase system is interference with the substrate NADPH. Janet Ludwig has evidence that zinc ions bind to NADPH. Elution patterns of NADPH from Sephadex G-10 columns equilibrated with various concentrations of Zn^{2+} ions indicated direct binding of zinc to NADPH in a molar ratio 2:1. Zinc did not bind to NADH. Analysis of ^{31}P n.m.r. spectra of NADPH and NADH with and without zinc ions indicated that NADPH binds zinc through the monophosphate on C-2 of the adenine-ribosyl portion. The u.v. and fluorescence

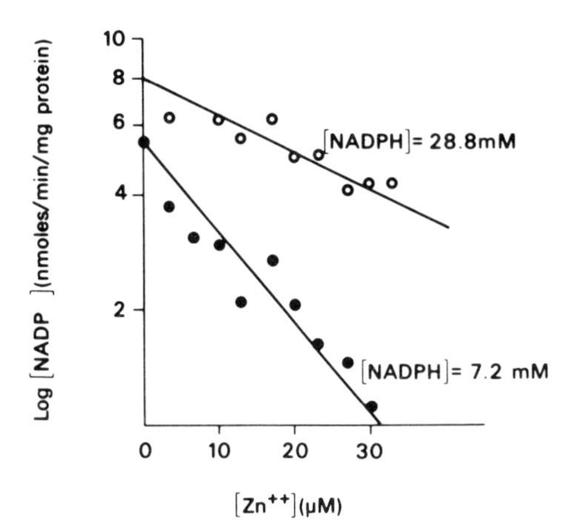

FIG. 2 (Chvapil). Effect of zinc ions on the activity of NADPH oxidase in liver microsomes at low and high substrate levels. Enzyme activity in the microsomal fraction was assayed at pH 5.5 in a system containing 8 μmol Na_2HPO_4, 54 μmol KH_2PO_4, 1 μmol $MnCl_2$, 340 μmol sucrose and 21.6 or 86.4 μmol NADPH in 3 ml final volume. Control samples did not contain a subcellular fraction. The rate of oxidation of the reduced coenzyme was scanned at 340 nm.

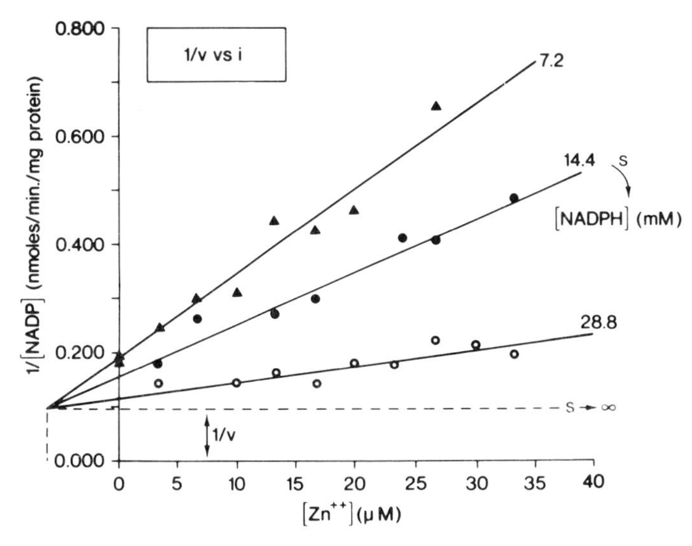

FIG. 3 (Chvapil). Lineweaver–Burk curve of inhibitor concentration, $[Zn^{2+}]$, against (1/velocity) based on the initial rate of enzyme activity at three different concentrations of substrate. NADPH oxidation was measured at 340 nm in a cuvette containing 22.5mM-KH_2PO_4, 13.3mM-Na_2HPO_4, 141.7mM-sucrose, and the amounts of NADPH and Zn^{2+} given in the figure; 2.0 mg of microsomal proteins were added to initiate the reaction at 32 °C. The $1/v_{max}$ value is 0.097 (nmol NADPH min^{-1} [mg of protein]$^{-1}$)$^{-1}$. The $1/\kappa_s$ value is 0.155 (mM-NADPH)$^{-1}$.

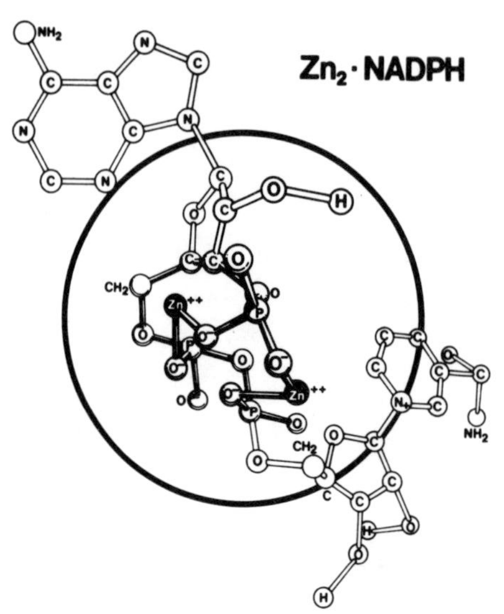

FIG. 4 (Chvapil). Schematic representation of binuclear complex Zn_2-NADPH.

spectra of NADPH did not change with zinc binding. Based on these results a Zn_2-NADPH model (Fig. 4) was proposed, in which the first zinc atom forms a linkage between the mono- and di-phosphate groups with $K_1 = 10^{3.76}$. The second zinc links the remaining oxygen atom of the monophosphate with the diphosphate to form a 10-membered ring structure with $K_2 = 10^{2.99}$. The overall formation constant of the Zn-NADPH complex is $10^{6.75}$. The binuclear complex, Zn_2-NADPH, may be a rigidly structured molecule in which the reduced nicotinamide ring is less available for enzymic oxidation than the uncomplexed NADPH.

Results of other studies have indicated interactions of zinc with related nucleotides. Zinc forms stable complexes with 5′-AMP, ADP and ATP. NADH did not bind zinc. All these nucleotides contain adenine and phosphate groups and all, except NADH, bind zinc.

Since the relative stability of some zinc-nucleotide complexes appears to increase with increasing numbers of phosphate groups in the nucleotide, we hypothesized that, although the zinc might not bind to NADH, it could bind NADPH as the latter compound contains an additional phosphate group. In this study we demonstrated that zinc does bind to NADPH and we suggest the structure shown in Fig. 4.

CAERULOPLASMIN AND SERUM ANTIOXIDANT ACTIVITY

Dormandy: Professor Slater outlined some of the critical antioxidant protective mechanisms which operate at the cellular and subcellular level. If one looks at organisms as a whole, in particular at humans, there are other tiers of protection. Human blood serum has been shown to be an exceedingly powerful antioxidant (Barber 1961; Vidlakova *et al.* 1972); and it was soon established that this property is unrelated to its vitamin E content. We have shown that serum antioxidant activity is the function of two well defined proteins. One is the iron-free fraction of transferrin: the other, the blue copper protein, caeruloplasmin (Stocks *et al.* 1974*a*,*b*). The antioxidant activity of iron-free transferrin was not altogether surprising since in most test systems 'free' iron is the most important lipid autoxidation catalyst; and free iron is readily mopped up and 'immobilized' by transferrin. But the more powerful antioxidant activity of caeruloplasmin was unexpected. We thought at first that it could be explained by analogy with the antioxidant activity of transferrin and that caeruloplasmin could mop up copper. But caeruloplasmin is not a copper-transporting protein in the sense that transferrin is iron-transporting (Scudder 1976): it is fully saturated with copper over a wide range of concentrations and plasma has no 'available copper-binding capacity' comparable to its available iron-binding capacity. More decisively, when copper is removed from caeruloplasmin *in vitro*, its antioxidant activity is abolished (Al-Timimi 1977; Al-Timimi & Dormandy 1977).

Over the past three years we have measured serum antioxidant activity, caeruloplasmin, copper, iron, and transferrin (as well as other related parameters) in over 300 subjects, in most cases on several occasions overy varying periods of time (Table 1) (Dormandy 1978; Cranfield *et al.* 1979). The groups were chosen because of the known association of various diseases and physiological states either with abnormal serum concentrations of caeruloplasmin or with an abnormal saturation of transferrin (Table 2). In all groups there was a close correlation between serum antioxidant activity and serum caeruloplasmin: indeed, serum antioxidant activity was virtually governed by the serum caeruloplasmin (Table 3). The correlation between serum antioxidant activity and

TABLE 1 (Dormandy)
Groups of subjects investigated

Groups	*Numbers (male + female)*	*(total)*	*Age (yr)*	
			range	*mean*
Healthy controls	62 + 58	(120)	20–65	37
Women on oral contraceptives	34	(34)	15–40	24
Pregnant women	40	(40)	17–41	26
Rheumatoid patients	27 + 48	(75)	24–86	54
Thalassaemic patients	20 + 32	(52)	5–18	10
Wilson's disease patients	12 + 8	(20)	14–36	22

TABLE 2 (Dormandy)

Serum antioxidant activity and related variables in various physiological and pathological states

Groups	*Antioxidant activity (% inhibition)**		*Caeruloplasmin (mg/l)*	*Copper (μmol/l)*	*Iron (μmol/l)*	*Iron-binding (μmol/l)*
	range	*mean* ± S.D.	*mean* ± S.D.	*mean* ± S.D.	*mean* ± S.D.	*mean* ± S.D.
Normal men	16.5–28.2	22.7 ± 3.7	328 ± 78	17.8 ± 3.8	21.1 ± 9.1	45.5 ± 6.2
Normal women	17.8–32.0	26.6 ± 2.4	357 ± 66	19.7 ± 2.8	17.1 ± 5.8	45.3 ± 7.2
Women on oral contraceptives	25.8–44.2	32.8 ± 4.6	570 ± 108	29.5 ± 5.8	21.9 ± 5.9	48.7 ± 10.2
Pregnant women	26.7–57.2	39.6 ± 10.1	552 ± 170	32.8 ± 7.8	19.1 ± 10.2	65.2 ± 22.0
Rheumatoid men	22.5–36.0	28.5 ± 3.8	419 ± 81	24.6 ± 4.6	10.9 ± 5.0	41.2 ± 6.3
Rheumatoid women	24.8–40.1	31.4 ± 3.0	489 ± 82	26.7 ± 4.2	12.9 ± 5.7	52.1 ± 12.8
Thalassaemic patients	7.2–28.1	20.2 ± 3.1	339 ± 63	19.4 ± 3.0	37.8 ± 7.2	6.0 ± 5.8
Wilson's disease patients	6.4–19.1	12.0 ± 4.2	86 ± 80	5.7 ± 2.9		

* Measured in a standard ox-brain homogenate assay system as described by Stocks *et al.* (1974*a*) and Al-Timimi & Dormandy (1977).

TABLE 3 (Dormandy)

Correlations between serum antioxidant activity, caeruloplasmin and available iron-binding capacity (Kendall rank correlation coefficients)

Groups	*Variables correlated (with significance)*		
	Antioxidant activity/ caeruloplasmin	*Antioxidant activity/ iron*	*Antioxidant activity/ available iron-binding capacity*
Normal controls	0.61 ($P < 0.001$)	0.25 ($P < 0.01$)	0.31 ($P < 0.01$)
Women on contraceptives	0.38 ($P < 0.01$)	0.18 (NS)*	0.22 (NS)
Pregnant women	0.24 ($P < 0.01$)	0.10 (NS)	0.18 (NS)
Rheumatoid patients	0.50 ($P < 0.001$)	0.12 (NS)	0.16 (NS)
Thalassaemic patients, pre-transfusion	0.54 ($P < 0.001$)	0.30 ($P < 0.01$)	0.31 ($P < 0.01$)
Thalassaemic patients, post-transfusion	0.55 ($P < 0.001$)	0.28 ($P < 0.01$)	0.26 ($P < 0.01$)
Wilson's disease patients	0.17 (NS)		
Whole series	0.68 ($P < 0.0005$)	0.28 (NS)	0.30 ($P < 0.01$)

* NS, not statistically significant

serum iron and iron-binding capacity was significant only in the normal control group and in thalassaemics.

These findings raise several questions of possible biological and even clinical relevance; but I should like to make only a few speculative comments. First, let me recall that caeruloplasmin is one of the acute-phase proteins — that is, one of the group of serum proteins whose concentration rises sharply after any form of tissue injury (even after minor planned surgical procedures). Most acute-phase proteins — as indeed most components of the inflammatory response — have fairly obvious survival value; but for long no such role could be assigned to caeruloplasmin. It now seems possible that the increased rate of caeruloplasmin synthesis after trauma relates to the greatly increased risk of free-radical damage associated with tissue destruction. Second, much has been said about iron and iron toxicity; and the evidence that the damage of iron overload depends on the role of iron as a free-radical oxidation catalyst is overwhelming. But there is equally strong evidence that the activity of iron is strongly influenced by caeruloplasmin and copper; and I do not think that either transitional metal should be studied in isolation.

Chvapil: We have heard about the importance of antioxidants, as represented by superoxide dismutase, glutathione peroxidase, catalase and probably vitamin E. I should like to propose that zinc ions have a vital part in a homeostatic mechanism, preserving the integrity of cell membrane, especially when the cell is exposed to some noxious agent. For instance, zinc deficiency leads to haemolysis of erythrocytes. Supplementation of zinc stabilizes the erythrocyte membranes.

INFLAMMATORY CELLS AND ZINC

Chvapil: In the fast few years I have been repeatedly amazed with the regulatory effect of zinc ions both *in vivo* and *in vitro* on various functions of cells, such as PMS, macrophages or mast cells.

Peritoneal inflammatory cells isolated three days after intraperitoneal injection of thioglycolate show a certain activation as measured by bioluminescence. Addition of yeast particles (1:50) to the incubation medium enriched with Mg^{2+} ions results in a several-fold increase in bioluminescence. When these cells are preincubated in a medium containing 80μM-zinc, the luminescence is decreased and, instead of showing an outburst after adding yeast particles, it continues for several minutes (Fig. 5). If we accept the view that bioluminescence reflects the formation of O_2^-, it follows that Zn^{2+} interfered with this reaction.

In another experiment we administered zinc intraperitoneally to rabbits and studied the reduction of NBT in peripheral blood leukocytes activated by

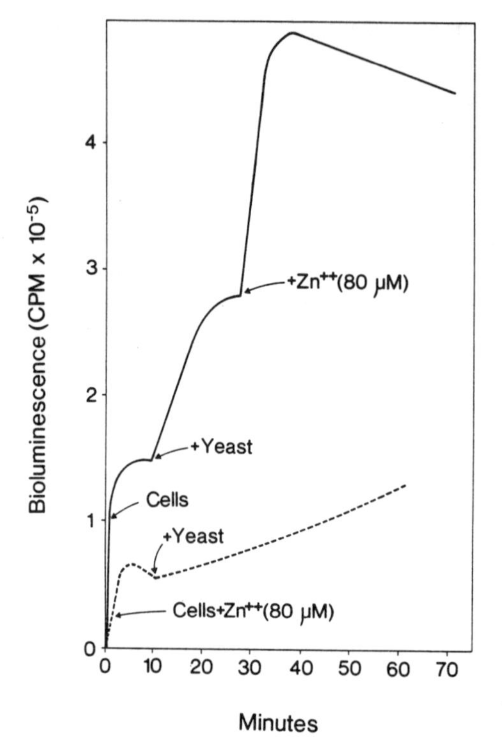

FIG. 5 (Chvapil). Bioluminescence studies with inflammatory cells harvested from peritoneal cavity of rats four days after Na caseinate injection. Cells (6×10^6; 0.05 ml) were added to 0.1M-PIPES (5.0 ml) pH 7.4 buffer containing: 3.1mM-KCl, 3.1mM-$MgSO_4$, 200 mg/100 ml glucose and 0.116M-NaCl. The cells were incubated at 25 °C (room temperature) and a background bioluminescence (c.p.m.) was established. At zero time, 0.1 ml of luminol (1 mg luminol per 100 ml of a 0.1% Me_2SO solution) was added and bioluminescence (c.p.m.; 0.1 s) was established at differing time periods. After 10 min 0.2 ml of yeast (280×10^6 opsinized) was added and the bioluminescence (c.p.m.) was established at various times. Finally, at 17.5 min, after the yeast was added, 79μM-$ZnSO_4$ was added and the bioluminescence was followed for various times.

The second set of data was determined by preincubating cells and 79μM-$ZnSO_4$ for 10–15 min. Luminol was added at zero time and the bioluminescence c.p.m. was determined at various time intervals up to 12.5 min. Yeast was then added and the bioluminescence (c.p.m.) was determined at various times after the yeast addition.

endotoxin. Zinc supplementation reduced the formation of formazan by leukocytes (Table 1). As it is assumed that reduction of NBT is mediated by O_2^-, we conclude that Zn^{2+} inhibited the formation of this highly reactive oxygen species.

Although we have no final proof on the molecular mechanism of the effect of zinc on leukocytes, some of our data favour a membrane-stabilizing effect of zinc. Multiple elongations and protrusions of intact peritoneal macrophages

TABLE 1 (Chvapil)

NBT test in rabbits treated with zinc

Sample	% NBT positive granulocytes		
	Control 0 time	*After Zn^{2+} injection*	
		2 h	*4 h*
Spontaneous reaction	70	7	12
Endotoxin stimulated	74	10	35

Three adult New Zealand White rabbit males were injected i.p. with $ZnSO_4.7H_2O$, pH 6.5, 2.9 mg/kg body weight. Venous blood was taken both before and 2 and 4 h after the treatment. Plasma zinc concentration increased from the original 0.7 μg/ml to 1.5 and 1.4 μg Zn/ml at 2 and 4 h, respectively. For spontaneous NBT test 0.1 ml of heparinized blood and 0.1 ml NBT dye were mixed and incubated at 37 °C for 25 min with gentle agitation. After incubation, thick blood smears were stained with 1% safranine. 100 granulocytes were counted to determine the percentage of NBT positive cells. Activation of granulocytes with endotoxin was done by mixing 1 ml whole blood either with 0.05 ml or 0.1 ml of endotoxin (J. Peacock and M. Chvapil, unpublished results).

disappear and the macrophages resume a rounded shape when Zn^{2+} from the medium by complexation with phosphate or cystine buffer restores the normal morphology of macrophages.

The functions of polymorphonuclear leukocytes or macrophages (e.g. phagocytosis, oxygen consumption, directional chemotaxis, bacterial killing, migration, etc.) are inhibited by Zn^{2+}, however, only in the presence of Mg^{2+}. The effect of Zn^{2+} is reversible and depends on the presence of zinc in the medium (Chvapil *et al.* 1977*a,b*).

We believe, therefore, that zinc ions interact with membrane constituents, either of lipid groups or sialoproteins, thereby changing the fluidity of the

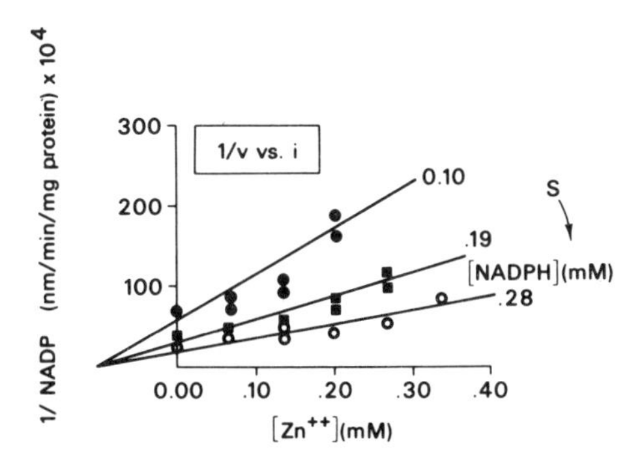

FIG. 6 (Chvapil). Effect of various doses of zinc on the oxidation of NADPH by mitochondrial fraction of pulmonary alveolar macrophages. The rate of oxidation was measured at three concentrations of the substrate and at three to five concentrations of zinc ions. Reciprocal values of the rate of oxidation of NADPH ($1/v$) are plotted against the concentration of zinc in the medium (i).

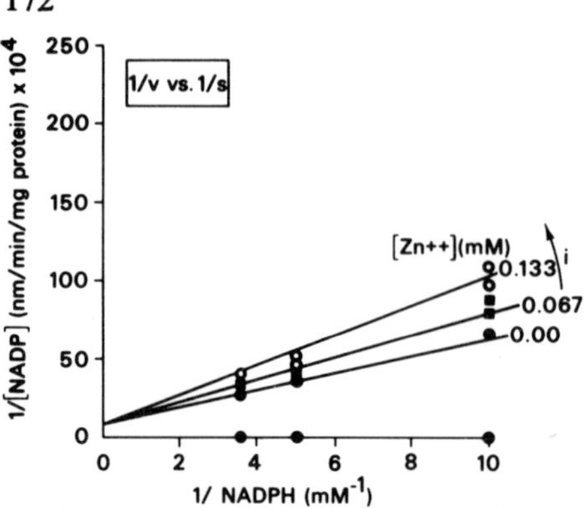

FIG. 7 (Chvapil). Double-reciprocal plot for the analysis of the effect of zinc on NADPH oxidase. The rate of NADP formation (*v*) was measured at three concentrations of the substrate (*s*) and at two concentrations of zinc (*i*).

membrane or affecting the communication pathway between surface sialic acid and O_2^- production of phagocytosing leukocytes, as proposed by Tsan & McIntyre (1976).

Another possible effect of Zn^{2+} on the production of unstable intermediate products of oxygen reduction refers to the demonstrated linkage of zinc to NADPH (Chvapil *et al.* 1975, 1976) and NADPH oxidase (Figs. 6 and 7). Janet Ludwig has shown that oxidation of NADPH by isolated cytoplasmic granules from macrophages is inhibited by zinc.

Slater: We can confirm the protection afforded against CCl_4 by zinc *in vitro*. Other factors, however, may contribute; intact animals have many ways of protecting the liver. For instance, diethyldithiocarbamate protects against CCl_4 by decreasing the uptake of CCl_4 from the stomach (when CCl_4 is administered by stomach tube). M. U. Dianzani *et al.* (personal communication) have confirmed this with zinc.

Hill: Since zinc has so many actions, surely one of the first things to look at ought to be protein synthesis and, in particular, the protective repair mechanisms?

Chvapil: Analysing the mechanisms of 'zinc effect' I see three major functions of this metal: (1) participation in the function of several enzymes; (2) involvement in protein and DNA synthesis; (3) stabilization of biomembranes or biostructures. This latter concept we proposed almost five years ago and we are still working on it. Another essential aspect of zinc refers to the established antagonism between zinc and copper or calcium.

Hill: A mechanism for zinc homeostasis must exist.

Williams: The calcium-binding sites on some enzymes are highly specific and would be little affected by the small amounts of zinc mentioned but millimolar concentrations will compete with calcium.

Chvapil: It has been shown that Ca^{2+}-dependent ATPase is inhibited by Zn^{2+}. It is also known that supplementation of zinc to rats with isoproterenol-induced myocardial infarction tends to reduce the amount of the necrosis possibly by interference with calcium flux into the damaged tissue. While calcium ions are increasing cardiac work, zinc ions inhibit the contraction of heart muscle thus protecting the myocardium against increasing oxygen deficiency (Chvapil & Owen 1977).

Willson: The loose binding of the zinc that Dr Chvapil stressed is in some ways important. For instance, iron will catalyse the oxidation of thiol groups. Addition of zinc (an inert and ubiquitous metal) can reduce the rate of catalysis considerably by binding competitively to the thiol: iron has to displace the zinc from the thiol before autoxidation can take place. I believe this is an important role for zinc *in vivo* but it is difficult to prove.

Crichton: Zinc overload has disastrous consequences on iron metabolism because in apoferritin it binds to the catalytic sites where iron is deposited. So, when iron is added, stimulation of apoferritin synthesis is normal but the apoferritin is poor in iron.

Hill: But how can one arrange for this zinc overload?

Crichton: Matrone *et al.* (1975) produced zinc toxicity by feeding a diet containing 0.75% zinc to rats and showed that the effects of zinc toxicity on iron metabolism *in vivo* were a reduced red cell lifespan and a decreased ferritin content with a lower percentage of iron in liver.

Chvapil: It is extremely difficult to overload adult rats by feeding them a high-zinc diet containing, for instance, 2000 p.p.m. zinc (normal diet contains 20–40 p.p.m. Zn). After administration of CCl_4 or after injury in general, the concentration of serum zinc goes down, supplementation of zinc only normalizes serum and possibly tissue content of zinc. In other words, there exists an efficient homeostatic regulation of zinc uptake.

Hill: That is reasonable; the control mechanism for zinc homeostasis is probably located in the liver.

Chvapil: Zinc metabolism has been shown to be regulated by a complex homeostatic mechanism. Although the whole picture of tissues and factors involved in the regulation of zinc body stores is not clear, it seems that intestinal mucosal epithelium is of great importance in the resorption of zinc. From there it goes to the plasma, where two major carrier proteins were identified. There seems to exist a certain dynamic equilibrium between plasma zinc and the content of zinc in various tissues. Liver is especially prone to accumulate zinc. It seems that the synthesis and the actual content of metallothionein is essential to the transport as well as disposition of zinc (and other metals as well).

References

AL-TIMIMI, D. J. (1977) *Serum Antioxidant Activity*, M. Phil. Thesis, London University

AL-TIMIMI, D. J. & DORMANDY, T. L. (1977) Inhibition of lipid autooxidation by human caeruloplasmin. *Biochem. J. 168*, 283–288

BARBER, A. A. (1961) Inhibition of lipid peroxide formation by vertebrate blood serum. *Arch. Biochem. Biophys. 92*, 38–43

BEAUCHAMP, C. & FRIDOVICH, I. (1971) Superoxide dismutase: improved assays and an assay applicable to acrylamide gels. *Anal. Biochem. 44*, 276–287

BENEDETTI, A., CASINI, A. F. & FERRALI, M. (1977*a*) Red cell lysis coupled to the peroxidation of liver microsomal lipids. Compartmentalization of the hemolytic system. *Res. Commun. Chem. Pathol. Pharmacol. 17*, 519–528

BENEDETTI, A., CASINI, A. F., FERRALI, M. & COMPORTI, M. (1977*b*) Studies on the relationships between carbon tetrachloride-induced alterations of liver microsomal lipids and impairment of glucose-6-phosphatase activity. *Exp. Mol. Pathol. 27*, 309–323

BRAUER, R. W. (1965) in *The Biliary System* (Taylor, W., ed.), p. 101, Blackwell, Oxford

CAMERON, G. R. & KARUNARATNE, W. A. E. (1936) *J. Pathol. Bacteriol. 42*, 1

CHIO, K. S. & TAPPEL, A. L. (1969*a*) Synthesis and characterization of the fluorescent products derived from malonaldehyde and amino acids. *Biochemistry 8*, 2821–2826

CHIO, K. S. & TAPPEL, A. L. (1969*b*) Inactivation of ribonuclease and other enzymes by peroxidizing lipids and by malonaldehyde. *Biochemistry 8*, 2827–2832

CHVAPIL, M. & OWEN, J. A. (1977) Effect of zinc on acute and chronic isoproterenol induced heart injury. *J. Mol. Cell. Cardiol. 9*, 151–159

CHVAPIL, M., RYAN, J. N., ELIAS, S. L. & PENG, Y. M. (1973) Protective effect of zinc on carbon tetrachloride-induced liver injury in rats. *Exp. Mol. Pathol. 19*, 186–196

CHVAPIL, M., SIPES, I. G., LUDWIG, J. C. & HALLADAY, S. C. (1975) Inhibition of NADPH oxidation and oxidative metabolism of drugs in liver microsomes by zinc. *Biochem. Pharmacol. 24*, 917–919

CHVAPIL, M., LUDWIG, J. C., SIPES, I. G. & MISIOROWSKI, R. L. (1976) Inhibition of NADPH oxidation and related drug oxidation in liver microsomes by zinc. *Biochem. Pharmacol. 25*, 1787–1791

CHVAPIL, M., STANKOVA, L., ZUKOSKI, C. IV & ZUKOSKI, C. III (1977*a*) Inhibition of some functions of polymorphonuclear leukocytes by *in vitro* zinc. *J. Lab. Clin. Med. 89*, 135–146

CHVAPIL, M., STANKOVA, L., BERNHARD, D. S., WELDY, P. L., CARLSON, E. C. & CAMPBELL, J. B. (1977*b*) Effect of zinc on peritoneal macrophages *in vitro*. *Infect. Immun. 16*, 367–373

CONROY, P. J., NODES, J. T., SLATER, T. F. & WHITE, G. W. (1977) The inhibitory effects of a 4-hydroxypentenal: cysteine adduct against sarcoma cells in mice. *Eur. J. Cancer 13*, 55–63

CRANFIELD, L. M., AL-TIMIMI, D. J., OWEN, M., MCMURRAY, W., GOLLAN, J. L., WHITE, A. G. & DORMANDY, T. L. (1979) The relation of serum antioxidant activity to serum caeruloplasmin and serum iron-binding capacity in normal and abnormal subjects. *Ann. Clin. Biochem.*, in press

DORMANDY, T. L. (1978) Free-radical oxidation and antioxidants. *Lancet 1*, 647–650

FISKE, C. H. & SUBBAROW, Y. (1925) The colorimetric determination of phosphorus. *J. Biol. Chem. 66*, 375–400

GOODING, P. E., CHAYEN, J., SAWYER, B. & SLATER, T. F. (1978) Cytochrome P450 distribution in rat liver and the effect of sodium phenobarbitone administration. *Chem.-Biol. Interact. 20*, 299–310

HIMSWORTH, H. P. (1950) in *The Liver and its Diseases*, p. 32, Blackwell, Oxford

HOCHSTEIN, P. & ERNSTER, L. (1963) ADP activated lipid peroxidation coupled to the TPNH system of microsomes. *Biochem. Biophys. Res. Commun. 12*, 388–394

INGALL, A., LOTT, K. A. K., STIER, A. & SLATER, T. F. (1979) in preparation

KING, M. M., LAI, E. K. & MCCAY, P. B. (1975) Singlet oxygen production associated with enzyme-catalyzed lipid peroxidation in liver microsomes. *J. Biol. Chem. 250*, 6496–6502

KWON, T. W., MENZEL, D. B. & ALCOTT, H. S. (1965). *J. Food Sci. 30*, 808

LAND, E. J. & SWALLOW, A. J. (1968) One-electron reactions in biochemical systems as studied by pulse radiolysis. I. Nicotinamide-adenine dinucleotide and related compounds. *Biochim. Biophys. Acta 162*, 327–337

MATRONE, G., TUGGLE, B. C. & RAMSAY, P. B. (1975) Studies on the elucidation of high levels of dietary zinc on synthesis of liver ferritin, in *Proteins of Iron Storage and Transport in Biochemistry and Medicine* (Crichton, R. R., ed.) pp. 337–342, North-Holland, Amsterdam

MCLEAN, A. E. M. & MCLEAN, E. K. (1966) The effect of diet and 1,1,1-trichloro-2,2-bis-(*p*-chlorophenyl)ethane (DDT) on microsomal hydroxylating enzymes and on sensitivity of rats to carbon tetrachloride poisoning. *Biochem. J. 100*, 564–571

NAKANO, M., NOGUCHI, T., SUGIOKA, K., FUKUYAMA, H. & SATO, M. (1975) Spectroscopic evidence for the generation of singlet oxygen in the reduced nicotinamide adenine dinucleotide phospate-dependent microsomal lipid peroxidation system. *J. Biol. Chem. 250*, 2404-2406

PONTI, V., DIANZANI, M. U., CHEESEMAN, K. & SLATER, T. F. (1978) Studies on the reduction of nitroblue tetrazolium chloride mediated through the action of NADH and phenazine methosulphate. *Chem.-Biol. Interact. 23*, 281–291

RECKNAGEL, R. O. & GLENDE, E. A. (1973). *C.R.C. Crit. Rev. Toxicol. 2*, 263

REYNOLDS, E. S., THIERS, R. E. & VALLEE, B. L. (1962) Mitochondrial function and metal content in carbon tetrachloride poisoning. *J. Biol. Chem. 237*, 3546–3551

RODERS, M. K., GLENDE, E. A. & RECKNAGEL, R. O. (1978) NADPH-dependent microsomal lipid peroxidation and the problem of pathological action at a distance. New data on induction of red cell damage. *Biochem. Pharmacol. 27*, 437–444

ROUBAL, W. T. & TAPPEL, A. L. (1966*a*) Damage to proteins, enzymes, and amino acids by peroxidizing lipids. *Arch. Biochem. 113*, 5–8

ROUBAL, W. T. & TAPPEL, A. L. (1966*b*) Polymerization of proteins induced by free-radical lipid peroxidation. *Arch. Biochem. 113*, 150–155

SCHAICH, K. M. & KAREL, M. (1976) Free radical reactions of peroxidizing lipids with amino acids and proteins: an ESR study. *Lipids 11*, 392–400

SCUDDER, P. R. (1976) *Copper Metabolism in Rheumatoid Arthritis and Related Disorders*, M. Phil. Thesis, London University

SLATER, T. F. (1972) in *Free Radical Mechanisms in Tissue Injury*, pp. 109–115 and 135–141, Pion, London

SLATER, T. F. (1976) Biochemical pathology in microtime. *Panminerva Med. 18*, 381–390

SLATER, T. F. & GREENBAUM, A. L. (1965) Changes in lysosomal enzymes in acute experimental liver injury. *Biochem. J. 96*, 484–491

SLATER, T. F. & JOSE, P. J. (1969) Destruction of reduced nicotinamide-adenine dinucleotide phosphate by bromotrichloromethane and by carbon tetrachloride *in vitro* and *in vivo*. *Biochem. J. 114*, 7P

SLATER, T. F. & SAWYER, B. C. (1977) The effects of CCl_4 on the content of nicotinamide adenine dinucleotide phosphate in rat liver. *Chem.-Biol. Interact. 16*, 359–364

SLATER, T. F., STRAULI, V. D. & SAWYER, B. C. (1964) Changes in liver nucleotide concentrations in experimental liver injury. *Biochem. J. 93*, 260–266

STOCKS, J., GUTTERIDGE, J. M. C., SHARP, R. J. & DORMANDY, T. L. (1974*a*) Assay using brain homogenate for measuring the antioxidant activity of biological fluids. *Clin. Sci. Mol. Med. 47*, 215–222

STOCKS, J., GUTTERIDGE, J. M. C., SHARP, R. J. & DORMANDY, T. L. (1974*b*) The inhibition of lipid autoxidation by human serum and its relation to serum proteins and α-tocopherol. *Clin. Sci. Mol. Med. 47*, 223–232

TSAN, M.-F. & MCINTYRE, P. A. (1976) The requirement for membrane sialic acid in the stimulation of superoxide production during phagocytosis by human polymorphonuclear leukocytes. *J. Exp. Med. 143*, 1308–1316

UGAZIO, G., TORRIELLI, M. V., BURDINO, E., SAWYER, B. C. & SLATER, T. F. (1976) Long-range effects of products of carbon tetrachloride-stimulated lipid peroxidation. *Biochem. Soc. Trans. 4*, 353–356

VIDLAKOVA, M., ERAZIMOVA, J., HORKI, J. & PLACER, Z. (1972) Relationship of serum antioxidative activity to tocopherol and serum inhibitor of lipid peroxidation. *Clin. Chim. Acta 36*, 61–66

WILLSON, R. L. & SLATER, T. F. (1975) in *Fast Processes in Radiation Chemistry* (Adams, G. E., Fielden, E. M. & Michael, B. F., eds.), pp. 147–161, Wiley, London

Lipid peroxidation: detection *in vivo* and *in vitro* through the formation of saturated hydrocarbon gases

GERALD COHEN

Department of Neurology, Mount Sinai School of Medicine, New York

Abstract Saturated, short-chain hydrocarbon gases (ethane, propane, etc. — but not methane) are evolved during the spontaneous lipid peroxidation of mouse tissue slices or homogenates. A relatively-selective increase in either ethane or pentane was observed when either linolenic acid or linoleic acid, respectively, was added to tissue homogenates. When mice were challenged by injection of carbon tetrachloride or cumene hydroperoxide (two agents that induce a lipid peroxidative attack on liver), they exhaled ethane. The gas chromatographic detection and monitoring of hydrocarbon gases can serve as an index of lipid peroxidation *in vivo*.

One problem facing biologists is the assessment of lipid peroxidation, particularly its occurrence *in vivo*. In this paper I am concerned with a novel approach: the measurement of saturated, short-chain hydrocarbon gases that are spontaneously formed during lipid peroxidation. The evolution of these gases can be readily monitored by gas chromatography. The exhalation of hydrocarbon gases by animals provides a means to detect lipid peroxidation *in vivo* as it is happening.

Lipid peroxidation (or oxidative rancification) represents one form of tissue damage (Plaa & Witschi 1976; see also Slater, in this volume). It can be initiated by oxygen, by oxygen radicals, and by hydrogen peroxide (e.g. Fong *et al.* 1973; King *et al.* 1975; Cohen 1975). Once initiated, the process takes the form of a free-radical chain reaction leading to the formation of organic peroxides and other products from unsaturated fatty acids. The accumulation of organic peroxides and the oxidation of membrane lipids places a stress on cellular vitality which can lead to an early demise. Protective mechanisms against lipid peroxidation include enzymes that remove H_2O_2, organic peroxides and oxygen radicals (see other articles in this volume, e.g. Flohé; Fridovich) and the presence of natural lipid antioxidants such as α-tocopherol (vitamin E). The controlled formation of organic endoperoxides from arachidonic acid can also serve a beneficial role in the synthesis of prostaglandins (see Flower, this volume).

There is general agreement that the peroxidation of membrane lipids is a major form of damage underlying the variety of tissue disorders seen in experimental vitamin E deficiency (e.g. Tappel 1962; Barber & Bernheim 1967). Although there is less of a consensus, lipid peroxidation is believed to play a role in the process of cellular ageing and in the cellular damage induced in target organs by various toxic agents (e.g. carbon tetrachloride, paraquat).

When experimental evidence for lipid peroxidation is found in excised tissues (e.g. formation of conjugated dienes, or the 2-thiobarbituric acid test), the question should be asked: were these products preformed *in vivo* or were they formed artifactually during the handling and work-up of tissues? A particular value of the measurement of exhaled hydrocarbon gases is that it provides a means for remote monitoring of lipid peroxidation *in vivo* without trauma to the experimental subjects.

STUDIES WITH TISSUE SLICES AND TISSUE HOMOGENATES

It has long been known that fruits and vegetables generate the unsaturated hydrocarbon gas, ethylene, which serves as a putative plant hormone (e.g. Yang 1974). The process appears to be enzymically controlled. Disruption of cellular integrity results in production of ethane instead of ethylene. More recently, the formation of ethane was noted from slices and homogenates of mammalian tissues (Cohen *et al.* 1968; Riely *et al.* 1974). This latter process was correlated with oxidative rancification as measured by the 2-thiobarbituric acid reaction (2-TBA test). Whereas sliced plant tissue (e.g. sliced apple or scallion incubated at ambient temperature in air) produced mainly ethylene, sliced animal tissue (e.g. sliced mouse liver or brain, incubated in isotonic saline at 37 °C or homogenized) produced ethane and much smaller amounts of ethylene. Animal tissues also produced other saturated hydrocarbons, such as propane, butane and pentane (but not methane), in addition to other compounds, as yet unidentified by gas chromatography.

The time course for the production of ethane by homogenates of mouse liver or brain and the correlation with the accumulation of 'lipid peroxides' are shown in Fig. 1. Ethane production by brain (Fig. 1A) proceeded linearly from zero time for up to two hours and paralleled the accumulation of material that gave a positive 2-thiobarbituric acid test. In liver (Fig. 1B), both the production of ethane and the accumulation of lipid peroxides showed a lag of about one hour but, thereafter, the two increased in parallel. α-Tocopherol, added to liver homogenates at zero time, blocked lipid peroxidation (as expected) and simul-

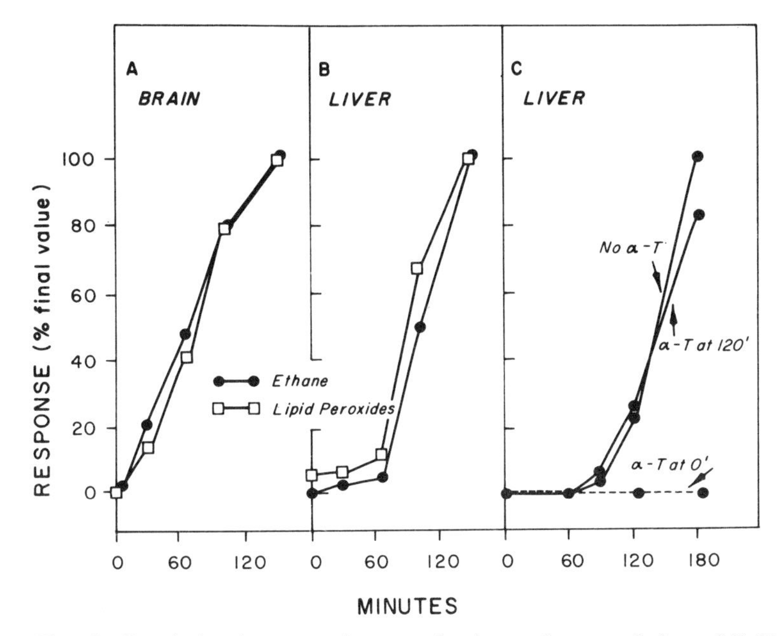

FIG. 1. Correlation between ethane production and accumulation of lipid peroxides in homogenates of mouse brain and liver: 10 ml of 10% (w/v) tissue homogenates in buffered (pH 7.4) isotonic saline were incubated at 37 °C with shaking in Erlenmeyer flasks that had been sealed with rubber septa. Lipid peroxides were measured with the 2-thiobarbituric acid test. Ethane in the atmosphere above the samples was measured by gas chromatography. α-Tocopherol (α-T) was added where indicated (panel C) to a concentration of 1 μg/ml. The final values in the 2-thiobarbituric acid test (100% response) were 23.8 and 14.6 nmol malonaldehyde/ml homogenate of brain (A) and liver (B), respectively. The final ethane concentrations were 0.14, 0.15 and 0.30 nmol/ml in the gas phase for brain (A), liver (B) and liver (C), respectively. (From Riely *et al.* 1974; copyright 1974 by the American Association for the Advancement of Science.)

taneously blocked the production of ethane (Fig. 1C, dashed line). It can be concluded that ethane is a by-product of the lipid peroxidative process.

In other *in vitro* studies (unpublished), CCl_4, a lipid pro-oxidant for liver, provoked ethane production by mouse liver slices from zero time. When animals were pretreated with α-tocopherol (25m g/kg) for several days, the production of ethane was suppressed.

The addition of unsaturated fatty acids to the incubated tissue homogenates enhanced the production of hydrocarbon gases (Table 1). Linolenic acid spurred a relatively selective production of ethane, but linoleic acid spurred a relatively selective production of pentane. Similar observations have been reported by others (Dumelin *et al.* 1977). This observation implies that specific hydrocarbons may be markers for the rancification of specific fatty acids.

TABLE 1

The effect of fatty acids on the production of ethane and pentane by homogenates of mouse liver

Sample	*Production (in nmol[g tissue]$^{-1}$[2 h]$^{-1}$) of hydrocarbon*	
	Ethane	*Pentane*
Control	0.5 ± 0.1	0.5 ± 0.0
Control + linolenic acid	23.7 ± 7.8	0.7 ± 0.2
Control + linoleic acid	0.9 ± 0.2	2.9 ± 0.8

2 ml of 10% (w/v) homogenates of mouse liver in isotonic saline were incubated at 37 °C in 18 ml screw-cap tubes that had been sealed with silicone septa. After two hours, the head space above the tissue homogenates was sampled by syringe and injected directly into a gas chromatograph. Linoleic or linolenic acid was added to a final concentration of 15 mmol/l. Results are the mean and standard error for three separate experiments.

IN VIVO STUDIES WITH MICE

We selected CCl_4 toxicity as an experimental model to evaluate the possibility that lipid peroxidation *in vivo* leads to the exhalation of hydrocarbon gases (Cohen *et al.* 1968; Riely *et al.* 1974). CCl_4 toxicity represents the best available model in a nutritionally-adequate state (that is, *not* a vitamin E-deficient state). Prior studies, primarily by Recknagel and his colleagues (Recknagel 1967; Hashimoto *et al.* 1968), had provided strong evidence for a lipid peroxidative phenomenon associated with the destructive attack by CCl_4 on liver. Moreover, CCl_4 provides an acute lesion that can be induced rapidly, encompassing large areas of a major organ. Some of the action of CCl_4 can be ameliorated by pretreatment with vitamin E (e.g., Gallagher 1961).

We observed (Fig. 2) that animals injected with CCl_4 exhaled ethane and that the rate of ethane production was significantly diminished by pretreatment with α-tocopherol. Moreover, phenobarbital, which induces liver enzymes that metabolize CCl_4, provoked a greater rate of production of ethane. The lipid peroxidation of liver is believed to be provoked by free radicals generated by the metabolism of CCl_4; it is known that the toxicity of CCl_4 is enhanced by phenobarbital treatment (e.g. Garner & McLean 1969). These studies (Fig. 2) showed that continuing lipid peroxidation in an internal organ can be monitored by following the exhalation of ethane.

In more recent studies (G. Cohen & P. J. O'Brien, unpublished observations) we investigated the effects of an organic peroxide, namely, cumene hydroperoxide, on liver. *In vitro* cumene hydroperoxide spurred both lipid peroxidation (as indicated by the 2-thiobarbituric acid test) and ethane evolution in homogenates of mouse liver. Both ethane production and the accumulation of material that gave a positive 2-thiobarbituric acid test could be blocked by

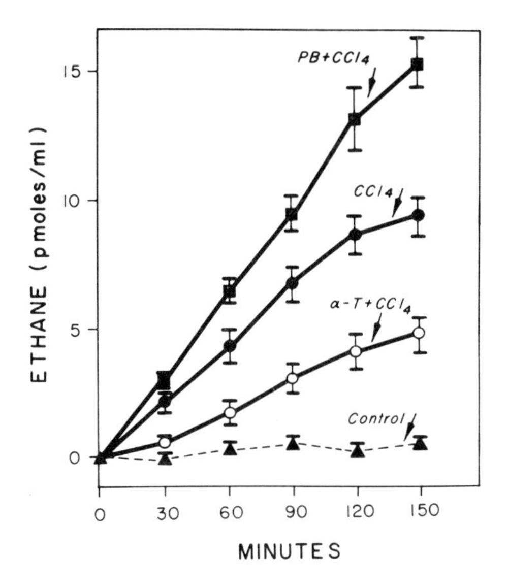

FIG. 2. Ethane exhalation by mice after intraperitoneal injection of carbon tetrachloride: groups of five Swiss–Webster mice were sealed into a chamber (2.4 l) in which access to the internal atmosphere was provided by a three-way stop-cock. Mice received i.p. injections of CCl_4 (3.2 g/kg) in light mineral oil. Control animals received the mineral oil alone. Some animals were pretreated with either α-tocopherol (α-T; 25 mg/kg, in isotonic saline containing 1% Tween 80, three times over two days) or phenobarbital (PB; 80 mg/kg, in isotonic saline, three times over three days). Control animals received an equivalent number of injections of isotonic saline alone with and without Tween 80. Data are the mean and standard error for the control ($n = 4$) and experimental ($n = 7$) groups. The chamber air was analysed by gas chromatography. (From Riely *et al.* 1974; copyright 1974 by the American Association for the Advancement of Science.)

TABLE 2

Lipid peroxidation provoked by injection of cumene hydroperoxide into mice

Time after injection (min)	*Production of ethane (nmol)*			
	1% Cumene hydroperoxide		*2% Cumene hydroperoxide*	
	(Exp't. 1)	*(Exp't. 2)*	*(Exp't. 3)*	*(Exp't. 4)*
5	1.5	2.3	9.2	8.2
30	3.8	5.5	18.8	16.1
60	3.8	5.0	24.1	14.9

Groups of six Swiss–Webster mice were sealed into a closed chamber (2.4 l) (Riely *et al.* 1974). Animals received i.p. injections of 0.5 ml of 1% or 2% (v/v) suspensions of cumene hydroperoxide in isotonic saline. The atmosphere in the chamber was sampled by syringe and injected directly into a gas chromatograph.

α-tocopherol. When cumene hydroperoxide was injected into mice, a dose-dependent exhalation of ethane (Table 2), and, simultaneously, a toxic action on the liver were noted.

Recent experiments by other investigators (Dillard *et al.* 1977; Hafeman & Hoekstra 1977; Koster *et al.* 1977) offer additional examples of the utility of studying the exhalation of hydrocarbon gases (ethane, pentane) in the evaluation of lipid peroxidation *in vivo.*

References

BARBER, A. A. & BERNHEIM, F. (1967) Lipid peroxidation: its measurement, occurrence and significance in animal tissues. *Adv. Gerontol. Res. 2*, 355–403

COHEN, G. (1975) Unusual defense mechanisms against H_2O_2-cytotoxicity in erythrocytes deficient in glucose-6-phosphate dehydrogenase or tocopherol. *Progr. Clin. Biol. Res. 1*, 685–698

COHEN, G., RIELY, C. & LIEBERMAN, M. (1968) Ethane formation from lipid peroxides in mammalian cells. *Fed. Proc. 27*, 648 (abstr. 2436)

DILLARD, C. J., DUMELIN, E. E. & TAPPEL, A. L. (1977) Effect of dietary vitamin E on expiration of pentane and ethane by the rat. *Lipids 12*, 109–114

DUMELIN, E. E., DILLARD, C. J., PURDY, R. E. & TAPPEL, A. L. (1977) Sensitive measurement of lipid peroxidation products *in vivo* and *in vitro. Fed. Proc. 36*, 1160 (abstr. 4705)

FLOHÉ, L. (1979) Glutathione peroxidase: fact and fiction, in *This Volume*, pp. 95–113

FLOWER, R. J. (1979) Biosynthesis of prostaglandins, in *This Volume*, pp. 123–139

FONG, K.-L., MCCAY, P. B., POYER, J. L., KEELE, B. B. & MISRA, H. (1973) Evidence that peroxidation of lysosomal membranes is initiated by hydroxyl free radicals produced during flavin enzyme activity. *J. Biol. Chem. 248*, 7792–7797

FRIDOVICH, I. (1979) Superoxide dismutases: defence against endogenous superoxide radical, in *This Volume*, pp. 77–86

GALLAGHER, C. H. (1961) Protection by antioxidants against lethal doses of carbon tetrachloride. *Nature (Lond.) 192*, 881–882

GARNER, R. C. & MCLEAN, A. E. M. (1969) Increased susceptibility to carbon tetrachloride poisoning in the rat after pretreatment with oral phenobarbitone. *Biochem. Pharmacol. 18*, 645–650

HAFEMAN, D. G. & HOEKSTRA, W. G. (1977) Lipid peroxidation *in vivo* during vitamin E and selenium deficiency in the rat as monitored by ethane evolution. *J. Nutr. 107*, 666–672

HASHIMOTO, S., GLENDE, E. A. JR. & RECKNAGEL, R. (1968) Hepatic lipid peroxidation in acute fatal human carbon tetrachloride poisoning. *N. Engl. J. Med. 279*, 1082–1085

KING, M. M., LAI, E. K. & MCCAY, P. B. (1975) Singlet oxygen production associated with enzyme-catalyzed lipid peroxidation in liver microsomes. *J. Biol. Chem. 250*, 6496–6502

KOSTER, U., ALBRECHT, D. & KAPPUS, H. (1977) Evidence for carbon tetrachloride- and ethanol-induced lipid peroxidation *in vivo* demonstrated by ethane production in mice and rats. *Toxicol. Appl. Pharmacol. 41*, 639–648

PLAA, G. L. & WITSCHI, H. (1976) Chemicals, drugs and lipid peroxidation. *Ann. Rev. Pharmacol. Toxicol. 16*, 125–141

RECKNAGEL, R. O. (1967) Carbon tetrachloride hepatotoxicity. *Pharmacol. Rev. 19*, 145–208

RIELY, C., COHEN, G. & LIEBERMAN, M. (1974) Ethane evolution: a new index of lipid peroxidation. *Science (Wash. D.C.) 183*, 208–210

SLATER, T. F. (1979) Mechanisms of protection against the tissue damage produced in biological systems by oxygen-derived radicals, in *This Volume*, pp. 143–159

TAPPEL, A. L. (1962) Vitamin E as the biological lipid antioxidant. *Vitam. Horm. 20*, 493–510

YANG, S. F. (1974) The biochemistry of ethylene: biogenesis and metabolism. *Rec. Adv. Phytochem. 7*, 131–164

Discussion

Crichton: How many mice do you need to collect sufficient ethane for analysis?

Cohen: After two hours, five mice produce enough to be detected by a good gas chromatograph (that is, 5–6 times the baseline value), although one is working at the limit of detection in the early phases. Other workers (Dillard *et al.* 1977) have now bound the hydrocarbon gases onto cold alumina columns in order to concentrate the material.

Crichton: Are you in a position to study single animals?

Cohen: Yes — if one accepts a doubling of the baseline value as a reasonable result. However, five mice give a good average figure.

Allison: One could, of course, use rats.

Cohen: This has been done (Lindstrom & Anders 1978).

Smith: The method does not locate the site of lipid peroxidation. Paraquat, for example, which selectively attacks lungs, would give similar results to CCl_4.

Cohen: In vitro studies with isolated liver slices reproduce these data; without doubt, the liver is the site for CCl_4 damage.

Slater: Your technique strikes me as a good and useful method for studying peroxidation. The results obtained with CCl_4 are consistent with what we know about peroxidation induced by CCl_4 and measured by diene conjugation or by the *in vivo* production of malonaldehyde. With ethanol, however, there is an apparent conflict of results. Many groups have tried to detect increased lipid peroxidation in liver after administering ethanol to rats, using either diene conjugation or malonaldehyde measurements as the experimental approach (summarized by Comporti 1978). In general, their results have been negative yet with the ethane procedure considerable lipid peroxidation appears to be going on (Koster *et al.* 1977).

Cohen: I cannot explain the apparent conflict in results. I find that liver slices from mice show an increase in production of both TBA-positive material and ethane when they are incubated with ethanol (unpublished observations). Litov *et al.* (1978) have recently shown enhanced exhalation of pentane when animals were fed a diet deficient in vitamin E and exposed to ethanol. Some of the debate revolves around the question of where the alcohol induces the damage; the mitochondria or the microsomes? DiLuzio (1973) claims that the major site of the damage is in the mitochondria.

Slater: The whole question of lipid peroxidation in ethanol intoxication is still in dispute.

Cohen: Yes; Hashimoto & Recknagel (1968) disputed that conclusion on the grounds that they could not detect lipid peroxides.

Dormandy: This method could become an important clinical diagnostic advance; but amongst the difficulties in interpreting the constituents of human exhaled gases are the autoxidative changes which may occur in the gut. In other words, could the peroxidation or autoxidation of fatty food in the gut interfere with the method by producing these breakdown products?

Cohen: Flatus (gastrointestinal gas) mainly contains methane and is not much contaminated with the other hydrocarbons (Gall 1968; Levitt & Inglefinger 1968). But a breakdown of lipids in the gut might pose a potential problem. Certainly it may complicate animal studies when diets rich in certain polyunsaturated fatty acids (e.g. corn oil, cod liver oil) are fed. Under such circumstances, one may wonder about the source of ethane or pentane.

Williams: Have you used this system to determine the source of the relative amounts of the different hydrocarbons? Knowledge of the origin of ethane, propane, butane, pentane and the isomeric alkanes would help our understanding of mechanisms.

Cohen: We were surprised that methane was not a product. Initially, we thought that the biosynthesis of the longer chain hydrocarbons was due to chain lengthening by methyl radicals. But that should have generated branched-chain hydrocarbons as well. So far, however, only straight-chain hydrocarbons have been reported.

Williams: Modern gas chromatographs should detect the branched isomers if they were there even in minute amounts.

Cohen: Yes, I agree. With regard to the limits of detection, in New York we are limited by automotive gas exhausts — this, of course, is a major problem in London, too. To avoid this problem, one should work with a source of purified air. The technique of concentrating the material by binding to columns seems practical, although contaminants in room air will be concentrated as well. It may also be necessary to prohibit the lighting of bunsen burners in a laboratory, as they are a source of hydrocarbons. Lack of such precautions can be serious: I remember distinctly when we traced one of our problems to a gas leak in the laboratory. The plumber claimed there was no leak since he could not ignite it: his definition of a leak was that it should be possible to ignite it. It was necessary to give him a short course in gas chromatography to show him the leak and get it repaired!

References

COMPORTI, M. (1978) in *Biochemical Mechanisms in Liver Injury* (Slater, T. F., ed.), Academic Press, London

DILLARD, C. J., DUMELIN, E. E. & TAPPEL, A. L. (1977) Effect of dietary vitamin E on expiration of ethane and pentane by the rat. *Lipids 12*, 109–114

DiLuzio, N. R. (1973) Antioxidants, lipid peroxidation and chemical induced liver injury. *Fed. Proc. 32*, 1875–1881

Gall. L. S. (1968) The role of intestinal flora in gas formation. *Ann. N.Y. Acad. Sci. 150*, 27–30

Hashimoto, S. & Recknagel, R. O. (1968) No chemical evidence of hepatic lipid peroxidation in acute ethanol toxicity. *Exp. Mol. Pathol. 8*, 225–242

Koster, U., Albrecht, D. & Kappus, H. (1977) Evidence for carbon tetrachloride- and ethanol-induced lipid peroxidation *in vivo* demonstrated by ethane production in mice and rats. *Toxicol. Appl. Pharmacol. 41*, 639

Levitt, M. D. & Inglefinger, F. J. (1968) Hydrogen and methane production in man. *Ann. N.Y. Acad. Sci. 150*, 75–81

Lindstrom, T. D. & Anders, M. W. (1978) Effect of agents known to alter carbon tetrachloride hepatotoxicity and cytochrome P-450 levels on carbon tetrachloride-stimulated lipid peroxidation and ethane expiration in the intact rat. *Biochem. Pharmacol. 27*, 563–567

Litov, R. E., Irving, D. H., Downey, J. E. & Tappel, A. L. (1978) Lipid peroxidation: a mechanism involved in acute ethanol toxicity as demonstrated by in vivo pentane production in the rat. *Lipids 13*, 305–307

Dioxygen and the vitamin K-dependent synthesis of prothrombin

M. PETER ESNOUF, MARTIN R. GREEN,* H. ALLEN O. HILL,* G. BRENT IRVINE and STEPHEN J. WALTER

*Nuffield Department of Clinical Biochemistry and *Inorganic Chemistry Laboratory, University of Oxford*

Abstract It has been shown that bovine erythrocyte superoxide dismutase inhibits the γ-carboxylation of glutamyl residues in the precursor of prothrombin and in a related synthetic peptide by a vitamin K-dependent rat liver microsomal carboxylase. In the same conditions a simple copper(II) tyrosinyl complex also inhibits the carboxylation. The formation of the vitamin K epoxide by the same systems is inhibited by both superoxide dismutase and catalase. It is suggested that the formation of the epoxide is a process distinct from carboxylation, representing perhaps a protective mechanism against the superoxide ion, or species derived therefrom, generated by the reaction of reduced vitamin K with dioxygen. Furthermore, the possibility that the carboxylating species is formed by the reaction of the superoxide ion, or species derived therefrom, with carbon dioxide, is proposed.

A considerable time elapsed (Olson & Suttie 1977) between the discovery of vitamin K and the identification of its metabolic role as an obligatory component of the microsomal particle responsible for the synthesis of prothrombin, a key constituent of the blood clotting cascade. Prothrombin is derived from its precursor by the carboxylation of 10 glutamyl (Glu) residues in the NH_2-terminal region and consequently contains 10 γ-carboxyglutamyl residues (Gla) (Fig. 1). These provide excellent chelating centres for the calcium ions which, along with factor Xa, factor V and a phospholipid, are required for the generation of thrombin. Vitamin K (Fig. 2) is part of the microsomal membrane-bound carboxylase system which requires, in addition to the vitamin, carbon dioxide, dioxygen, NADH, a vitamin K reductase and, of course, the substrate which is either the prothrombin precursor or synthetic peptides based on residues 5–9 of bovine or rat prothrombin. The carboxylation of either the protein or the peptide substrate can be achieved by adding vitamin K as the hydroquinone and omitting the NADH. This suggests that the vitamin K hydroquinone is simply

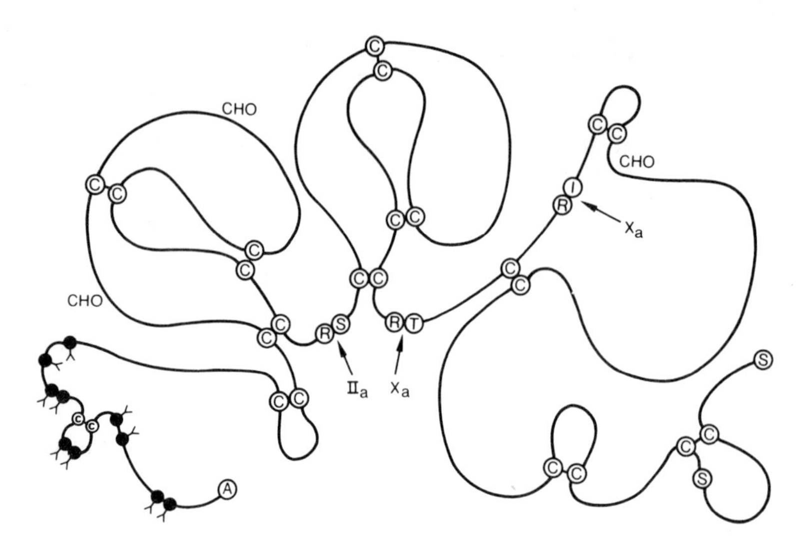

FIG. 1. A schematic representation of the structure of prothrombin, based on the known sequence, showing the sites of carboxylation (solid circles). The one-letter notation for amino acids is used. (From Esnouf 1977.)

reacting with some other component, most likely dioxygen, to form the semiquinone and the quinone. There is an additional complication since the vitamin K epoxide is also formed, especially in animals administered warfarin (Fig. 3), which inhibits the reduction of the epoxide to the quinone. However, the most recent evidence (Sadowski *et al.* 1977) suggests that epoxidation can occur in the absence of carboxylation.

The presence in any reaction of both dioxygen and a reductant warrants the presumption of the involvement of the superoxide ion and/or peroxide and reactants derived therefrom. Patel & Willson (1973) had previously shown that vitamin K semiquinone reacts with dioxygen to form superoxide and that 2-methyl-1,4-naphthoquinol yields hydrogen peroxide (Misra & Fridovich 1972). In addition, rat liver microsomes produce superoxide by an NADPH-dependent pathway (Bartoli *et al.* 1977). We have extended these results (Esnouf *et al.* 1978) (Table 1) and shown that the rate of production of superoxide, as shown by its reaction with adrenalin at neutral pH, is considerably enhanced by the addition of vitamin K. The latter reaction is subject to pronounced inhibition by bovine superoxide dismutase (Table 2). The carboxylation of a synthetic peptide, e.g. L-phenylalanyl-L-leucyl-L-glutamyl-L-glutamyl-L-leucine, by a microsomal fraction from vitamin K-deficient rat livers was investigated by studying the incorporation of ^{14}C from $H^{14}CO_3$. The incorporation, in the apparent absence of dioxygen, was only 12% of that when dioxygen was present.

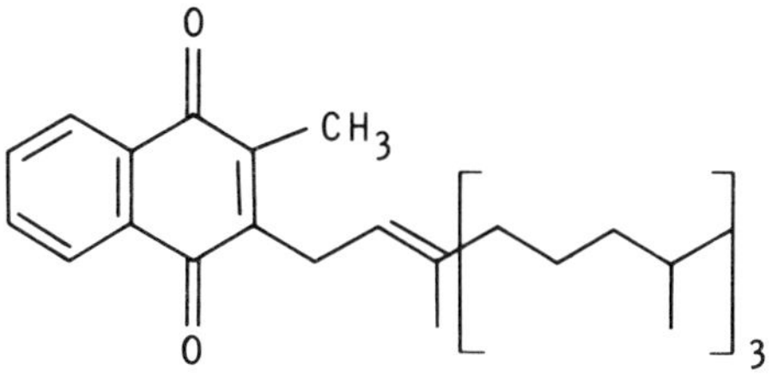

Vitamin K_1

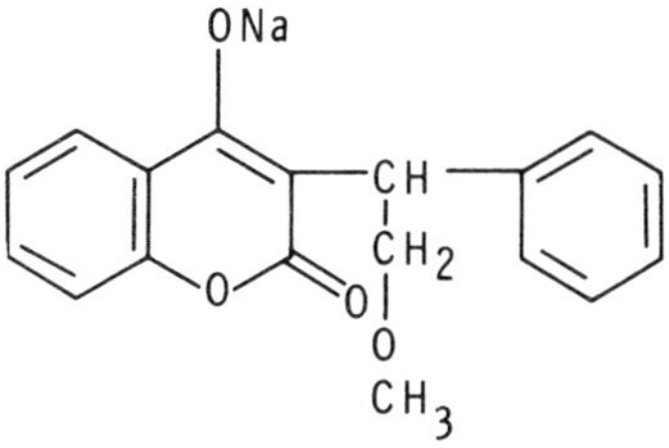

Warfarin

FIG. 2. The structures of warfarin and vitamin K.

The ^{14}C is also incorporated into the trichloroacetic acid-insoluble fraction of the microsomes, which contains the prothrombin precursor.

Both reactions are inhibited by superoxide dismutase, but not by bovine serum albumin. The inhibition is not complete and this suggests that in these nearly intact microsomes the carboxylase is still membrane-bound and consequently the access of superoxide dismutase is impeded, as Strobel & Coon (1971) found in studies on other microsomal systems. No such impediment obviously exists for the copper(II)-tyrosine complex known to be effective in catalysing the disproportionation of the superoxide ion (Richter *et al.* 1977). The carboxylation is inhibited at concentrations similar to those of the bovine superoxide dismutase. In contrast to the inhibition of carboxylation, both superoxide dismutase and catalase inhibit the formation of the vitamin K epoxide (Table 3). This suggests that the vitamin K epoxide is formed as a

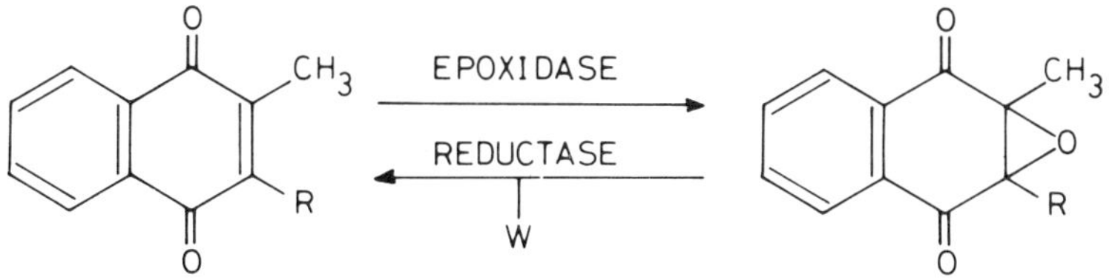

EPOXIDASE

REDUCTASE

W

FIG. 3. The formation of vitamin K epoxide and its warfarin (W) -inhibited reduction.

TABLE 1

The effect of superoxide dismutase (SOD) on the incorporation of ^{14}C into the pentapeptide Phe-Leu-Glu-Glu-Leu and the trichloroacetic acid-insoluble fraction of liver microsomes from vitamin K-deficient rats

System	*Peptide incorporation*		*Protein incorporation*	
	c.p.m./ml	*% of control*	*c.p.m./ml*	*% of control*
Microsomes − vitamin K	1440		2000	
Microsomes	27 600	100	4840	100
Microsomes + 20μM-SOD	19 530	68	3930	82
Microsomes + 69μM-SOD			3470	71
Microsomes + 138μM-SOD	11 320	36	3160	65
Microsomes + 276μM-SOD	12 290	39		
Microsomes + 276μM-SOD + 4μM-catalase	8870	26		
Microsomes + 50μM-Cu(tyr)$_2$	2890	0[a]		

Vitamin K was present in all experiments except the first listed. It was added as the hydroquinone in the experiment with the pentapeptide and as the quinone for the protein incorporation. The figures were obtained from duplicate incubations. The leucine pentapeptide was present at a final concentration of 2 mmol/l and was omitted from the protein incorporation experiment.

[a] The incorporation in this experiment was indistinguishable from that with microsomes alone.

consequence of the generation of superoxide and species derived therefrom but that the formation is not directly connected with the carboxylation.

The results suggest, therefore, that the superoxide ion is required for the formation of the carboxylating agent. It seems unlikely that it would function as a Brønsted base to remove the γ-proton of the glutamyl residue though preliminary results (M. P. Esnouf & S. J. Walter, unpublished work) show that solvent protons exchange readily with tritium in this position in the presence of the carboxylase. (The peroxide di-anion, O_2^{2-}, would be sufficiently basic but its formation in an aprotic environment would require considerable energy.) As with many reactions in which the superoxide ion is involved, it is difficult to

TABLE 2

The production of superoxide by normal rat liver microsomes

Microsomal system	*Production of superoxide (pmol min^{-1} ml^{-1})*
Microsomes	0
Microsomes + NADH	60
Microsomes + NADH + vitamin K	230
Microsomes + NADH + vitamin K + SOD	0

The incubation mixture contains O_2-saturated buffer (2 ml), microsomal suspension (0.1 ml), adrenalin (0.1 ml; 4 mg/ml), NADH (0.05 ml; 10 mg/ml), vitamin K (0.1 ml; 10 mg/ml), superoxide dismutase (SOD; 10 μg/ml).

TABLE 3

Effect of catalase and superoxide dismutase (SOD) on the formation of vitamin K epoxide by liver microsomes from vitamin K-deficient rats

System	*Molar ratio (vitamin K epoxide/vitamin K)*
Control	1.89
Control + SOD (10 mg/ml)	0.88
Control + catalase (1 mg/ml)	0.50
Control + SOD (10 mg/ml) + catalase (1 mg/ml)	0.48

The incubations were carried out in the same conditions as for the experiment described in Table 1, except that warfarin (40 μg/ml) was added and $NaH^{14}CO_3$ was omitted. Vitamin K was present at a final concentration of 5 μg/ml.

identify the reacting species because of the possible intervention of reactions such as (1) or (2).

$$O_2^- + O_2^- \xrightleftharpoons{2H^+} H_2O_2 + O_2 \quad (1)$$

$$H_2O_2 + O_2^- \rightleftharpoons OH^{\bullet} + OH^- + O_2 \quad (\text{or } {}^1O_2) \quad (2)$$

It has previously been observed (Stauff *et al.* 1973; Hodgson & Fridovich 1976; Puget & Michelson 1976; Michelson & Durosay 1977) that carbon dioxide or hydrogen carbonate influences other reactions in which the superoxide ion is involved, and the species CO_3^- and CO_2^- have been invoked as intermediates. However, superoxide, probably *via* peroxide, reacts with CO_2, to give either the peroxobicarbonate anion HCO_4^- or the peroxodicarbonate anion $C_2O_6^{2-}$ and these, or species derived from them, may be the carboxylating species (Mel'nikov *et al.* 1962; Firsova *et al.* 1963). It is interesting that chemiluminescence has been observed in some of the above reactions; we have shown that potassium peroxodicarbonate $K_2C_2O_6$ and peroxocarbonate $KHCO_4$ chemiluminesce when dissolved in ethanol/water mixtures, in which they decompose, perhaps with the concomitant formation of some of the active species referred to above. Thus the carboxylation may be described by a reaction scheme such as that in Fig. 4. The generation of the superoxide ion yields, by reaction with CO_2, a carboxylating species; the formation of the epoxide simply represents an unavoidable consequence of the generation of the O_2^- or species derived therefrom and might thus represent a protective device. It will be interesting to discover whether other vitamin K-dependent carboxylations such as those necessary for the synthesis of other blood-clotting factors, the bone protein osteocalcin, and γ-carboxyglutamyl-containing proteins found in plasma and kidneys are subject to the same inhibition. The relationship, if any, to other physiological processes which require or release superoxide or to intrinsic or extrinsic inhibitors of superoxide should be explored.

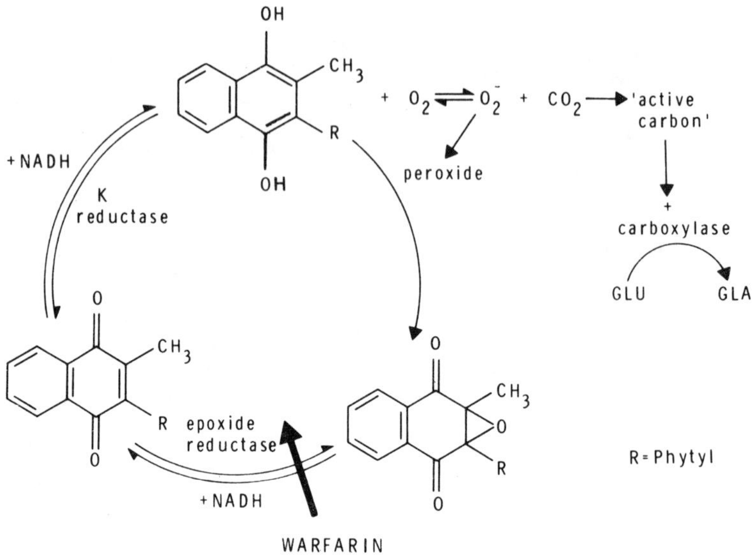

FIG. 4. A scheme proposed to account for the vitamin K-dependent carboxylation and the formation of vitamin K epoxide.

ACKNOWLEDGEMENTS

We are grateful to the Science Research Council, the Medical Research Council and the British Heart Foundation for support. This is a contribution from the Oxford Enzyme Group of which two of us, M.P.E. and H.A.O.H. are members.

References

BARTOLI, G. M., GALEOTTI, T., PALOMBINI, G., PARISI, G. & AZZI, A. (1977) Different contributions of rat liver microsomal pigments in the formation of superoxide anions and hydrogen peroxide during development. *Arch. Biochem. Biophys. 184*, 276–281

ESNOUF, M. P. (1977) *Br. Med. Bull. 33*, 213–219

ESNOUF, M. P., GREEN, M. R., HILL, H. A. O., IRVINE, G. B. & WALTER, S. J. (1978) Involvement of superoxide in the vitamin K dependent carboxylation of glutamyl residues. *Biochem. J. 174*, 345–348

FIRSOVA, T. P., MOLODKINA, A. W., MOROZOVA, T. G. & AKSENOVA, I. V. (1963) Preparation of sodium peroxocarbonates. *Russ. J. Inorg. Chem. 8*, 140–144

HODGSON, E. K. & FRIDOVICH, I. (1976) The mechanism of the activity-dependent luminescence of xanthine oxidase. *Arch. Biochem. Biophys. 172*, 202–205

MEL'NIKOV, A. KH., FIRSOVA, T. P. & MOLODKINA, A. N. (1962) Preparation of pure potassium peroxodicarbonate and some of its properties. *Russ. J. Inorg. Chem. 7*, 637–640

MICHELSON, A. M. & DUROSAY, P. (1977) Hemolysis of human erythrocyte by activated oxygen species. *Photochem. Photobiol. 25*, 55–63

MISRA, H. P. & FRIDOVICH, I. (1972) The role of superoxide anion in the autoxidation of epinephrine and a simple assay for superoxide dismutase. *J. Biol. Chem. 247*, 3170–3175

OLSON, R. E. & SUTTIE, J. W. (1977) Vitamin K and γ-carboxyglutamate biosynthesis. *Vitam. Horm. 35*, 59–108

PATEL, K. B. & WILLSON, R. L. (1973) Semiquinone free radicals and oxygen. *J. Chem. Soc. Faraday I 69*, 814–825
PUGET, K. & MICHELSON, A. M. (1976) Oxidation of luminol by the xanthine oxidase system in the presence of carbonate anions. *Photochem. Photobiol. 24*, 499–501
RICHTER, C., AZZI, A., WESER, U. & WENDEL, A. (1977) The action of Cu(tyrosine)$_2$ as a superoxide dismutase model on hepatic microsomal demethylation, in *Superoxide and Superoxide Dismutases* (Michelson, A. M., McCord, J. M. & Fridovich, I., eds.), pp. 375–386, Academic Press, London & New York
SADOWSKI, J. A., ESMON, C. T. & SUTTIE, J. W. (1977) Vitamin K epoxidase: properties and relationship to prothrombin synthesis. *Biochemistry 16*, 3865–3868
STAUFF, J., SANDER, U. & JAESCHKE, W. (1973) Chemiluminescence of perhydroxyl- and carbonate-radicals, in *Chemiluminescence and Bioluminescence* (Cormur, M. J., Hercules, D. M. & Lee, J., eds.), pp. 131–140, Plenum Press, New York
STROBEL, H. W. & COON, M. J. (1971) Effect of superoxide generation and dismutation on hydroxylation reactions catalysed by liver microsomal cytochrome P-450. *J. Biol. Chem. 246*, 7826–7829

Discussion

Slater: Is the vitamin K epoxide a substrate for epoxide hydrolase?

Hill: That is not known.

Slater: What is known about the warfarin-sensitive enzyme step? The reduction of the epoxide as shown in the scheme is unusual in my experience.

Hill: The epoxide is formed readily from the vitamin by reaction with hydrogen peroxide at high pH. Again, we do not know whether peroxide is the reactive material or a progenitor of the reactive species. The epoxide reductase has not been isolated. So one cannot say whether the reduction is direct or whether there are intermediates. However, in vitamin K-deficient rats one can easily observe a rapid accumulation of the epoxide when one administers warfarin (which is the basis of the use of this treatment).

Fridovich: If the scheme in Fig. 4 were correct, would you not expect an alternative source of O_2^- and CO_2 to carry out the carboxylation?

Hill: Yes; we are trying that, but the difficulty we face is generating the species close to the intact microsome. We are at present using the peroxocarbonate as the carboxylating agent.

Fridovich: Superoxide dismutase, at low concentrations, effectively scavenges O_2^-; you were using much higher concentrations. Since your system is microsomal, there will be a problem of getting the dismutase close to the membrane. Dr McCord has circumvented this, using porcine superoxide dismutase which has a much higher isoelectric point than the corresponding bovine enzyme.

McCord: We isolated naturally-occurring superoxide dismutases from many sources and use enzymes that have isoelectric points that differ from the bovine enzyme: the dismutases from pig and sheep are much less negatively charged. In protecting phagocytosing leukocytes these enzymes from other species were up to two orders of magnitude more effective than the bovine enzyme. We attributed the difference to the charge-charge interaction; the negatively charged bovine enzyme may be repelled by the negatively charged leukocyte surface (McCord & Salin 1977).

Hill: The use of superoxide dismutases with different isoelectric points could be useful. [*Note added in proof:* In slightly different conditions we have found superoxide dismutase to be a much more effective inhibitor.] Although a copper(II)-tyrosine complex inhibits, that is no guarantee of its specificity.

McCord: There could be a steric problem as well.

Hill: The same problem was encountered in the study of the inhibition of P450 from microsomes. Keeping the system intact so that these complex reactions can proceed makes ingress of materials the size of the dismutase very difficult.

Fridovich: Another possibly useful scavenger might be tiron, the catechol disulphonate, which effectively scavenges O_2^-.

Reiter: What is the problem with higher concentrations of superoxide dismutase?

Fridovich: There may be effects due to the protein itself or, with the copper-zinc-enzyme, to the copper. At concentrations of a few mg/ml, the enzyme can act as a weak peroxidase. To eliminate these effects one should use the enzyme, whenever possible, at low levels.

Willson: Dr Hill, I am not happy with the reaction of O_2^- with CO_2, although it may go in a lipid enviroment. In water the reverse reaction (1) occurs with a rate constant of 2.4×10^9 l mol^{-1} s^{-1} (Adams & Willson 1969).

$$CO_2^- + O_2 \rightarrow CO_2 + O_2^- \quad (1)$$

Hill: Certainly, but the product of the reaction of superoxide with carbon dioxide has not been characterized (see p. 191).

Willson: May I suggest another mechanism, nevertheless? I am reminded of some work by Scholes *et al.* (1960) on the sites of radical formation in irradiated macromolecules using labelled CO_2 in conditions in which the solvated electron, a strong reducing agent, reacts to give CO_2^- (reaction 2) and the hydroxyl

$$e_{aq}^- + CO_2 \rightarrow CO_2^- \quad (2)$$

radical reacts with the organic molecule to form a radical — in your case this could be the glutamic acid. They showed that these radicals would combine with CO_2^- to give the carboxylate derivative (reactions 3 and 4). This simple

$$RH + HO^{\bullet} \rightarrow R^{\bullet} + H_2O \quad (3)$$

$$R^{\bullet} + CO_2^- \rightarrow RCO_2^- \quad (4)$$

mechanism could explain the carboxylation. Questions remain as to how the $OH^{\bullet}$ and CO_2^- radicals are generated. I should have thought that the cell would contain other nucleophiles which would be better one-electron donors than O_2^-.

Hill: As with all the actions of O_2^- we have difficulty in identifying the reactive species but that mechanism is possible. Some of the species that Professor Michelson implicated in cell damage may also be present and, since CO_2 may lead to CO_2^-, CO_3^-, CO_4^-, HCO_4^- and $C_2O_6^{2-}$, I was deliberately ambiguous when I referred to the reactive material; all are possible. Although in aprotic media peroxocarbonates are formed, in protic media they decompose, presumably through one or more of these species and with chemiluminescence. By rejecting the idea that carbon dioxide directly reacts with the quinone we have eliminated the requirement for ATP, which was previously assumed.

Fridovich: If carbon dioxide does react with O_2^-, it must do so in a reversible way; if not, the reduced carbon dioxide will reduce cytochrome *c*. I say that because SOD-inhibitable reduction of cytochrome *c*, caused by O_2^-, is not affected by carbonate buffer.

Hill: No, I did not say carbonate; carbon dioxide is the reactive material in the carboxylation.

Fridovich: But some CO_2 will be in equilibrium with the carbonate in the buffer.

Hill: Even so, I am convinced (Jones *et al.* 1977) that carbon dioxide is the reactive species. As I mentioned (p. 15), one has to define the medium before one can specify reactions. Furthermore, protons, in particular, crucially influence all these equilibria.

Michelson: Isn't the formate radical anion a strong reducing agent?

Hill: Yes.

Michelson: It is interesting that ribulosediphosphate carboxylase can act as an oxygenase in different conditions.

Fridovich: But that enzyme seems to have two separate sites.

Bielski: How does the semiquinone react with CO_2? The reaction could be checked in anaerobic conditions by oxidizing the hydroquinone to the semiquinone with cerium(IV) in the presence of CO_2. Through its resonance structures the semiquinone has several sites on the ring available for reaction with CO_2.

Hill: Possibly. Among the variables that often are not controlled in studies of reactions of superoxide, particularly with intact systems, are carbonate and CO_2. I expect more evidence will soon emerge that superoxide and its derivatives form highly reactive species by reaction with carbonate or CO_2.

Bielski: Since the absorption spectra of CO_2^- and CO_3^- are known (Keene *et al.* 1965) it should not be difficult to identify these species in your system.

Willson: The mechanism I suggested would only work in anaerobic conditions, since the radical so produced will react with oxygen to give the peroxyl radical. Also CO_2^- will react with oxygen to give O_2^-; this, of course, assumes that the reaction environment is aqueous.

Williams: In enzymes that require O_2^- is superoxide postulated as a free species at any time, reaching a steady-state concentration? If so, what sort of steady-state concentrations of superoxide might be reached in biological systems?

Hill: In truly intact systems superoxide may never be a free species. That is probably true for cytochrome P450 systems as well. In the intact system all this takes place in the locus of the carboxylase. To investigate it, by prizing it open, we interfere with it.

Fridovich: When Professor Hayaishi first described the indoleamine dioxygenase as a superoxygenase I caused him some considerable aggravation by my scepticism.

Hayaishi: The indoleamine dioxygenase reacts in the iron(III) state with O_2^- to form an oxygenated species, which might be called the EO_2^- complex and which decomposes with any of the substrates (see also pp. 199–203). We tried to demonstrate this in intact cell suspensions of enterocytes in two ways. In one case we added substrates for xanthine oxidase to see if O_2^- would be generated *in situ* and, therefore, accelerate the indoleamine dioxygenase activity. With about 5μM-inosine or any other purine derivative that can serve as substrate for xanthine oxidase, the dioxygenase activity was increased about 5–10-fold. This effect could be completely abolished by adding about 1mM-allopurinol, a specific inhibitor of xanthine oxidase. Secondly, in the presence of dietnyl-dithiocarbamate, which inhibits superoxide dismutase *in vivo*, the indoleamine dioxygenase activity increased about 10-fold. The highly purified indoleamine dioxygenase activity is not affected at all by either inosine, allopurinol or diethyldithiocarbamate *in vitro*.

Fridovich: One has to conclude, from these data and the abundance of superoxide dismutase in the cell, that there is a steady-state level of O_2^- and it is extremely low — less than 1 pmol/l — in the cell sap. There may be privileged environments with less dismutase and more O_2^- — enough to carry out this kind of reaction.

Hill: The species that destroy superoxide *in vivo* could interfere with the processes such as these and others that require superoxide.

Williams: Professor Hayaishi believes that the O_2^- is a genuine free mobile substrate for the dioxygenase. You have to examine the same point by the same techniques as he used for only then would you know whether O_2^- was a true intermediate. It is hard to prove what the bound species are.

Hill; I agree; the quinol, oxygen and carbon dioxide could give the reduced carbon dioxide species directly. One might suppose that that would be the best way to generate the reactive carboxylating species. We have recently shown that vitamin K quinol, carbon dioxide and oxygen together give a highly chemiluminescent system — that is, those three reagents may form a reactive species derived from carbon dioxide.

References

ADAMS, G. E. & WILLSON, R. L. (1969) Pulse radiolysis studies on the oxidation of organic radicals in aqueous solution. *Faraday Soc. Trans. 65*, 2981–2987

JONES, J. P., GARDNER, E. J., COOPER, T. G. & OLSON, R. E. (1977) Vitamin K-dependent carboxylation of peptide-bound glutamate. *J. Biol. Chem. 252*, 7738–7742

KEENE, J. P., RAEF, Y. & SWALLOW, A. J. (1965) Pulse radiolysis studies of carboxyl and related radicals, in *Pulse Radiolysis* (Ebert, M., Keene, J. P., Swallow, A. J. & Baxendale, J. H., eds.), pp. 99–106, Academic Press, London & New York

McCORD, J. M. & SALIN, M. L. (1977) Self-directed cytotoxicity of phagocyte-generated superoxide free radical, in *Movement, Metabolism and Bactericidal Mechanisms of Phagocytes* (Rossi, F., Patriarca, P. L. & Romeo, D., eds.), pp. 257–264, Piccin, Padova

SCHOLES, G., SIMIC, M. & WEISS, J. J. (1960) Radiation-induced carboxylation of organic compounds: formation and reaction of the carboxyl radical ion. *Nature (Lond.) 188*, 1019–1020

Specific induction of pulmonary indoleamine 2,3-dioxygenase by bacterial lipopolysaccharide

OSAMU HAYAISHI and RYOTARO YOSHIDA

Department of Medical Chemistry, Kyoto University Faculty of Medicine, Kyoto, Japan

Abstract Indoleamine 2,3-dioxygenase (molecular weight about 42 000) has been purified from rabbit intestines and contains one mole of protohaem IX as the sole prosthetic group. It catalyses the oxidative cleavage of the pyrrole ring of various indoleamines with a much broader specificity of substrate than tryptophan 2,3-dioxygenase. The enzyme has an absolute requirement for superoxide anion for catalytic activity.

The enzyme was induced specifically in the lungs of mice for 24 h after administration of the lipopolysaccharide fraction of *E. coli*. This increase is due to synthesis of enzyme protein and is specific for the lipopolysaccharide fraction.

These results are interpreted to mean that indoleamine dioxygenase is induced in pulmonary inflammatory processes in response to an increase in production of superoxide anion, 5-hydroxytryptamine or other indoleamines in the lung as a consequence of inflammation. The dioxygenase reaction is a more innocuous way of disposing of superoxide than dismutation.

Indoleamine 2,3-dioxygenase, a haemoprotein, is widely distributed in various tissues and organs of mammals (Hayaishi 1976). It catalyses the oxidative cleavage of the pyrrole ring of various indoleamine derivatives, including tryptophan, to yield corresponding anthraniloylamines (Fig. 1). The enzyme has been purified extensively from rabbit intestine. The molecular weight has been estimated to be about 42 000. One mole of protohaem IX has been found to be

Indoleamine **Anthraniloylamine**

FIG. 1. Catalytic action of indoleamine 2,3-dioxygenase: conversion of indoleamines into the corresponding anthraniloylamines.

contained as a sole prosthetic group per mole of enzyme (Shimizu *et al.* 1978). The reaction catalysed by this enzyme is essentially identical to that catalysed by the well known hepatic tryptophan 2,3-dioxygenase (tryptophan pyrrolase) with two important exceptions. First, the substrate specificity of the enzyme is much broader than that of tryptophan pyrrolase; the enzyme catalyses the oxidative cleavage of not only the parent compound tryptophan but also 5-hydroxytryptophan, which is a precursor of 5-hydroxytryptamine, and other indoleamine derivatives such as 5-hydroxytryptamine itself, tryptamine, melatonin and so forth. Secondly, the enzyme has an absolute requirement for superoxide anion for catalytic activity. Superoxide anion labelled with oxygen-18 is incorporated into the product of the reaction (Hayaishi *et al.* 1977). Consequently, the enzyme might be termed a 'superoxygenase'.

In an attempt to find out the true substrate of this enzyme *in vivo* and to study its biological function, we recently determined levels of the indoleamine dioxygenase activity in various tissues in a variety of physiological and pathological conditions.

When a bacterial endotoxin, for example, the lipopolysaccharide fraction of *E. coli* (LPS), was administered to mice intraperitoneally, a remarkable increase in the enzyme activity was observed in the lung within 24 h (Fig. 2). The specific activity in the high-speed supernatant fraction of the lung increased almost linearly for about 24 h; after about 48 h it gradually decreased and reached a normal value within 6–7 days. A single injection of about 20 μg/mouse of LPS was sufficient to induce a 30–50-fold, sometimes nearly 100-fold, increase in enzyme activity.

This increase of the indoleamine dioxygenase activity appears to be specific for the lung, because in all other organs so far examined, the increment was usually negligible or, at most, less than 2–3-fold, but in the lung the indoleamine dioxygenase activity increased about 50-fold (Fig. 3).

The increase in the enzyme activity appears to be due to the net synthesis of

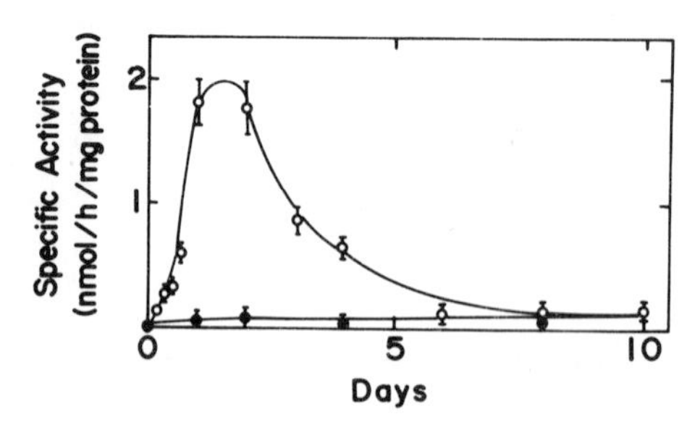

FIG. 2. Induction of indoleamine 2,3-dioxygenase in mice lungs by the lipopolysaccharide fraction of *E. coli*.

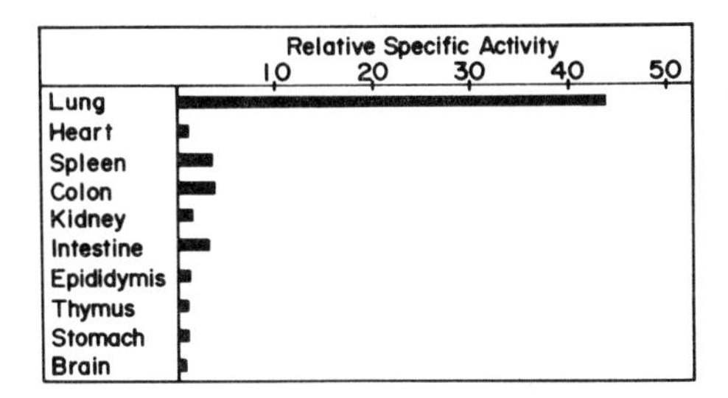

FIG. 3. Induction of indoleamine 2,3-dioxygenase by the lipopolysaccharide fraction of *E. coli* in various organs.

enzyme protein, rather than to the removal of inhibitors, an increase in activators or to other reasons. The simultaneous administration of actinomycin D or cycloheximide abolished the increase in enzyme activity, at least for 4 h, as shown in Fig. 4. Thereafter, as a consequence of the metabolism or excretion of the inhibitor, the induction was gradually established.

The induction of the indoleamine dioxygenase activity in the mouse lung appears to be specific for the lipopolysaccharide fraction, because other inflammatory agents or stimulants, such as glycogen or zymosan (a protein polysaccharide complex of yeast cell wall) had no similar effect, but the LPS fractions from *E. coli* and *Salmonella* were both effective (Fig. 5).

To determine whether the induction is specific for the indoleamine dioxygenase activity, we tested various other enzyme activities. For example, lysosomal enzymes such as β-glucuronidase and acid phosphatase, or enzymes reportedly involved in the inflammation such as monoamine oxidase were determined. None of these enzymes was induced to any significant extent (Fig. 6).

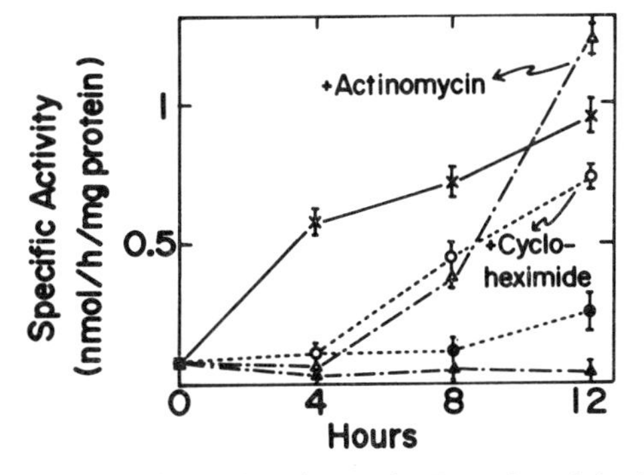

FIG. 4. Effect of actinomycin D and cycloheximide on the activity of indoleamine 2,3-dioxygenase.

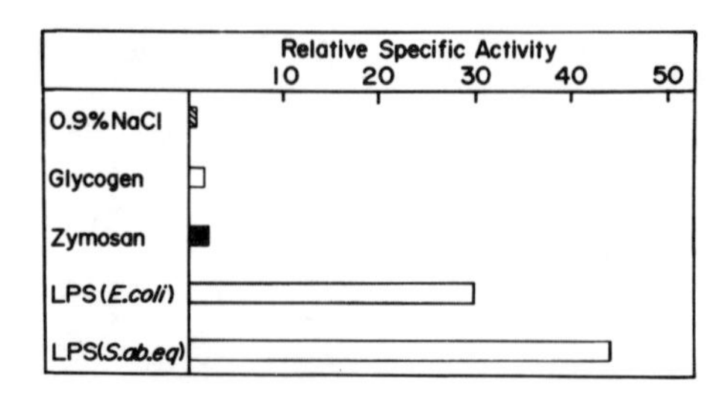

FIG. 5. Induction of indoleamine 2,3-dioxygenase in mice lungs by the lipopolysaccharide fraction of *E. coli* and *Salmonella*.

These results, although rather preliminary, suggest several interesting possibilities. The indoleamine dioxygenase activity may be related to the inflammatory process of the lung. This induction may be due to the increase in the production of the superoxide anion or 5-hydroxytryptamine or some other indoleamines in the lung as a consequence of inflammation. The indoleamine dioxygenase-catalysed reaction may be a more innocuous way to obliterate the superoxide anion, because it does not produce hydrogen peroxide and it also eliminates 5-hydroxytryptamine from the area of inflammation — so, the enzyme may be killing two birds with one stone. Obviously, further studies are required to elucidate the mechanism of the specific induction of the indoleamine dioxygenase activity in the lung by bacterial endotoxin, but it may open up a new area of research related to this unique enzyme, lung inflammation and/or the mode of action of bacterial endotoxin.

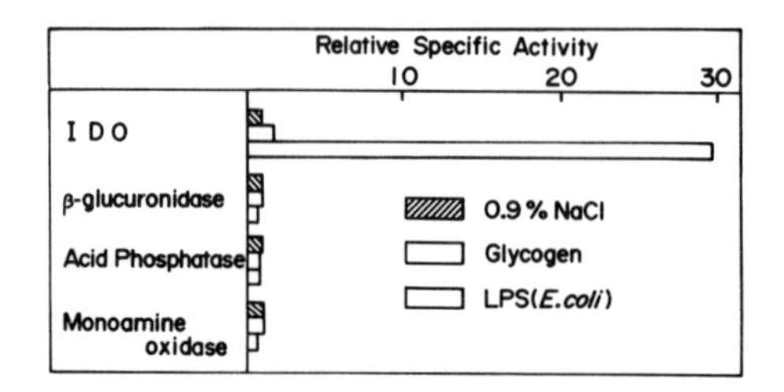

FIG. 6. Induction of various enzymes in lung: IDO, indoleamine 2,3-dioxygenase.

References

HAYAISHI, O. (1976) Properties and function of indoleamine 2,3-dioxygenase. *J. Biochem.* *79*, 13–21

HAYAISHI, O., HIRATA, F., OHNISHI, T., HENRY, J.-P., ROSENTHAL, I. & KATOH, A. (1977) Indoleamine 2,3-dioxygenase. Incorporation of $^{18}O_2^-$ and $^{18}O_2$ into the reaction products. *J. Biol. Chem. 252*, 3548–3550

SHIMIZU, T., NOMIYAMA, S., HIRATA, F. & HAYAISHI, O. (1978) Indoleamine 2,3-dioxygenase. Purification and some properties. *J. Biol. Chem. 253*, 4700–4706

Discussion

McCord: How did you generate the [^{18}O]superoxide and prevent it from exchanging with molecular oxygen?

Hayaishi: The details have been published (Hayaishi *et al.* 1977). A solution of ^{18}O-labelled potassium superoxide (synthesized in the Weizmann Institute in Israel) in anhydrous dimethyl sulphoxide was infused into the reaction mixture in anaerobic and aerobic conditions. We also did this with cold KO_2 and ^{18}O-labelled molecular oxygen. Consequently, we could calculate exactly how much oxygen is derived from O_2^- and from O_2, respectively. In our conditions about 67% of the oxygen incorporated in the product was estimated to be derived from the superoxide anion whereas about 33% came from molecular oxygen. This proportion varies with the conditions.

Fridovich: Your results, Professor Hayaishi, call to mind a report that an injection of endotoxin provided resistance to oxygen toxicity in rats. Now I begin to understand that report (Frank *et al.* 1978).

Allison: Are there macrophages in the lung that you can get out?

Hayaishi: That is what we are studying now.

Cohen: Why did you suppose that monoamine oxidase is involved in the inflammatory process?

Hayaishi: We were under the impression that some indoleamines are involved in inflammation.

Reiter: There have been reports of an unexplained increase of resistance to infection in mice and cows when lipopolysaccharides were injected (Hill *et al.* 1973; Hibbitt *et al.* 1977). These authors ruled out activation of macrophages.

References

Frank, L., Yam, J. & Roberts, R. J. (1978) The role of endotoxin in protection of adult rats from oxygen-induced lung toxicity. *J. Clin. Invest. 61*, 269–275

Hayaishi, O., Hirata, F., Ohnishi, T., Henry, J.-P., Rosenthal, I. & Katoh, A. (1977) Indoleamine 2,3-dioxygenase. Incorporation of $^{18}O_2^-$ and $^{18}O_2$ into the reaction products. *J. Biol. Chem. 252*, 3548–3550

Hibbitt, R. G., Hill, A. W., Young, J. L. & Sheers, A. L. (1977) The effect of endotoxin tolerance on some serum constituents of cows. *J. Comp. Pathol. 87*, 195–203

Hill, A. W., Hibbitt, R. G. & Sheers, A. (1973) The stimulation by endotoxin of the nonspecific resistance of mice to bacterial infections. *Br. J. Exp. Pathol. 55*, 194–201

Oxygen consumption by stimulated human neutrophils

A. W. SEGAL and A. C. ALLISON

Division of Cell Pathology, Clinical Research Centre, Harrow, Middlesex

Abstract Oxygen consumption by stimulated human neutrophils has been studied. An initial lag after stimulation of about 20 s is followed by a linear phase of oxygen consumption which lasts about 60 s and then declines exponentially. The duration of the respiratory burst is thus much shorter than has been generally recognized. The cessation of the linear phase of oxygen consumption occurs at a constant time after stimulation, is not due to depletion of the substrate of the oxidase enzyme (since linear oxygen consumption can be re-initiated), and is due to termination or saturation of particle uptake. The curtailment of oxygen consumption probably marks the end of the oxygen-dependent microbial role of the phagocytic vacuole, preparing it for the secondary and independent function of digestion.

A novel cytochrome *b* becomes associated with the phagosomes and probably forms a component of a complex electron-transport chain. Superoxide may be an intermediate in this system, but it is unlikely to be released free from the cell or to form the final product of the system.

Neutrophil granulocytes comprise the largest proportion of leucocytes in humans and the primary defence against bacterial infection. Phagocytosis of the invading organism is associated with a burst of oxygen consumption (Baldridge & Gerard 1933) which is not due to enhanced mitochondrial respiration, as it is not inhibited by classical mitochondrial inhibitors such as cyanide or azide (Sbarra & Karnovsky 1959). The oxygen metabolism is associated with mechanisms which appear to play an important bactericidal role (Mandell 1974).

The importance of oxidative processes in the killing of bacteria is clearly demonstrated in anaerobic conditions and by the syndrome of chronic granulomatous disease (CGD) (Berendes *et al.* 1957). In this condition the burst of oxygen consumption that is normally associated with phagocytosis is absent (Holmes *et al.* 1967). The children with this disease are extremely susceptible to infection by organisms such as staphylococci that produce catalase and so can catabolize H_2O_2 but not by bacteria that themselves produce hydrogen peroxide (Mandell & Hook 1969). *In vitro* studies show that the particular organisms that demonstrate a characteristic and abnormal virulence in these patients are killed slowly, if at all, by their neutrophils (Holmes *et al.* 1966).

The burst of oxidative metabolism has generated considerable interest and controversy primarily because its biochemical basis still remains unknown 40 years after it was first described, despite intensive investigation. In view of this controversy, and the great variety of results obtained to date, we decided to re-examine the whole process of oxygen consumption in association with phagocytosis, and a brief survey of our preliminary observations is presented here.

THE FEATURES OF THE RESPIRATORY BURST OF HUMAN NEUTROPHILS

We used latex (0.81 μm in diameter) as the object of phagocytosis and opsonized it with human IgG (Singer & Plotz 1956) in a concentration of about 10^7 molecules/particle. This particle was chosen as it is readily phagocytosed, metabolically inert, and because it is light, thus allowing the phagocytic vacuoles themselves to be separated from other subcellular organelles by density centrifugation. Oxygen consumption was studied at 37 °C in the chamber of a Clark oxygen electrode.

Unstimulated neutrophils consume oxygen slowly, at a rate of about 3 nmol

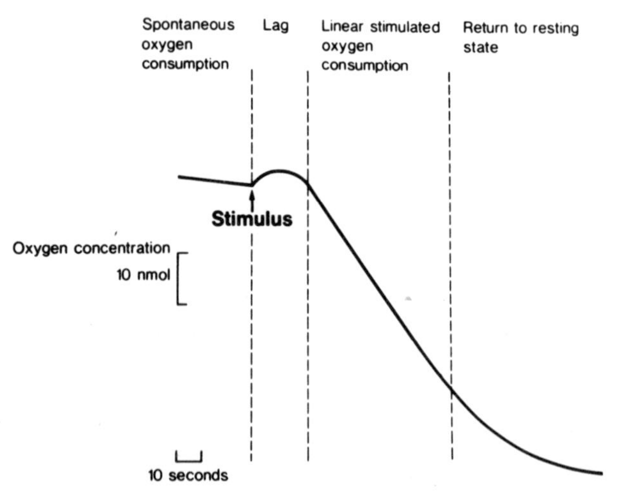

FIG. 1. Polarographic measurement of oxygen consumption by human neutrophils (2 × 10^7) after the addition of latex particles (50/cell) opsonized with IgG.

min^{-1} (1×10^7 cells)$^{-1}$. After the particles are added to the rapidly stirred cells, there is a lag of about 10 s before oxygen consumption commences after which it rapidly increases until it reaches a linear rate after about 25 s. This linear consumption is maintained for about 60 s and then the rate of oxygen consumption rapidly tails off (Fig. 1).

The interesting and surprising feature of this process is that, after a submaximal dose of latex particles is originally given, oxygen consumption can be induced to return to a rapid linear rate by the addition of further particles to the medium (Fig. 2). Thus the oxygen consumption does not stop because the oxygen-consuming process is depleted of substrate. As the number of particles in the initial stimulus is increased, the rate of oxygen consumption increases and the rate of oxygen consumption induced by a subsequent maximal dose of particles is reduced. The sum of the rates induced by a submaximal and then by a maximal stimulus generally equals that of the maximal stimulus alone. This indicates that the process is saturable, an observation that was supported by the results of determining the effect of changing the incubation temperature. As the temperature is raised from 20 to 50 °C there is a striking rise in the rate of oxygen consumption but, as this rate increases, there is a corresponding reduction in the duration of linear consumption, so that the total linear oxygen consumption remains constant (Fig. 3).

The effect of varying the ratio of particles to cells is informative. There is a linear relationship between the maximal rate of oxygen consumption and the logarithm (to the base 10) of the number of particles (from 1 to 50) (Fig. 4). This was at first surprising, but was shown to be due to a similar logarithmic

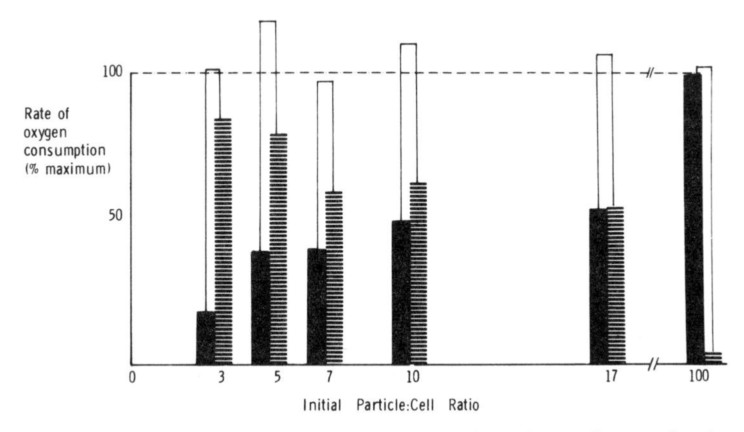

FIG. 2. Rate of oxygen consumption after addition of a maximum stimulus of latex (striped columns) to human neutrophils previously submaximally stimulated with latex (solid columns). Open columns represent the sum of these rates.

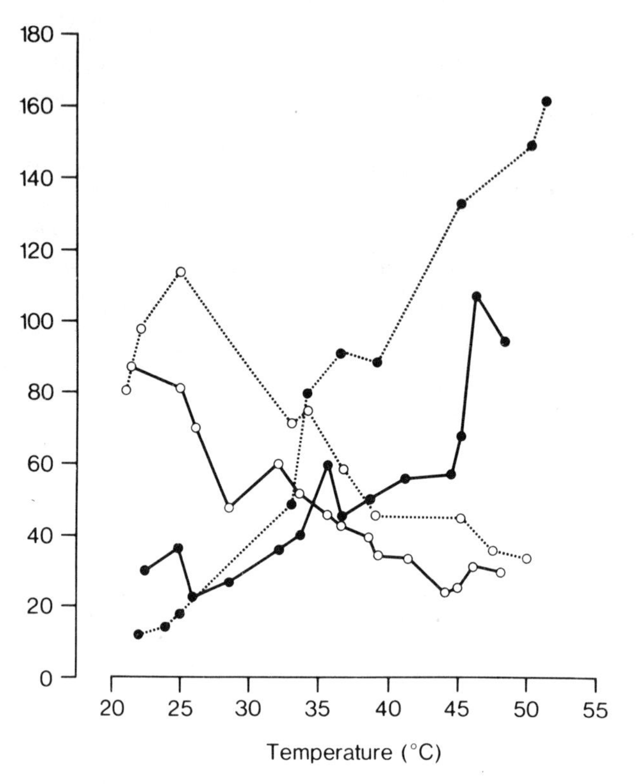

FIG. 3. Effect of temperature on linear oxygen consumption by phagocytosing neutrophils: ●, rate (nmol/min); ○, linearity (seconds); . . ., 6 × 10^7 cells/ml; ——, 4 × 10^7 cells/ml.

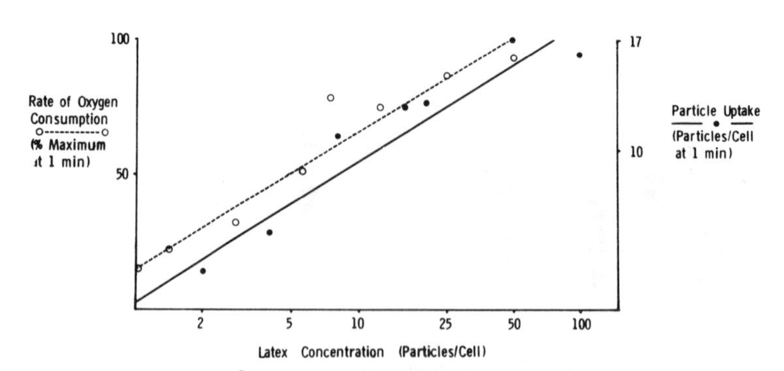

FIG. 4. Relationship between particle uptake and rate of oxygen uptake by human neutrophils exposed to various concentrations of latex. (Note log scale for abscissa.)

relationship between the number of particles administered and the number taken up by the cells (Fig. 4). Thus, as expected, the rate of oxygen consumption is proportional to the number of particles taken up by the cell.

When a saturating dose of latex particles (greater than 50 per cell) is administered, about 20 particles are taken up per cell, which corresponds to oxygen consumption of about 40 nmol/(1 $\times$ 10^7 cells), or ± 0.2 fmol/vacuole.

The rate of particle uptake closely parallels that of the burst of oxygen consumption after the initial lag period (Fig. 5). These data indicate that: (1) oxygen consumption is directly related to uptake of particles by neutrophils; (2) uptake of particles starts soon after addition to the cell suspension. Oxygen consumption parallels particle uptake, but only after a lag period of about 20 s, indicating that the oxidase process is distinct from but associated with phagocytosis and that the triggering of the process takes about 20 s. (3) Oxygen consumption ceases. The fact that the kinetics of particle uptake and oxygen consumption are similar indicates that in any particular vacuole oxygen consumption must be brief or else an additive effect would have been observed.

THE OXIDASE PROCESS

There has been no convincing evidence as to the mechanisms of the oxidase

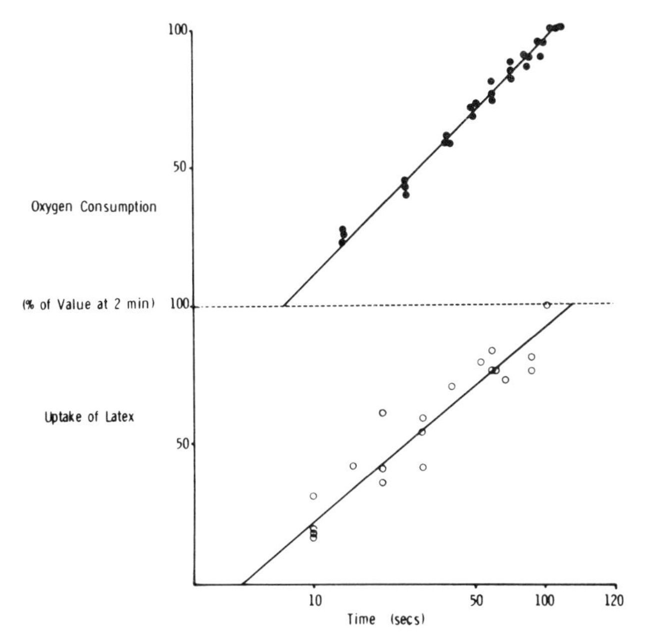

FIG. 5. Relationship of latex uptake and oxygen consumption to time: ○, uptake of latex; ●, oxygen consumption after lag period of 23 s.

system, whether it is enzymic or non-enzymic and, if enzymic, the identity of the natural substrate of the system.

The visible spectra of neutrophils stimulated with IgG-coated latex particles and with phorbol myristate acetate were examined in a split-beam spectrophotometer. Dramatic but different changes were found with the two stimuli (Fig. 6).

Dithionite and CO difference spectra of neutrophil homogenates were then examined and, in addition to the myeloperoxidase absorption peaks at 475 nm, there were also distinct peaks at 560 nm and 429 nm (Figs. 7 and 8). Absorption at these wavelengths is characteristic of a *b* cytochrome. Phagocytic vacuoles containing latex particles were then prepared by a combination of flotation on sucrose gradients and differential centrifugation (Segal & Jones 1978) (Fig. 9) and these vacuole preparations were examined for the presence of this *b* cytochrome. Not only was it clearly present in the vacuoles (Fig. 10) but it was spectroscopically free of contamination by cytochrome P450 and cytochrome oxidase, confirming that it is not a contaminant from the endoplasmic reticulum or mitochondria (Segal & Jones 1978).

Homogenates of neutrophils from patients with CGD were examined. The normal 429 nm cytochrome *b* peak was missing in all five patients studied and

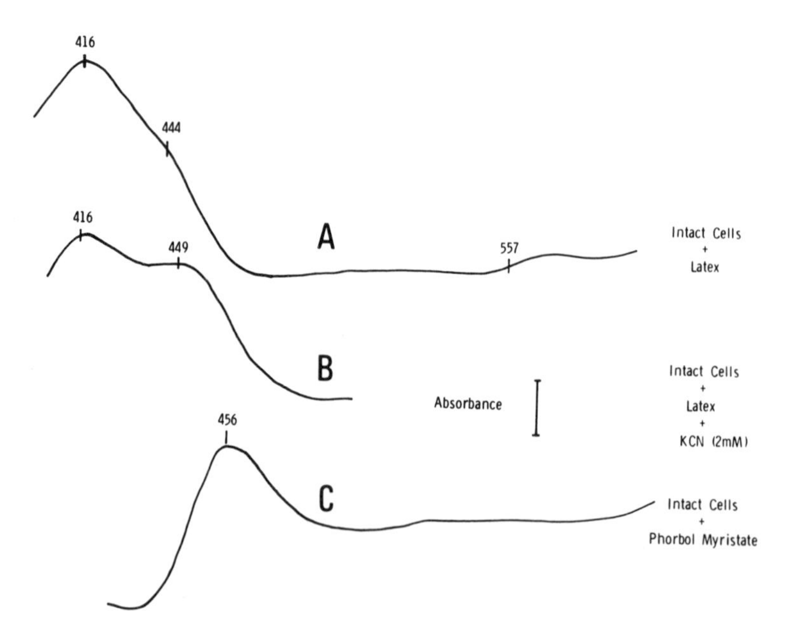

FIG. 6. Difference spectra of intact human neutrophils stimulated with IgG-opsonized latex in the presence and absence of KCN, and with phorbol myristate acetate.

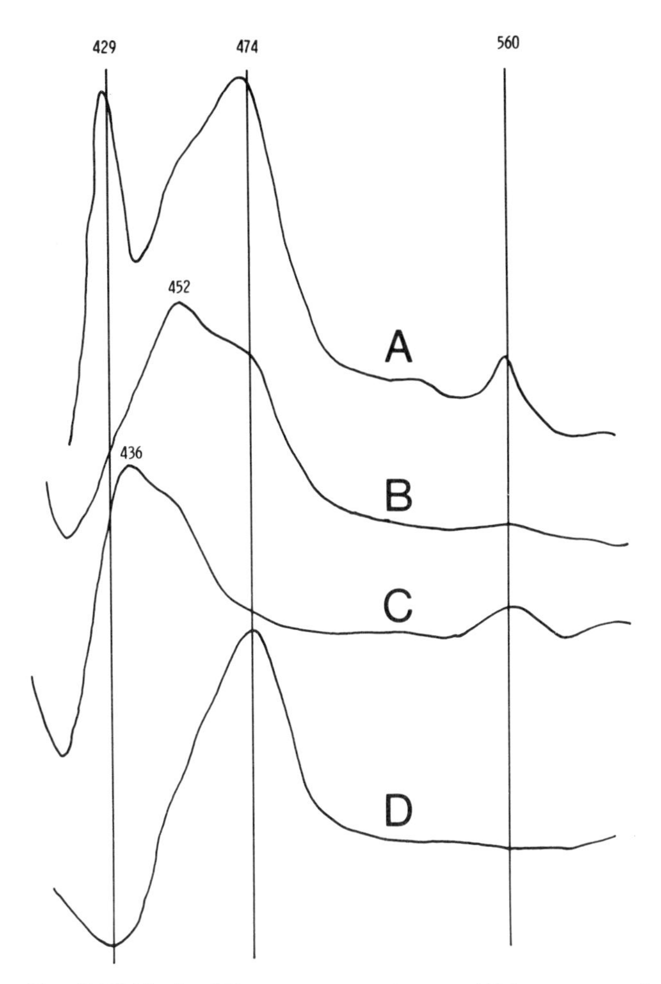

FIG. 7. Dithionite difference spectra of neutrophil homogenates from a normal subject (A), and a male (B) and female (C) patient with chronic granulomatous disease. The spectrum obtained with pure myeloperoxidase is also shown (D).

unusually low levels were observed in both of the patients' mothers who were obligate heterozygotes of the X-linked form of the disease.

Thus, a unique cytochrome *b* system has been demonstrated in human neutrophils (Segal & Jones 1978). The observation that it becomes associated with phagocytic vacuoles, uncontaminated by endoplasmic reticulum or mitochondria, indicates that it is a novel cytochrome system in a previously undescribed situation, possibly the plasma membrane. Absence or abnormalities of this cytochrome in CGD directly implicate it in the microbicidal oxidase system.

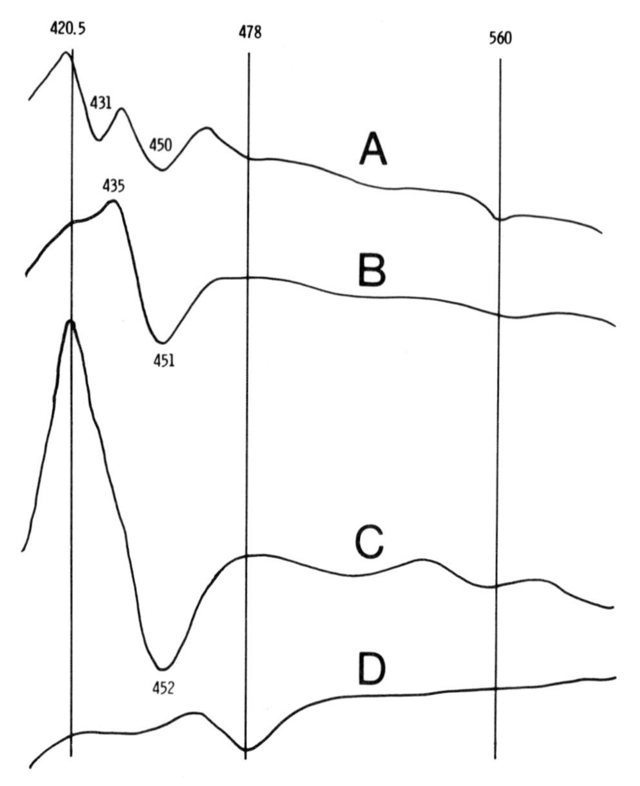

FIG. 8. CO difference spectra of the neutrophil homogenates described in Fig. 7.

THE PRODUCT OF THE OXIDASE SYSTEM

Hydrogen peroxide

It is generally believed that hydrogen peroxide is the natural product of the oxidase system. This was first considered after [^{14}C]formate was shown to be oxidized to $^{14}CO_2$ (Iyer *et al.* 1961). Supportive evidence comes from the fact that myeloperoxidase can use H_2O_2 as substrate, that the bacteria that particularly cause infections in patients with CGD are catalase-positive, whereas bacteria that themselves produce H_2O_2 are easily killed (Mandell & Hook 1969), and that bacterial killing is promoted by the addition of an H_2O_2-generating system to their cells (Baehner *et al.* 1970*b*). When H_2O_2 is directly assayed in phagocytosing neutrophils, the recovery accounts for only a small percentage of the oxygen consumption (Iyer *et al.* 1961; Baehner *et al.* 1970*a*). In the presence of high concentrations of cyanide or azide, the recovery of H_2O_2 is greatly increased — this increase has been assumed to be due to the inhibition of the

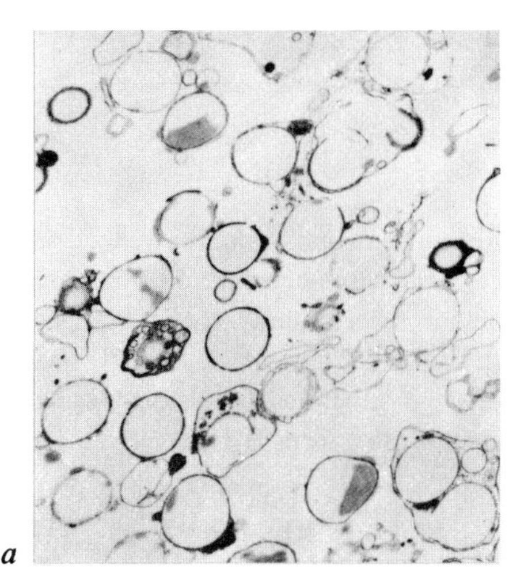

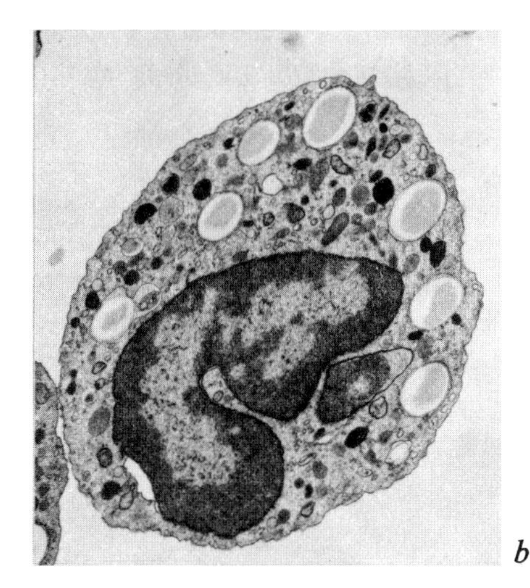

FIG. 9. Electron micrographs of phagocytic vacuoles (*a*; ×8500) prepared from neutrophils (*b*; ×8500) after exposure to IgG-opsonized latex particles for 1 min. (Electron microscopy by J. Dorling and J. Webb.)

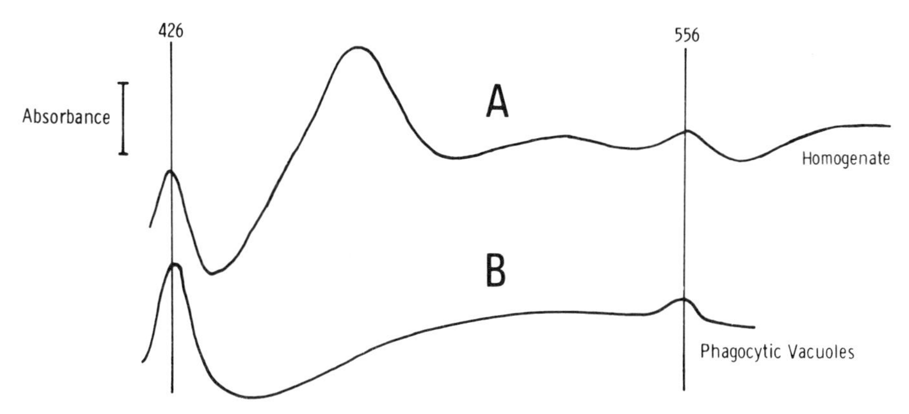

FIG. 10. Dithionite difference spectra of (A) neutrophil homogenates and (B) phagocytic vacuoles, at 77 K. Absorbance scale indicator = 0.0028 of an optical density unit.

normal degradative enzyme systems including the enzymic action of myeloperoxidase (Homan-Müller *et al.* 1975), which assumption is supported by the observation that H_2O_2 production is greatly increased in myeloperoxidase-deficient neutrophils (Klebanoff & Pincus 1971).

Most of the oxygen that is consumed eventually forms water and little, if any, is incorporated into the organic composition of the cell or ingested bacteria (Segal *et al.* 1978). Experiments with $^{15}O_2$ have shown that all the metabolized $^{15}O_2$ can be recovered in the aqueous medium and that the product is water rather than a water-soluble metabolite because its specific activity does not change after distillation (Segal *et al.* 1978).

Superoxide

Ferricytochrome *c* is reduced by phagocytosing neutrophils and this reduction can be inhibited by superoxide dismutase (SOD) (Babior *et al.* 1973). Also, SOD-inhibitable reduction of ferricytochrome in association with phagocytosis is defective in neutrophils from patients with CGD (Curnutte *et al.* 1974). It was thus proposed that the superoxide anion, O_2^-, is generated by phagocytosing neutrophils and that this spontaneously dismutates to form H_2O_2 (reaction 1).

$$4O_2 \xrightarrow{\text{oxidase}} 4O_2^- \xrightarrow[+2H^+]{\text{spontaneous dismutation}} 2H_2O_2 + 2O_2 \qquad (1)$$

The effect of cytochrome *c* on oxygen consumption does not seem to have been previously examined. We determined the reduction of cytochrome *c* by neutrophils exposed to IgG-coated latex particles in the chamber of an oxygen electrode. As previously described by Babior *et al.* (1973) and Curnutte *et al.* (1974), cytochrome *c* is reduced by phagocytosing neutrophils; this process can be inhibited by SOD, but not by boiled SOD, and reduction does not take place in anaerobic conditions. However, it made no difference whatsoever to the rate or extent of oxygen consumption (Fig. 11) whether the cytochrome *c* was reduced, or whether this reduction was inhibited by SOD. In addition, the time course of oxygen consumption and cytochrome *c* reduction are different: whereas linear oxygen consumption only commences after 25 s, cytochrome *c* reduction is largely complete 10 s after the addition of the stimulus.

If, as has been previously thought, O_2^- is the intermediate in the production of H_2O_2 (Babior 1978), then reduction of cytochrome *c* by the free electron on O_2^- should regenerate O_2. However, as the reduction of cytochrome *c* does not affect O_2 consumption, we must conclude that the previous assumptions are erroneous and that superoxide is unlikely to be released freely from the cell and is probably an intermediate in, or by-product of, the oxidase system.

THE FUNCTION OF THE OXIDASE SYSTEM

It seems clear that the function of the cytochrome *b* is to carry electrons. This cytochrome also seems to bind CO which could give it the characteristics of a cytochrome O (Lemberg & Barrett 1973), previously only described in bacteria. It is feasible that the main function of this system is not the production of H_2O_2

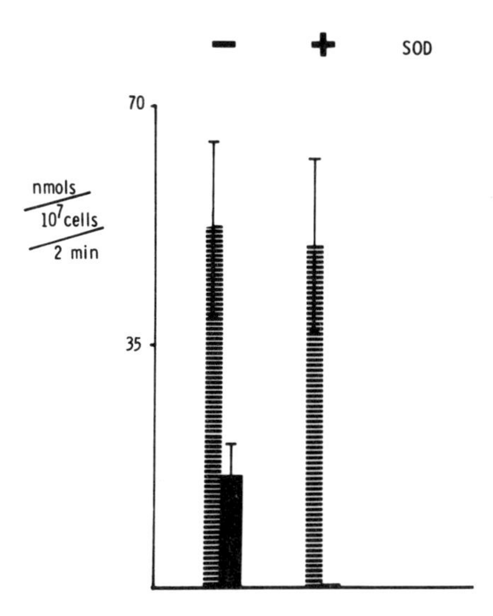

FIG. 11. Oxygen consumption (striped columns) by neutrophils stimulated with latex particles (50/cell) in the presence of cytochrome *c*. The mean ± S.E.M. of four studies is shown. Superoxide dismutase strongly inhibits cytochrome *c* reduction (solid columns) but does not influence oxygen consumption.

but to act as a proton pump, in accord with the chemiosmotic theory of Mitchell (1962), to reduce the intravacuolar pH to the optimal for lysosomal degradative enzymes. It is possible that myeloperoxidase forms the terminal part of this electron transporting system onto oxygen to form water, and that when myeloperoxidase is missing or inactivated H_2O_2 is produced as an alternative.

The mild bactericidal defect in the cells of myeloperoxidase-deficient neutrophils could be as much due to oxidative damage to the lysosomal digestive enzymes by H_2O_2 as to the lack of a positive killing mechanism.

It is, of course, also possible that the oxidase system works in the generally accepted way to produce H_2O_2 which then forms the substrate for a myeloperoxidase-dependent microbicidal system (Klebanoff 1975).

THE CHARACTERISTICS OF THE OXIDASE REACTION

In any particular vacuole, the oxidase reaction is activated after a lag period of about 25 s, is very brief and appears to be irreversible. In order to determine the reason for this cessation of oxygen consumption, we allowed cells to phagocytose zymosan or latex particles in an atmosphere of nitrogen and then exposed them to oxygen in the chamber of an oxygen electrode. No oxygen debt was built up by intact cells, cell homogenates or a 27 000 *g* preparation containing granules and phagocytic vacuoles, or by highly purified vacuoles themselves. Neither intact cells nor isolated granule preparations consumed oxygen when lysed by sonication in anaerobic conditions and then exposed to oxygen. Thus it does not seem to be a simple autoxidation process of a substrate which might have been previously reduced by the cytochrome system, because if this were the case oxygen should be consumed immediately on contact with the substrate. The process appears to be closely associated with electron transport by the cytochrome *b*. It cannot be reactivated after exposure of stimulated anaerobic cells to oxygen. Thus, either there must be an alternative to O_2 as the electron acceptor, or the process can only consume oxygen at a well defined stage in the ordered sequence of endocytosis and degranulation.

THE NADPH OXIDASE

Granule preparations from normal neutrophils contain an NADPH oxidase that is activated by phagocytosis (Patriarca *et al.* 1971, 1977). A similar activation of an NADPH oxidase is not observed in whole cells or neutrophil preparations from patients with CGD (Hohn & Lehrer 1975). We, in common with other investigators (Takanaka & O'Brien 1975), have found that purified human myeloperoxidase is itself an NADPH oxidase with similar properties to those of the NADPH oxidase of neutrophils. It has been thought that the NADPH oxidase is distinct from myeloperoxidase because oxidase activity was determined in neutrophils from a patient in whom these cells lacked detectable peroxidase activity (Patriarca *et al.* 1975) and because the oxidase activity is resistant to cyanide whereas the peroxidatic activity is inhibited.

Myeloperoxidase is a unique protein, having two haem subunits with different properties (Odajima & Yamazaki 1972). It is not clear which of the subunits is responsible for the NADPH oxidase and which for the peroxidase activity, if these activities are subgroup-specific. These activities are definitely different, as cyanide will stimulate the former and yet inhibit the latter (Takanaka & O'Brien 1975).

If myeloperoxidase is an NADPH oxidase, why is this activity defective in patients with CGD whose neutrophils contain myeloperoxidase (Segal & Peters 1978)? It is possible that the peroxidase is different in these patients. The dithionite difference spectra of the neutrophils from the patients we have studied all showed a high peak of absorption at 452 nm which is the classical peak of absorption of myeloperoxidase compound III (Odajima & Yamazaki 1972), a relatively inactive form of the molecule. It seems possible that because the cytochrome *b* is absent or defective in these patients the granule contents cannot be maintained in a reduced state, and thus the myeloperoxidase forms inactive compound III, and that the lack of 'NADPH oxidase' activity is a result rather than the cause of the basic defect in this disease. A similar mechanism could explain the apparently inappropriate association of glutathione-peroxidase deficiency with chronic granulomatous disease (Holmes *et al.* 1970).

CONCLUSIONS

The process of phagocytosis-induced oxygen consumption has been re-examined using IgG-coated latex particles as the phagocytic stimulus. After an activation period of 20 s, oxygen consumption is closely related to particle uptake, and in the individual vacuole it is brief, self-limited and amounts to about 0.2 fmol.

A previously undescribed cytochrome *b* in a unique situation has been discovered in human neutrophils. It was absent in one patient with CGD and abnormal in another. This cytochrome appears to be the first clearly identified component of the neutrophil–oxidase enzyme system.

The product of this oxidase system is uncertain. Myeloperoxidase could be the terminal electron-transporting or -receiving compound and H_2O_2 may only be produced when its function is absent, defective or inhibited. Contrary to popular belief, O_2^- does not seem to be an important final product of the oxidase system.

ACKNOWLEDGEMENTS

We thank Mr S. Coade and Miss T. Meshulam for technical assistance, Dr O. T. G. Jones and Professors J. B. Chappell and H. Baum for helpful advice, and Drs D. Webster and D. I. K. Evans and Professors J. Soothill and M. Greaves for allowing us to study patients under their care.

References

BABIOR, B. M. (1978) Oxygen-dependent microbial killing by phagocytes. *N. Engl. J. Med. 298*, 659–668

BABIOR, B. M., KIPNES, R. S. & CURNUTTE, J. T. (1973) The production by leukocytes of superoxide, a potential bactericidal agent. *J. Clin. Invest. 52*, 741–744

BAEHNER, R. L., GILMAN, N. & KARNOVSKY, M. L. (1970*a*) Respiration and glucose oxidation in human and guinea pig leukocytes. Comparative studies. *J. Clin. Invest. 49*, 692–700

BAEHNER, R. L., NATHAN, D. G. & KARNOVSKY, M. L. (1970*b*) Correction of metabolic deficiencies in the leukocytes of patients with chronic granulomatous disease. *J. Clin. Invest. 49*, 865–870

BALDRIDGE, C. W. & GERARD, R. W. (1933) The extra respiration of phagocytosis. *Am. J. Physiol. 103*, 235–236

BERENDES, H., BRIDGES, R. A. & GOOD, R. A. (1957) A fatal granulomatosis of childhood. The clinical study of a new syndrome. *Minnesota Med. 40*, 309–312

CURNUTTE, J. T., WHITTEN, D. M. & BABIOR, B. M. (1974) Defective superoxide production by granulocytes from patients with chronic granulomatous disease. *N. Engl. J. Med. 290*, 593–597

HOHN, D. C. & LEHRER, R. I. (1975) NADPH oxidase deficiency in X-linked chronic granulomatous disease. *J. Clin. Invest. 55*, 707–713

HOLMES, B., QUIE, P. G., WINDHORST, D. B. & GOOD, R. A. (1966) Fatal granulomatous disease of childhood. An inborn abnormality of phagocytic function. *Lancet 1*, 1225–1228

HOLMES, B., PAGE, A. R. & GOOD, R. A. (1967) Studies of the metabolic activity of leukocytes from patients with a genetic abnormality of phagocytic function. *J. Clin. Invest. 46*, 1422–1432

HOLMES, B., PARK, B. H., MALAWISTA, S. E., QUIE, P. G., NELSON, D. L. & GOOD, R. A. (1970) Chronic granulomatous disease in females. A deficiency of leukocyte glutathione peroxidase. *N. Engl. J. Med. 283*, 211–217

HOMAN-MÜLLER, J. W. T., WEENING, R. S. & ROOS, D. (1975) Production of hydrogen peroxide by phagocytizing human granulocytes. *J. Lab. Clin. Med. 85*, 198–207

IYER, G. Y. N., ISLAM, M. F. & QUASTEL, J. H. (1961) Biochemical aspects of phagocytosis. *Nature (Lond.) 192*, 535–541

KLEBANOFF, S. J. (1975) Antimicrobial mechanisms in neutrophilic polymorphonuclear leucocytes. *Semin. Hematol. 12*, 117–142

KLEBANOFF, S. J. & PINCUS, S. H. (1971) Hydrogen peroxide utilisation in myeloperoxidase-deficient leukocytes: a possible microbicidal control mechanism. *J. Clin. Invest. 50*, 2226–2229

LEMBERG, R. & BARRETT, J. (1973) *Cytochromes*, pp. 225–233, Academic Press, London & New York

MANDELL, G. L. (1974) Bactericidal activity of aerobic and anaerobic polymorphonuclear neutrophils. *Infect. Immun. 9*, 337–341

MANDELL, G. L. & HOOK, E. W. (1969) Leukocyte bactericidal activity in chronic granulomatous disease: correlation of bacterial hydrogen peroxide production and susceptibility to intracellular killing. *J. Bacteriol. 100*, 531–532

MITCHELL, P. (1962) Metabolism, transport, and morphogenesis: which drives which. *J. Gen. Microbiol. 29*, 25–37

ODAJIMA, T. & YAMAZAKI, I. (1972) Myeloperoxidase of the leukocyte of normal blood. IV. Some physicochemical properties. *Biochim. Biophys. Acta 284*, 360–367

PATRIARCA, P., CRAMER, R., MONCALVO, S., ROSSI, F. & ROMEO, D. (1971) Enzymic basis of metabolic stimulation in leukocytes during phagocytosis: the role of activated NADPH oxidase. *Arch. Biochem. Biophys. 145*, 255–262

PATRIARCA, P., CRAMER, R., TEDESCO, F. & KAKINUMA, K. (1975) Studies on the mechanism of metabolic stimulation in polymorphonuclear leukocytes during phagocytosis. II. Presence of the $NADPH_2$ oxidising activity in a myeloperoxidase-deficient subject. *Biochim. Biophys. Acta 385*, 387–393

PATRIARCA, P. L., CRAMER, R. & DRI, P. (1977) The present status of the subcellular localization of the NAD(P)H oxidase in polymorphonuclear leucocytes, in *Movement, Metabolism and Bactericidal Mechanisms of Phagocytes* (Rossi, F., Patriarca, P. L. & Romeo, D., eds.), Piccin Editore, Padua & London

SBARRA, A. J. & KARNOVSKY, M. L. (1959) The biochemical basis of phagocytosis. I. Metabolic changes during the ingestion of particles by polymorphonuclear leukocytes. *J. Biol. Chem. 234*, 1355–1362

SEGAL, A. W. & JONES, O. T. G. (1978) A novel cytochrome *b* system in phagocytic vacuoles from human granulocytes. *Nature (Lond.) 276*, 515–517

SEGAL, A. W. & PETERS, T. J. (1978) Analytical subcellular fractionation of neutrophils from patients with chronic granulomatous disease. *Q. J. Med. 47*, 213–220
SEGAL, A. W., CLARK, J. & ALLISON, A. C. (1978) Tracing the fate of oxygen consumed during phagocytosis by human neutrophils with $^{15}O_2$. *Clin. Sci. Mol. Med. 55*, 413–415
SINGER, J. M. & PLOTZ, C. M. (1956) The latex fixation test. I. Application to the serologic diagnosis of rheumatoid arthritis. *Am. J. Med. 21*, 888–892
TAKANAKA, K. & O'BRIEN, P. J. (1975) Mechanisms of hydrogen peroxide formation in leucocytes: the NAD(P)H oxidase activity of myeloperoxidase. *Biochem. Biophys. Res. Commun. 62*, 966–971

Discussion

Klebanoff: How pure were the neutrophil preparations?

Segal: Generally, about 97%.

Klebanoff: Were they Ficoll-Hypaque preparations?

Segal: Yes.

Klebanoff: What were the contaminating cells?

Segal: A few mononuclear cells, eosinophils and basophils.

Klebanoff: Your claim that isolated vacuoles do not contain peroxidase runs contrary to the cytochemical data.

Segal: The vacuoles are prepared after one minute of exposure to latex particles. They show no *spectroscopic* evidence of myeloperoxidase.

Reiter: So the lysosomes have not yet migrated?

Segal: Some have, because in some vacuoles one can find degranulated lysosomal contents. Biochemical assays indicate small amounts of myeloperoxidase. My point is that spectroscopy of the homogenate differs completely from that of the phagocytic vacuole preparation in which the cytochrome *b* peak is unaccompanied by a myeloperoxidase peak.

Fridovich: Addition of (ferri)cytochrome *c* to a superoxide-producing system should return the O_2^- to the gas phase or the oxygen 'pool' — i.e. it should inhibit or diminish the oxygen consumption (cf. Fig. 11). SOD should overcome that effect. Is that what you failed to observe?

Segal: Since the effect of adding cytochrome *c* to the incubation mixture is not known, we incubated the cells with cytochrome *c* and measured the effect of the addition of superoxide dismutase. The dismutase inhibited cytochrome *c* reduction but had no effect on oxygen consumption — thus the expected increase in oxygen consumption was not observed.

McCord: I cannot accept that there was no effect on oxygen consumption.

Michelson: The changes in oxygen consumption need not be great. Consider the reactions (1)–(3). Consequently, with SOD 1.5 molecules of oxygen will be liberated instead of two — i.e. an oxygen consumption about 25% lower than

$$2O_2^- + 2\text{cytochrome } c \text{ (oxidized)} \longrightarrow 2O_2 + 2\text{cytochrome } c \text{ (reduced)} \quad (1)$$

$$2O_2^- \xrightarrow[2H^+]{SOD} H_2O_2 + O_2 \quad (2)$$

$$H_2O_2 \longrightarrow \tfrac{1}{2}O_2 + H_2O \quad (3)$$

with cytochrome *c* reduction (not 25% of the total) and that is the maximum difference.

Segal: We do not observe that; the polarographic recordings of experiments with the same cell preparations measured in the presence and absence of SOD can be superimposed on each other.

McCord: Wouldn't the standard errors shown obscure such a difference?

Segal: No. We expressed the results in this way to illustrate the relative amount of oxygen consumption in comparison with cytochrome *c* reduction in different specimens. The errors refer to composites from different preparations. The variation between repeated studies on the same cell preparation is very small.

Roos: How do you know that the cytochrome *c* gets to the place where most of the superoxide is formed? In phagocytosing leucocytes, most of the oxygen is consumed after the particle has been taken up, in the phagosome. Not much space is left there for cytochrome *c*; once it is reduced, it stays reduced. The only effect you can observe is in the medium around the cells. Root & Metcalf (1977) found that granulocytes treated with cytochalasin B, which do not take up the particles but still show metabolic activation, show a small effect only, and they concluded that hydrogen peroxide was still formed with superoxide as an intermediate.

Segal: I understand your point about the uptake of particles removing the site of O_2^- generation from the cytochrome *c* in the suspending medium. The geography of the site of reduction is not important in relationship to our observations, because wherever the cytochrome *c* is being reduced, the reduction is inhibited by SOD and this is unaccompanied by an effect on O_2 consumption.

Roos: One parameter is missing in your data — the product, H_2O_2. Root & Metcalf (1977) were able to measure it in the presence of cytochrome *c*. You should compare your data with theirs.

McCord: The kinetics of superoxide production when the cells are stimulated are at variance with those reported by other groups. Most other investigators add 5 or 10% serum to the medium. Curnutte & Babior (1974) have shown that the serum is intimately involved, at least when bacteria are the stimulus. Complement factors are also engaged in the turning-on process. Might the very short burst you see be explained by a lack of proper stimulation and a consequently abortive turn on?

Segal: IgG is a much better opsonin than serum is for latex.

McCord: But it also fixes complement.

Allison: There was no complement in the system.

McCord: I know; I am suggesting that, if there were complement, there might be a more traditional response.

Segal: The traditional response results from inadequate opsonization of the particles. Consequently, both uptake and consumption of oxygen are slow and continuous. Jandl *et al.* (1978) discuss the termination of oxygen consumption and completely miss the point: there is a close relationship between particle uptake and oxygen consumption and without measuring particle uptake one doesn't know what one is dealing with. They claim that the process is switched off after several minutes, as a result of an inactivation during this time of an NADPH oxidase. I agree that oxygen consumption is terminated; but I believe that in any individual vacuole it is rapidly switched off, probably within seconds or fractions thereof, and that the reason that this process appears to continue for much longer is that phagocytosis continues for this time.

Roos: In our work with IgG-coated latex without complement we observed oxygen consumption for about 10 min. Uptake of the particle runs parallel — it takes about 10 min before the process comes to an end (Weening *et al.*, unpublished work, 1974). But uptake of particles doesn't have much to do with the activation of the superoxide-generating system because binding to the cell is sufficient to start this process (Goldstein *et al.* 1975).

Segal: Oxygen consumption does not stop after one minute; it continues but slows down. I was referring to the linear phase of oxygen consumption. On stimulation of neutrophils by latex particles (see Fig. 1) uptake proceeds with the same *rate* as oxygen consumption but without the 20 s lag. If there were a cumulative effect of uptake on oxygen consumption, the two responses would not show such parallelism, rather the particles should continue to consume oxygen and there should be an increase in the rate. Instead, the rate slows down; therefore, it must be related to uptake and it must be brief in each individual phagocytic vacuole.

Fridovich: Unless the entire membrane is activated at once.

Segal: But then the dose-response curve would not be the same. We have shown a direct relationship between particle uptake by cells and oxygen consumption. If the entire membrane were activated at once one would expect a sigmoidal relationship between particle uptake and oxygen consumption. There would be little response until the stimulus was great enough to evoke a response, then there would be a dramatic response as the entire membrane of the cell was activated, and then the cell would be refractory to further stimuli. This is unlike the linear relationship between particle uptake and oxygen consumption that we observe.

Goldstein: The conclusion from your data is heretical! This burst of oxygen consumption is now accepted as representing a final common pathway occurring after many different insults to the cell. But the results with phorbol myristate acetate imply that within the white blood cell there are different cytochrome systems which react in different ways to different stimuli.

Segal: You interpret the data correctly. Rather than a single enzyme we believe that an electron-transport system with several components is responsible, of which myeloperoxidase may be the last one. Stimulation with phorbol may only activate the early part of the chain as it is a soluble stimulus and does not appear to result in phagosome formation and degranulation in the same way as a particulate stimulus.

Crichton: What is the direct evidence for the involvement of a *b*-type cytochrome in the process of oxygen consumption?

Segal: The difference spectra we have described are typical of a cytochrome *b*. Its absence from (or abnormality in) patients with chronic granulomatous disease is extremely good evidence because everything else appears to be normal in that disease.

Crichton: Is that such good evidence? Is this cytochrome missing from every other plasma membrane of every other cell type? That raises another question: how pure is the plasma membrane preparation?

Segal: This cytochrome system appears to be unique; it is not associated with any of the normal accompaniments of the two major cytochrome systems. It is not associated with mitochondrial cytochrome oxidase, nor does it appear to be associated with cytochrome P450 of the endoplasmic reticulum. It is very difficult to purify plasma membranes from human neutrophils because of the lack of a specific marker for this organelle. Phagocytic vacuole preparations include plasma membrane which invaginates to form the wall of the vacuole. It is difficult to separate phagosomes from granule contents as the granules fuse with and discharge their contents into the phagosome. The point we wanted to establish was that the cytochrome *b* becomes associated with the phagosome, where the oxidative metabolism is probably located, and that in this situation it is free of measurable contamination by cytochrome oxidase and P450.

Crichton: Babior (1977) implicates a flavoprotein.

Segal: Cytochromes are often reduced by flavoprotein enzymes.

Dormandy: What happens during the 20 s delay?

Segal: I don't know; during this time the phagocytic vacuole is being formed, and the lag period may be related to closure of the vacuole, passage of the vacuole into the cytosol, or degranulation of one of the cytosolic granules into the vacuole. We are investigating these possibilities.

Hill: Have you used anything other than cytochrome *c* to react with the supposed superoxide anion?

Segal: No. Unfortunately, NBT is toxic to the cells.

Hill: As you said, the action of cytochrome *c* on the cell is unknown; it might make the cell leakier or divert the superoxide into an abnormal route.

Segal: This is just my point: superoxide dismutase-inhibitable cytochrome *c* reduction is the major evidence that has been used to establish superoxide as the product of the oxidase process, and this evidence is shaky.

Hill: Given time one might be able to reconcile the oxygen consumption data but it would be comforting to know the results with reagents in addition to cytochrome *c*.

Fridovich: What might the reductant of cytochrome *c* be?

Segal: I don't know; the reduction seems to be initiated by the presence of cytochrome *c*. The cytochrome *c* appears to be reduced by O_2^- because it is inhibited by SOD, but not by heat-inactivated SOD, and does not occur in anaerobic conditions. However, O_2^- does not appear to be released from the cell in the absence of cytochrome *c* which seems to initiate O_2^- production!

Fridovich: But if cytochrome *c* is being directly reduced at some point why does the reduction need oxygen?

Segal: Cytochrome *c* seems to be reduced by O_2^-. The oxidative process probably involves a complex electron-transport chain, and oxygen is probably reduced to O_2^- somewhere along this chain. The O_2^- may simply be a transiently reduced intermediate in this chain. Under these circumstances, the addition of cytochrome *c* may provide an alternative substrate for reduction. This does not explain why oxygen consumption is not increased in the presence of SOD which inhibits cytochrome *c* reduction. This suggests that cytochrome *c* initiates an autocatalytic reduction involving O_2^- as an intermediate or that SOD blocks cytochrome *c* by some mechanism other than dismutation. [*See also Discussion after Dr Roos's paper, pp. 254–262.*]

References

BABIOR, B. M. (1977) Recent studies on oxygen metabolism in human neutrophils: superoxide and chemiluminescence, in *Superoxide and Superoxide Dismutases* (Michelson, A. M., McCord, J. M. & Fridovich, I., eds.), pp. 271–281, Academic Press, London

CURNUTTE, J. T. & BABIOR, B. M. (1974) Biological defense mechanisms — the effect of bacteria and serum on superoxide production by granulocytes. *J. Clin. Invest. 53*, 1662–1672

GOLDSTEIN, I. M., ROOS, D., KAPLAN, H. B. & WEISSMANN, G. (1975) Complement and immunoglobulins stimulate superoxide production by human leukocytes independently of phagocytosis. *J. Clin. Invest. 56*, 1155–1163

JANDL, R. C., ANDRÉ-SCHWARTZ, J., BORGES-DUBOIS, L., KIPNES, R. S., MCMURRICH, B. J. & BABIOR, B. M. (1978) Termination of the respiratory burst in human neutrophils. *J. Clin. Invest. 62*, 1176–1185

ROOT, R. K. & METCALF, J. A. (1977) H_2O_2 release from human granulocytes during phagocytosis. Relationships to superoxide anion formation and cellular catabolism of H_2O_2: studies with normal and cytochalasin B-treated cells. *J. Clin. Invest. 60*, 1266–1279

Defects in the oxidative killing of micro-organisms by phagocytic leukocytes

DIRK ROOS* and RON S. WEENING*†

* *Central Laboratory of the Netherlands Red Cross Blood Transfusion Service and Laboratory of Experimental and Clinical Immunology of the University of Amsterdam, and † Pediatric Clinic, Binnengasthuis, University of Amsterdam*

Abstract One of the most important mechanisms of phagocytic killing of ingested microorganisms by leukocytes is the generation of toxic oxygen products. During phagocytosis, neutrophils, as well as monocytes and macrophages, display a strongly increased cell respiration. Quantitatively the most important product of this reaction is hydrogen peroxide. Superoxide is also generated in large amounts, probably as an intermediate in the formation of hydrogen peroxide. Indications exist that singlet oxygen and hydroxyl radicals are also formed in this process. Some of these oxygen products have microbicidal properties by themselves. The effect of hydrogen peroxide is greatly enhanced by the enzyme myeloperoxidase.

Several dysfunctions of this system are known. In chronic granulomatous disease the enzyme system that produces superoxide is not operative. Thus, no superoxide or hydrogen peroxide is generated, leading to a severely decreased bacterial killing capacity. The exact molecular defects in the X-linked and the autosomal form are as yet undefined. Two variants are also known: lipochrome histiocytosis, with different clinical and histological manifestations, and a 'triggering defect' where only strongly opsonized particles trigger the respiratory burst. Myeloperoxidase deficiency leads to slightly decreased killing capacity, especially for yeasts. In glucose-6-phosphate dehydrogenase deficiency no oxygen radicals or hydrogen peroxide are produced because no equivalents for oxygen reduction can be generated in the hexose-monophosphate shunt. Deficiencies in the glutathione redox system also result in impaired phagocyte function, probably because the cells have to be protected against their own toxic oxygen products.

The host-defence against invading microorganisms is executed by a series of humoral and cellular mechanisms. First, the production of lactic and fatty acids by the skin and the mucous secretions of the mucous tissues offer effective barriers against these pathogens. In the secretory fluids, lysozyme, phospholipases and immunoglobulins of the IgA class constitute another line of defence. Next, in the tissues, the lymph fluid and the blood, specific antibodies of the IgM and the IgG class, and components of the complement system form an efficient humoral microbicidal system. Moreover, in the same location, antibody-secreting and cytotoxic lymphocytes comprise the cellular immunological defence system. And, finally, the phagocytic cells of the blood and tissues form a last, but certainly not least, line of resistance.

The phagocytic cells are extremely well equipped for killing and removing microorganisms. The polymorphonuclear leukocytes (neutrophilic granulocytes) and the monocytes in the blood are able to leave the circulation and sense their way to a site of infection, where they can accumulate in large numbers. The macrophages are located exclusively in the tissues, either fixed between other tissue cells or suspended in tissue fluids. Each of these cell types can ingest the foreign material, generate microbicidal products and release pre-synthesized hydrolytic enzymes (and other antimicrobial proteins) into the vicinity of the ingested pathogens. Finally, some of these cells, especially the macrophages, can digest the ingested material and release or reutilize the small molecular products.

From the time of Metchnikoff, at the end of the last century, people have wondered which molecular mechanisms are responsible for the killing of microorganisms by phagocytic cells. The past 10–15 years have witnessed a large increase in our knowledge about these processes, and it appears that many different reactions are involved. As is so often the case, much has been learned about the normal process from the study of hereditary defects in this function. The study of phagocytic leukocytes from patients with chronic granulomatous disease, in particular, has proven the important role of oxygen products in the antimicrobial defence. In this paper we shall focus on the generation, the use and the protection against these toxic oxygen products and some well characterized disturbances in these reactions.

THE RESPIRATORY BURST

During ingestion of particulate material, phagocytic leukocytes display a sudden increase in the oxidative metabolism which is called the respiratory burst. First, the consumption of oxygen by neutrophils is enhanced 10–15-fold within a few seconds after contact with the stimulating substance (Baldridge & Gerard 1933; Sbarra & Karnovsky 1959; Weening *et al.* 1974). As this increase in oxygen consumption is not inhibited by cyanide (Sbarra & Karnovsky 1959), the oxygen is not used in the mitochondrial process of oxidative phosphorylation. This reaction and all other reactions of the respiratory burst have been demonstrated not only in neutrophils, but also in monocytes and macrophages (for a review see Roos & Balm 1979).

The first product of the increased oxidative metabolism to be recognized was hydrogen peroxide. H_2O_2 is produced in large amounts and can be measured in the medium surrounding the phagocytes (Iyer *et al.* 1961; Paul & Sbarra 1968; Root *et al.* 1975), especially when its enzymic degradation is inhibited (Homan-Müller *et al.* 1975). With cytochemical methods, the presence of H_2O_2 has also been demonstrated in the phagosomes, in close proximity to the ingested microorganisms (Briggs *et al.* 1975).

Phagocytic leukocytes also generate large amounts of superoxide anion radicals during particle ingestion (Babior *et al.* 1973; Weening *et al.* 1975; Johnston *et al.* 1975). This product, too, is released in large amounts into the cell medium, where it can be detected by the reduction of cytochrome *c* or nitroblue tetrazolium. The superoxide is initially generated at the point of contact between a particle and the phagocytic cell membrane; subsequently, nitroblue tetrazolium dye reduction can be detected around the phagocytic vacuole after the particle has been ingested (Nathan *et al.* 1969). Most likely, superoxide is an intermediate in the formation of hydrogen peroxide, and probably all hydrogen peroxide produced by neutrophils is formed in this way (Root & Metcalf 1977).

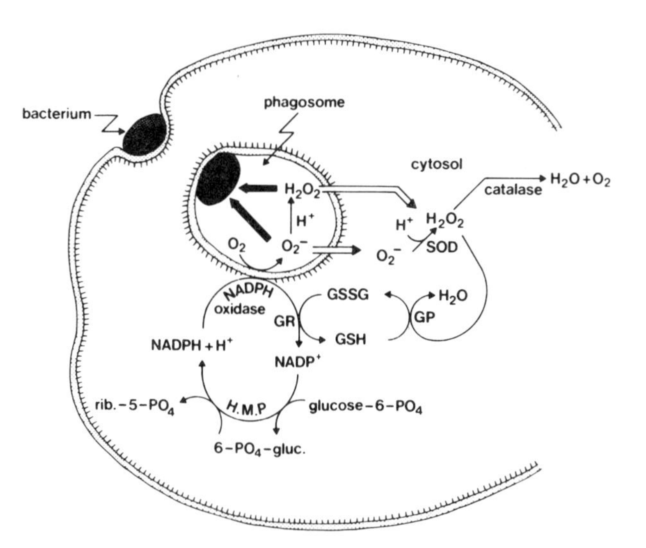

FIG. 1. Formation of bactericidal oxygen products in the phagosomes of phagocytic leukocytes and protection of the cytosol against oxidative injury. After attachment to the cells, bacteria are encircled by part of the plasma membrane and taken up into the cells. The inside of the phagosomal membrane consists of the original outside of the plasma membrane. The NADPH oxidase in this membrane generates superoxide into the phagosomes. The cytosol of the leukocytes is protected by superoxide dismutase, catalase and the glutathione redox system. →, reaction; ➡, attack; ⇨, diffusion; GR, glutathione reductase; GP, glutathione peroxidase; SOD, superoxide dismutase; HMP, hexose-monophosphate shunt (From Roos *et al.* 1979*b* with permission).

The reducing equivalents for the production of superoxide from oxygen are derived from glucose (see Fig. 1). During the respiratory burst, the percentage of glucose metabolized by the hexose-monophosphate shunt is increased several-fold (Sbarra & Karnovsky 1959; Stjernholm & Manak 1970). In this way, an increased amount of NADPH is generated, which can reduce oxygen, either directly by an NADPH oxidase or indirectly by reducing substrates for other oxidases (e.g. NADH oxidase). This reaction is described in more detail later.

Indications exist that two more reactive oxygen species may be formed by phagocytic leukocytes: singlet oxygen and hydroxyl radicals. The chemi-luminescence of phagocytosing neutrophils has been interpreted as resulting from the return of singlet oxygen to the triplet ground state (Allen *et al.* 1972). In principle, singlet oxygen may be generated in phagocytic leukocytes in the following reactions:

(1) the reaction of the myeloperoxidase-generated hypochlorite ion with hydrogen peroxide (Rosen & Klebanoff 1976):

$$H_2O_2 + Cl^- \xrightarrow{\text{myeloperoxidase}} H_2O + OCl^-$$

$$OCl^- + H_2O_2 \longrightarrow H_2O + {}^1O_2 + Cl^-$$

(2) the spontaneous dismutation of superoxide anions (Allen *et al.* 1972):

$$2O_2^- + 2H^+ \rightarrow H_2O_2 + {}^1O_2$$

(3) the Haber–Weiss reaction between superoxide and hydrogen peroxide (Kellogg & Fridovich 1975):

$$O_2^- + H_2O_2 \rightarrow {}^1O_2 + OH^{\bullet} + OH^-$$

As a consequence, the light emission is inhibited by catalase as well as by superoxide dismutase (Webb *et al.* 1974), and is also subnormal in cells lacking myeloperoxidase activity (Rosen & Klebanoff 1976). The chemiluminescence of phagocytosing leukocytes is by no means absolute proof for the existence of ${}^1O_2$ in these cells, however, since this process is not specific for the relaxation of singlet oxygen to triplet oxygen. The light emission depends to a large extent on the oxidation of constituents from the ingested particles (Cheson *et al.* 1976; Nelson *et al.* 1977). Moreover, when neutrophils are stimulated to increased superoxide production by fluoride, they do not chemiluminesce. Thus, the chemiluminescence in phagocytosing leukocytes is probably derived for a large part from the interaction of O_2^- or H_2O_2 with particle constituents. The results of experiments with scavengers of ${}^1O_2$ (Johnston *et al.* 1975; Rosen & Klebanoff

1976) are hard to interpret, owing to the lack of specificity of such agents (Held & Hurst 1978).

Hydroxyl radicals are also formed in the Haber–Weiss reaction, but (as mentioned above) the occurrence of this reaction in phagocytic leukocytes is doubtful. During phagocytosis, these cells liberate ethylene from methional (3-methylthiopropionaldehyde) (Tauber & Babior 1977; Weiss *et al.* 1977) and from 2-oxo-4-thiomethylbutyric acid (Weiss *et al.* 1978) but again, the specificity of these reactions for the presence of $OH^{\bullet}$ radicals is questionable (Bors *et al.* 1976). The fact that they are inhibited by catalase and by superoxide dismutase (Weiss *et al.* 1978) is compatible with the formation of $OH^{\bullet}$ in the Haber–Weiss reaction. After ingestion of bacteria or paraffin-oil droplets by neutrophils, unsaturated fatty acids are degraded (Shohet *et al.* 1974; Stossel *et al.* 1974), a process initiated by hydroxyl radicals (Fong *et al.* 1973). This process as such and the resulting formation of the potentially bactericidal malonaldehyde may contribute to the bacterial death.

None of these reactions can be observed in the phagocytosing leukocytes from patients with chronic granulomatous disease (CGD), a syndrome characterized by frequent recurrent infections of life-threatening severity (Holmes *et al.* 1967). The only abnormality found is in the leukocytes of these patients: a strongly depressed microbicidal activity correlated with a lack of oxidative stimulation (see below). Therefore, the oxidative reactions are regarded as extremely important for the host-defence of these cells against certain microorganisms.

The exact mechanism(s) of the process that kills microorganisms is not known. Probably, the leukocytes contain a whole array of microbicidal systems to cope with as many different microorganisms as possible. One of the systems that has been studied extensively is the peroxidase-mediated reaction (see Klebanoff & Rosen, this volume). In short, this system uses hydrogen peroxide and a halide and is catalysed by (myelo)peroxidase. During phagocytosis, these components are brought together in the phagosome, in direct contact with the ingested microorganisms. As a result, the halide is fixed to the microorganisms, and, with chloride, the amino acids of the microbial membrane proteins undergo oxidative decarboxylation. Moreover, toxic aldehydes and chloroamines are formed in this reaction, and, as mentioned above, possibly also singlet oxygen. Any one, or a combination of several, of these reactions may cause the death of the microorganisms. This system is particularly important in the killing of certain yeast species (Lehrer 1975), but probably less so in the reaction against other microorganisms. The effect of myeloperoxidase deficiency on the metabolism and function of phagocytic leukocytes is also described by Klebanoff & Rosen (pp. 264–284).

Other microbicidal agents in phagocytic leukocytes include hydrogen peroxide — alone or in combination with certain metal ions and ascorbate (Drath & Karnovsky 1974) or with lysozyme (Miller 1969). These reactions probably proceed through interactions of free radicals with the microbial cell wall, since they can be inhibited by free-radical scavengers. Evidence for direct participation of singlet oxygen, superoxide or hydroxyl radicals in the microbicidal reactions is scarce, however. The inhibition of the bactericidal activity of neutrophils by superoxide dismutase as well as by catalase may indicate that O_2^- and H_2O_2 react with each other to form hydroxyl radicals and singlet oxygen as the lethal compounds (Johnston *et al.* 1975). Also, the fact that a carotenoid-missing mutant of *Sarcina lutea* is killed by neutrophils much more rapidly than the wild-type bacteria is consistent with the involvement of singlet oxygen in this process (Krinsky 1974). Superoxide radicals are not microbicidal in themselves, probably because nearly every organism that can live in air contains superoxide dismutase (Fridovich 1974).

Finally, several oxygen-independent microbicidal or microbistatic systems are known in phagocytic leukocytes, including the drop in the intraphagosomal pH (Jensen & Bainton 1973) and several digestive (lysozyme, phospholipase), iron-binding (lactoferrin), cationic and fungicidal proteins (see Elsbach 1973; Klebanoff 1975).

THE SUPEROXIDE-GENERATING SYSTEM

The identification of the system in phagocytic leukocytes responsible for the one-electron reduction of oxygen to superoxide has not yet been completed. Because the respiratory burst is accompanied by a concomitant increase in hexose-monophosphate shunt activity, this system almost certainly uses NADPH or NADH as a substrate. In the case of NADH, a reaction must exist either to transfer electrons from NADPH to NAD or to save NADH from re-use in the lactate dehydrogenase reaction. An NAD–NADP transhydrogenase could catalyse the direct transfer of electrons between the two pyridine nucleotides, but neutrophils appear to lack this enzyme (Evans & Karnovsky 1961). Moreover, in cells with 5% of normal glucose-6-phosphate dehydrogenase activity, a strongly decreased ratio of $NADPH/NADP^+$ has been found, together with a practically normal $NADH/NAD^+$ ratio (Baehner *et al.* 1972). Thus, little exchange of reducing equivalents takes place between the different nucleotide pools.

Another suggestion has been the involvement of an NADPH-utilizing lactate dehydrogenase (Evans & Karnovsky 1961). According to this idea, glucose degradation to pyruvate reduces NAD^+ to NADH, which serves for the reduction of oxygen. The NADPH produced in the hexose-monophosphate shunt is used to convert pyruvate into lactate. However, the amount of lactate produced by phagocytosing neutrophils is insufficient to account for the amount of $NADP^+$ required for the shunt activity (Rossi & Zatti 1966; Rossi *et al.* 1972). Moreover, an NADPH-utilizing lactate dehydrogenase has not been detected in human (Baehner *et al.* 1970) or rat neutrophils (Reed & Tepperman 1969; Reed 1969). Thus, in case superoxide is generated by an NADH oxidase, the substrate availability and/or the HMP shunt activity remain to be explained.

At least two different NADH oxidases have been reported to exist in neutrophils. One is a flavoprotein, absolutely specific for NADH, with a pH optimum of 4.8 and not activated by phagocytosis (Evans & Karnovsky 1961; Cagan & Karnovsky 1964). Its supposed localization in the cytosol raises a problem as to its function (any O_2^- produced by such an enzyme would be catabolized immediately) but may be an artefact due to its easy elution from the plasma membrane (Segal 1978). Neutrophils from patients with CGD have been reported to contain about 25% of the normal activity of this enzyme (Baehner & Karnovsky 1968), but other investigators have been unable to confirm these results (Holmes *et al.* 1967, 1968; Hohn & Lehrer 1975; DeChatelet *et al.* 1975; Babior *et al.* 1975; Curnutte *et al.* 1975; Biggar *et al.* 1976). The explanation might be, according to Segal (1979), that two different NADH oxidases are located in the plasma membrane: one specific for neutrophils and inactive in CGD, and another present in the plasma membranes of all blood leukocytes, still active in CGD, and not related to the respiratory burst.

Another, or possibly the same, NADH oxidase was demonstrated cytochemically by Briggs *et al.* (1975) as an ecto-enzyme in the plasma membrane, specific for NADH, activated by phagocytosis and partially missing in CGD (Briggs *et al.* 1977). Whether this enzyme is identical with the non-specific enzyme mentioned by Segal (1979) remains to be determined.

Finally, Segal & Peters (1976, 1977) have also reported the existence of an NADH oxidase in the plasma membranes of neutrophils with a much higher affinity for NADH than for NADPH and with very low activity in CGD patients. The same fraction that contained the NADH oxidase activity appeared to produce superoxide radicals. No measurement was made on preparations from phagocytosing cells and neither pH optimum nor cyanide sensitivity was determined. It is unknown whether this enzyme is identical with the enzyme described by Evans & Karnovsky (1961) and Cagan & Karnovsky (1964).

With regard to NADPH oxidase there is also substantial evidence for the existence of such activity in neutrophils. This enzyme was first described by Iyer *et al.* (1961) in whole homogenates of neutrophils and subsequently shown to be stimulated by flavin compounds (Iyer & Quastel 1963). The activity of this preparation was substantially increased by manganese ions. This latter property, however, has recently been shown by Curnutte *et al.* (1976) to be due to the catalysis of a free-radical chain reaction, resulting in a non-enzymic amplification of the NADPH oxidation. Rossi and his co-workers (1964, 1972) demonstrated the insensitivity of this enzyme to cyanide, the preferential use of NADPH as a substrate, a pH optimum of 5.5 and the activation by phagocytosis. A strongly decreased NADPH oxidase activity in CGD patients was described practically simultaneously by Hohn & Lehrer (1975), DeChatelet *et al.* (1975) and Curnutte *et al.* (1975).

Since then, Babior and his colleagues have characterized this enzyme (in the absence of Mn^{2+}) in more detail and shown that superoxide is produced by the same particulate fraction of neutrophils that also displays the NADPH oxidase activity, that similar fractions from resting cells are inactive, and that the use of NADPH as a substrate yields six times more O_2^- than NADH does (Curnutte *et al.* 1975). The pH optimum of the O_2^- production is 5.5–6.0. At this pH the system has an optimal affinity for NADPH whereas the K_m for NADH remains much higher at any pH tested (Babior *et al.* 1976). The superoxide-generating activity is insensitive to cyanide (Baehner 1975; Babior *et al.* 1976) and accounts for about 35% of the oxygen consumed in the respiratory burst (Babior *et al.* 1976). Finally, the superoxide-generating system requires flavin for optimal activity (Babior & Kipnes 1977).

Thus, two enzyme activities are candidates for the superoxide-generating system in phagocytic leukocytes: NADH oxidase and NADPH oxidase, both of which have been shown to be missing in CGD. Several observations seem to indicate that the NADPH oxidase is the physiologically-important enzyme. First, as mentioned at the beginning of this section, additional enzyme activities are required (but have not been found) for NADH to act as the substrate. Second, when neutrophils are stimulated, the NADPH/$NADP^+$ ratio decreases, whereas the NADH/NAD^+ ratio remains unchanged (Selvaraj & Sbarra 1967). And third, several patients are known (see p. 242) in whom a severe deficiency of glucose-6-phosphate dehydrogenase is correlated with a defect in the respiratory burst of the neutrophils. Since a deficiency of glucose-6-phosphate dehydrogenase directly affects the reduction of NADP but not of NAD, this observation is only understandable if the primary oxygen-consuming, superoxide-generating enzyme were an NADPH oxidase.

How are the findings of Segal & Peters (1976, 1977) to be explained in this connection? Their observation that, at low (25 μmol/l) concentrations of pyridine nucleotides (claimed to be more physiological than the 2.5 mmol/l used by others), a much higher affinity for NADH is found than for NADPH is very puzzling when compared to the opposite results of Babior *et al.* (1976).

It must be remembered, however, that Segal & Peters used preparations from *resting* neutrophils and not, as Babior *et al.* did, from *phagocytosing* cells. Several authors have shown, however, that preparations from phagocytosing cells are much more active than those from resting cells (Rossi & Zatti 1964; Hohn & Lehrer 1975; DeChatelet *et al.* 1975; Curnutte *et al.* 1975). Since neutrophils from CGD patients differ from normal cells only in the stimulated state, any relevant enzymic defect should be demonstrated in activated cells. Moreover, the physiological concentrations of NADH and NADPH in intact neutrophils, at the site of the oxidase, are not known.

Better agreement exists about the localization of the oxidase system. On theoretical grounds, a plasma membrane localization had been predicted for some years already. This idea is based on the observation that up to 90% of the consumed oxygen can be recovered as O_2^- or H_2O_2 from the medium outside the cells, despite the presence of catabolizing systems for these products in the cells (Roos *et al.* 1977; Root & Metcalf 1977). Direct cytochemical evidence exists for the plasma and phagosomal membrane as the site of production of O_2^- (Nathan *et al.* 1969) and of H_2O_2 (Briggs *et al.* 1975). In addition, the oxygen consumption and superoxide generation by intact neutrophils is strongly decreased by inhibitors which do not penetrate the plasma membrane (Takanaka & O'Brien 1975; Goldstein *et al.* 1977). From these observations a model has been designed of an enzyme system located in the plasma membrane which is accessible to NAD(P)H from the cytosolic side and to oxygen from the outside. This system releases superoxide on the outside of the plasma membrane and is activated when particles attach to the cells (Fig. 1). Since the phagosomes consist of plasma membranes with the outside in, the superoxide will be released into the phagosomes in close contact with ingested microorganisms.

Recently, some direct biochemical evidence for this model has been published. Baehner (1975) found NBT-reducing activity with either NADH or NADPH as substrate in a subcellular fraction which contained microsomes and plasma membranes, but not lysosomes. Segal & Peters (1977) described a similar distribution for NADH oxidase and superoxide-generating activity. Dewald *et al.* (1979) also obtained evidence for the plasma membrane localization of the NADPH-dependent superoxide-forming system. This immediately raises two questions. Could it be that there is only one pyridine nucleotide-utilizing enzyme in the plasma membranes of neutrophils, which can use both NADPH and NADH? Iverson *et al.* (1977) have described a preparation which uses both NADH and NADPH; unfortunately, direct competition between the two substrates was not demonstrated. This activity, however, was localized in a fraction even heavier than the lysosomes and, therefore, was certainly not in the plasma membrane. Probably, this enzyme activity is identical with that observed by Segal & Peters (1977) in the azurophil granules, which uses NADPH but does not generate superoxide.

The second question is: can the NAD(P)H oxidase activity be localized in inactive form in the lysosomes and, after cell stimulation and lysosome-phagosome fusion, can the oxidase be incorporated and activated in the phagosomal membrane? Some (circumstantial) evidence pleads against this concept. First, Goldstein *et al.* (1975) have shown that, in certain conditions, neutrophils can be stimulated to generate superoxide without the release of lysosomal enzymes, indicating that oxidative stimulation can take place without plasma membrane–lysosome fusion. Second, Segal & Peters (1977) have worked with preparations from resting cells and found the oxidase activity in the plasma membrane, but, as indicated earlier, the physiological significance of their observations remains to be proven. Nevertheless, the original idea of Roberts & Quastel (1964) that the NAD(P)H oxidase activity might be identical with myeloperoxidase, confirmed by Patriarca *et al.* (1973), seems not tenable. Therefore, as judged from the present knowledge, the primary site of superoxide generation most probably is an NADPH oxidase located in the plasma membrane of the phagocytic leukocyte. For clarification of several controversies, it might be advisable for certain investigators in this area to come together and do some joint experiments.

CHRONIC GRANULOMATOUS DISEASE

Chronic granulomatous disease (CGD) is characterized by the manifestation of recurrent purulent infections of the skin, reticuloendothelial organs and lungs (Johnston & Newman 1977). Lymphadenopathy is frequently present; hepatomegaly and splenomegaly are common. Pneumonitis, subcutaneous abscesses and furunculosis are found sooner or later in almost all patients. The inability of the patients' phagocytic leukocytes to kill invading microorganisms leads to chronic infections, sometimes hemmed in by granuloma formation.

The microorganisms from which these patients suffer are mostly those that contain catalase and, therefore, do not release hydrogen peroxide. Pneumococcus and various strains of streptococci are killed normally by CGD leukocytes (Mandell & Hook 1969), probably because these microorganisms release H_2O_2 and in this way aid their own destruction.

As indicated in the previous section the molecular defect in this syndrome seems to be an absent or non-functional oxidase in the plasma membrane of the phagocytic leukocytes. Although elucidation of the precise defect will have to wait for purification of this enzyme, it is almost certain that several different deficiencies will be found. Genetically, for instance, two forms can be distinguished with a different mode of inheritance. The predominant form of CGD is X-linked: within most CGD families only boys are affected and a heterozygous state can be detected in the mothers, some sisters and some female maternal relatives. Mozaicism is found in the phagocytes of these individuals (Windhorst *et al.* 1968). Families with female patients are also known, though, as well as a few cases of affected brother–sister pairs and male CGD patients without detectable heterozygotes (Johnston & Newman 1977). Thus, a recessive autosomal form may also exist, but it cannot be excluded that the female patients are heterozygotes for the X-linked defect with an extremely low percentage of normal cells. The fact that no heterozygote has been detected in the families with female CGD patients supports the existence of a separate entity. On a molecular level, no difference has been found: neutrophils from male as well as from female CGD patients lack NADPH oxidase activity (McPhail *et al.* 1977).

In the past five years, we have diagnosed 15 CGD patients: 10 males and five females. Such a diagnosis is based on the clinical symptoms and on the inability of the patients' neutrophils to respond with a respiratory burst to challenge with phagocytosable particles, e.g. latex (in the presence of fresh human serum) or serum-opsonized zymosan particles (yeast membranes). Table 1 shows the

TABLE 1

Oxidative metabolism of neutrophils from CGD patients during phagocytosis of latex

Patient	O_2 *consumption*[a]	H_2O_2 *production*[a]
Male CGD		
K.B.	1.3	0
D.D.	0.1	0
J.G.	0.7	
K.H.	1.5	0.5
R.L.	2.2	
G.Q.	0.7	0
H.V.	0.7	3.0
M.Va.	2.0	0
G.Ve.	0.4	0
M.Z.	0.8	0
Female CGD		
M.D.	0	0
M.J.	0.3	2.8
G.M.	1.1	
M.K.	0.8	0
J.P.	0.3	0
10 normals (mean ± S.D.)	23 ± 4.3	16 ± 2.8

[a] Values are given in nmol $(10^7 \text{ cells})^{-1} \text{ min}^{-1}$.

TABLE 2

Oxidative metabolism of neutrophils from family Q. during latex-induced phagocytosis

Member of family	*O_2 consumption*[a]	*O_2^- production*[a]	*H_2O_2 production*[a]	*[$1-^{14}C$]-Glucose oxidation*[a]	*Iodination*[a]	*NBT test*[b]
Patient Q.	0.7	0.5	0	0.1	0	0.04
Mother	8.7	3.2	4.2	1.9	0.74	0.08
Father	22.7	8.6	13.8	2.3	0.75	0.12
Sister	6.8	1.7	3.7	0.7	0.37	0.06
Brother	23.8	6.9	15.3	3.6	0.82	0.10
10 Normals (mean ± S.D.)	23 ± 4.3	7 ± 1.0	16 ± 2.8	3 ± 0.7	0.6 ± 0.1	0.2 ± 0.04

[a] Values are given in nmol (10^7 cells)$^{-1}$ min^{-1}.
[b] Values are given as the change in light absorbance at 560 nm (2 × 10^7 cells)$^{-1}$ (15 min)$^{-1}$.

results of the oxygen consumption and H_2O_2 generation by the patients' neutrophils and by normal cells; the CGD cells are grossly defective in this respect. Moreover, the killing *in vitro* of *Staphylococcus aureus* was also well below normal. When tested, similar defects were found in the patients' monocytes. To exclude specific enzyme defects in other metabolic pathways as the cause of this defect (see pp. 242–248), we have also measured the activity of the following enzymes in the neutrophils: glucose-6-phosphate dehydrogenase, 6-phosphogluconate dehydrogenase, glutathione peroxidase and glutathione reductase. We have found no deficiency in any of these enzymes. Therefore, we conclude that these patients suffer from CGD.

Heterozygotes were also looked for. This can be done with a variety of techniques, the most widely used being the bacterial killing test *in vitro* and the nitroblue tetrazolium (NBT) test. In the latter assay, the cells are incubated with NBT dye and the degree of formazan production is measured either qualitatively as blue-stained cells under the light microscope (Nathan *et al.* 1969) or quantitatively as the increase in light absorption at 560 nm after pyridine extraction (Baehner & Nathan 1968). NBT is reduced to formazan by superoxide (Baehner *et al.* 1976). As Table 2 shows, other sensitive parameters of the oxidative metabolism can also be used for heterozygote detection. In the family shown, only the iodination technique (fixation of radioactive iodide to trichloroacetic acid-precipitable material by H_2O_2 and myeloperoxidase; Pincus & Klebanoff 1971) failed to indicate the mother as a heterozygote, although her cells produced only about one fourth of the normal amount of H_2O_2. Probably, hydrogen peroxide is normally generated in large excess over what is needed for bacterial killing.

In the families of the female CGD patients we detected no heterozygotes. We did find heterozygotes in five out of eight families from male CGD patients, but, as Table 3 shows, even then obligate heterozygotes were not always detectable (e.g. the maternal grandmother in family Z). This may be due to preferential inactivation of the defective X-chromosome in the somatic cells of this carrier early in embryonic life or to a spontaneous mutation in the mother of the patient.

Clinically, the five female CGD patients showed a more benign form of the disease: the symptoms manifested themselves at a later age and were not as life-threatening as those seen in the 10 male CGD patients. In contrast, Johnston & Newman (1977) conclude from a literature study of 140 published cases that no difference in the age of onset of the disease between male and female patients is apparent.

Our results confirm the idea that the syndrome of CGD consists of at least two different entities.

TABLE 3

Oxidative metabolism of neutrophils from family members of male CGD patients (see Table 1)

Member of family	*O_2 consumption*[a]	*H_2O_2 consumption*[a]
Family D.		
Mother	15.3	8.2
Father	19.7	11.7
Sister	12.5	7.2
Family H.		
Mother	15.7	7.1
Father	20.4	12.8
Sister	24.8	9.2
Family Q.		
Mother	8.7	4.2
Father	22.7	13.8
Sister	6.8	3.7
Brother	23.8	15.3
Family Ve.		
Mother		3.8
Father		11.9
Sister 1		13.2
Sister 2		13.9
Brother 1		12.6
Brother 2	17.0	
Family Z.		
Mother	8.5	9.3
Sister	15.0	13.3
Maternal grandmother	26.5	19.2
Maternal aunt 1		13.7
Maternal aunt 2		16.0
Maternal aunt 3		20.9
10 Normals (mean ± S.D.)	23 ± 4.3	16 ± 2.8

No heterozygous values were detected in families B., G. and V.
[a] Values are given in nmol $(10^7 \text{ cells})^{-1} \text{ min}^{-1}$.

VARIANTS OF CHRONIC GRANULOMATOUS DISEASE

Two variants of CGD are known: lipochrome histiocytosis, which can be distinguished from CGD by some clinical features, and a 'triggering defect', which can be distinguished on a cellular basis.

Lipochrome histiocytosis

The syndrome of recurrent infections and infiltration of the viscera by pigmented lipid histiocytes was first described in two unrelated boys by Landing & Shirkey (1957). Foci of granulomatous inflammation were noted in the liver of one of them. Later, Ford *et al.* (1962) described the same symptoms in three sisters. All three girls displayed also hyperglobulinaemia, pulmonary infiltrations and splenomegaly. Transient episodes of arthritis were observed in one girl and juvenile rheumatoid arthritis in another. Rodey *et al.* (1970*a*), studying the blood leukocytes of two of the same three sisters, found an impaired stimulation of the oxidative metabolism after phagocytosis of latex particles. The staphylococcicidal activity of these cells was also diminished. Moreover, the iodination capacity of the neutrophils of one of these girls was also defective (Pincus & Klebanoff 1971). Thus, the observed cellular abnormalities were identical to those found in patients with chronic granulomatous disease.

The syndrome of lipochrome histiocytosis can be distinguished from the syndrome of CGD by the occurrence of rheumatoid arthritis and the absence of suppurative lymphadenopathy in lipochrome histiocytosis. Moreover, lipochrome-pigmented histiocytes can be observed morphologically and the predominance of granuloma, as found in CGD, is absent.

Recently, we had the opportunity to study a family consisting of two parents (first cousins) and five children, of whom one boy has died with CGD-like symptoms (Corbeel *et al.* 1978). The leukocytes of this boy have not been investigated. Three of the four remaining children (one boy, two girls) still show the CGD-like symptoms. In contrast to CGD, however, dwarfism was observed in all four affected children, and lipochrome-loaded histiocytes were seen in at least one biopsy of each of the three affected children that were examined. One affected boy and his mother had chronic polyarthritis; the two girls suffer from chronic enterocolitis. Granulomata were only found in one of the affected children, after 13 biopsies with negative findings. Thus, these symptoms clearly remind one of the syndrome described by Ford *et al.* (1962).

The levels of immunoglobulins in the blood were high; the complement values and the lymphocyte functions were normal. The chemotactic responsiveness of the neutrophils and the rate of phagocytosis were normal, as was the opsonic activity of the serum. However, as Table 4 shows, the killing capacity towards *Staphylococcus aureus* and *Candida albicans* was strongly impaired, whereas *Lactobacillus acidophilus* (which releases H_2O_2) was killed normally. Moreover, those parameters of the respiratory burst that were tested were not increased by phagocytosis. Similar results were obtained with the patients' monocytes. The activity of glucose-6-phosphate and 6-phosphogluconate dehydrogenase, of glutathione peroxidase and reductase, and of myeloperoxidase was normal in each case. Heterozygotes were not found in this family.

Our results confirm, therefore, the identical cellular defect in lipochrome histiocytosis and CGD. Whether lipochrome histiocytosis is a truly different variant from CGD can as yet not be decided: the clinical and morphological differences between the two syndromes are not mutually exclusive (Dilworth

TABLE 5

Neutrophil functions in the 'triggering defect'[a]

Parameter	*Additions*	*Controls (n)*	*Patient*		*CGD*
			T.B. (♂)	*S.B.* (♀)[b]	(♂♂+♀♀)
Oxygen consumption	None	0.6–3.1 (24)	0–0.6 (2)	0 (3)	0–1.3 (11)
	Latex + serum	15.1–29.7 (24)	1.1–1.6 (2)	0.8–2.0 (3)	0–1.5 (10)
	IgG–latex	19.6–20.7 (4)	20.9–21.1 (3)	6.6 (1)	1.5 (1)
Superoxide generation	None	0.7–1.3 (13)	0–0.2 (3)	0 (2)	0–0.2 (7)
	Latex + serum	4.2–11.1 (13)	0–0.3 (3)	0–0.1 (2)	0–0.5 (6)
	IgG–latex	6.2–17.9 (4)	9.3–12.2 (2)	8.7 (1)	1.6 (1)
Hydrogen peroxide production	None	0–0.2 (10)	0–0.4 (3)	0.2–0.3 (2)	0 (10)
	IgG–latex	13.1–21.0 (10)	6.9–24.7 (3)	10–12.3 (2)	0–0.5 (10)
HMP shunt activity	None	0.1–0.2 (5)	0 (1)	0.2 (1)	0–0.2 (7)
	Latex + serum	1.3–3.7 (8)	0.4 (1)	0.3 (1)	0–0.2 (6)
	IgG–latex	1.0–5.8 (5)	1.5–4.5 (2)	2.4 (1)	0.03 (1)
Iodination	None	0–0.01 (11)	0 (3)	0 (2)	0 (3)
	Zymosan + serum	0.32–0.87 (11)	0–0.03 (3)	0.02–0.03 (2)	0–0.1 (3)

[a] The results are expressed in nmol (10^6 neutrophils)$^{-1}$ (10 min)$^{-1}$. The range of several different tests is given (number in parentheses). For comparison, the results with CGD neutrophils are also given.

[b] Patient S.B. received corticosteroids during these investigations; this may have decreased her neutrophil responsiveness.

TABLE 4

Neutrophil functions in lipochrome histiocytosis during zymosan phagocytosis

Parameter	*Member of family*						*Normals*
	Be ♀	*Aa* ♂	*Nu* ♀	*Bi* ♀	*father*	*mother*	*(n = 10)*
Killing of *S. aureus*[a]	0	0	0	70	63	75	52–95
Killing of *C. albicans*[a]	0	0	0	27	27	28	20–30
Killing of *L. acidophilus*[a]	75	71	n.t.[b]	n.t.	n.t.	n.t.	63–85
O_2 consumption[c]	0.4	0.2	0.3	3.0	3.4	4.0	1.7–7.7
O_2^- generation[c]	0.7	0.5	0.3	2.9	4.8	3.0	2.2–7.8
H_2O_2 release[d]	0	0	0	104	85	89	79–187

[a] % killed after 60 min incubation.
[b] n.t., not tested.
[c] in nmol $(10^6 \text{ cells})^{-1} \text{ min}^{-1}$.
[d] in nmol $(10^6 \text{ cells})^{-1} (30 \text{ min})^{-1}$.

& Mandell 1977; Johnston & Newman 1977). A definite conclusion has to await determination of the activity and the properties of the NAD(P)H oxidase activity in lipochrome histiocytosis.

The 'triggering defect'

A second phagocyte dysfunction closely related to CGD has been described by our group (Weening *et al.* 1976). It concerns two brothers and a sister, children of unrelated, healthy parents, with recurrent severe bacterial and fungal infections (Van der Meer *et al.* 1975). One boy died at the age of 20 years from pneumonia; his leukocyte functions have not been measured. The two surviving affected children had low to normal levels of immunoglobulins and a normal amount of complement components in the blood. The lymphocyte reactivity was also unaffected. The neutrophils of these patients ingested particles and bacteria normally, but showed a subnormal killing of *Staphylococcus aureus*. The chemotactic responsiveness of the patients' neutrophils was normal.

The oxidative metabolism of the patients' neutrophils was tested during incubation with two different kinds of particles: immunoglobulin G-coated latex particles and latex particles in the presence of 10% human serum. The serum is needed for the coating of the particles with serum factors to obtain optimal ingestion. Table 5 shows that the patients' cells were not stimulated by latex + serum, in contrast to normal cells. Moreover, zymosan particles were not iodinated. Heterozygosity in both parents and in one sister was observed only with the iodination test.

So far, the observed defects were consistent with the diagnosis of CGD. However, in complete contrast to the findings in CGD, a normal stimulation of the oxidative metabolism was found when the patients' neutrophils were stimulated with IgG-coated latex or IgG aggregates. Nevertheless, serum-opsonized and IgG-coated latex particles were equally well ingested, indicating that the defective metabolic stimulation by serum-opsonized latex was not due to a decreased binding or uptake of these particles. Moreover, the enzymes of the HMP shunt, the glutathione redox system and myeloperoxidase had a normal activity. We conclude that the decreased metabolic response of the patients' neutrophils to serum-opsonized particles must be due to a defect in the initial steps of cell activation.

The only measurable difference between serum-incubated and IgG-coated latex is the amount of IgG on the particles. It follows, therefore, that only heavily opsonized particles can stimulate the patients' cells. This abnormality may be caused either by a decreased number of binding places for IgG on the patients' neutrophils or to an increased binding threshold for metabolic activation. Apparently, enough binding between serum-opsonized latex and the patients' cells occurs to initiate phagocytosis, but not enough to trigger the NAD(P)H oxidase. Hence, we have called this syndrome the 'triggering defect'. From these results it follows that particle binding and phagocytosis is not sufficient to start the respiratory burst. It is also known that particle binding, but not phagocytosis, is a prerequisite for the metabolic stimulation (Goldstein *et al.* 1975). Thus, particle binding is a necessary, but not efficient, event in the stimulatory process.

The defects observed in this family and the absence of the Kx antigen on CGD cells (Marsh *et al.* 1976) are consistent with an important function of the plasma membrane in the initiation of the respiratory burst. Probably, the binding of phagocytosable material or soluble stimulators to the plasma membrane of phagocytic leukocytes activates the superoxide-generating system. In the family with the 'triggering defect', the system as such is present and can be stimulated to a normal activity. The defect must lie in the activation mechanism of the plasma membrane oxidase. Curnutte *et al.* (1975) and McPhail *et al.* (1977) have suggested that this might be one cause for the lack of metabolic response in CGD. From our results, it seems that this may indeed result in neutrophil dysfunctions.

GLUCOSE-6-PHOSPHATE DEHYDROGENASE DEFICIENCY

Glucose-6-phosphate dehydrogenase (G6PD) is the first enzyme of the hexose-monophosphate shunt pathway (Fig. 1). Deficiencies in red cell G6PD are common and may give rise to chronic non-spherocytic haemolytic anaemia. Usually, the rest activity of this enzyme in the leukocytes is higher than in the erythrocytes because leukocytes have a much higher turn-over rate. A few patients are known, however, with leukocytic G6PD values of less than 5% of normal. In these patients an increased susceptibility to infections is correlated with a decreased antibacterial capacity *in vitro* of the neutrophils (Cooper *et al.* 1972; Baehner *et al.* 1972; Gray *et al.* 1973). With G6PD values of 20% of normal or more, this increased incidence of infections and the bactericidal defect in the neutrophils is not found (Rodey *et al.* 1970*b*; Baehner *et al.* 1972). We have confirmed these last observations with a patient with 25% of normal G6PD levels in his neutrophils (erythrocytes: 5% of normal).

The G6PD-deficient neutrophils (less than 5% of normal activity) are defective in their capacity to kill catalase-positive bacteria *in vitro*. The ingestion of the bacteria and the degranulation are normal, but the generation of hydrogen peroxide is severely depressed. Thus, the defect is similar to that found in CGD neutrophils.

The simplest explanation for the effect of the G6PD deficiency on the neutrophil function is the inability to generate NADPH as a substrate for the NADPH oxidase. Alternatively, the reduction of NADH by NADPH may be impaired, thus causing a lack of substrate for an NADH oxidase. The latter explanation is unsatisfactory, however, since Baehner *et al.* (1972) have found a strongly decreased ratio of $NADPH/NADP^+$ in cells with 5% of normal G6PD activity, but a practically normal $NADH/NAD^+$ ratio. Thus, it seems that in G6PD deficiency the substrate (NADPH) is missing for the same enzyme (NADPH oxidase) which is inactive in CGD. Therefore, both defects lead to similar cellular dysfunctions and clinical symptoms.

THE GLUTATHIONE REDOX SYSTEM

As shown in Fig. 1 the glutathione system consists of reduced glutathione (γ-glutamylcysteinylglycine, GSH) which can react with peroxides in the glutathione peroxidase (GP) reaction. As a result, the oxidized dimer of glutathione (GSSG) is formed which, in its turn, can react with NADPH in the glutathione reductase (GR) reaction to reconstitute GSH. Thus, this redox cycle is able to detoxify peroxides. Not shown in Fig. 1 is the reaction (1) of glu-

$$2\,GSH + 2A^{\bullet} \rightarrow GSSG + 2AH \qquad (1)$$

tathione with free radicals (see also Slater, pp. 143–159). Finally, the glutathione system is also able to reduce oxidized thiol groups in proteins. Therefore, this system is very important in the protection of cells against oxidative and radical damage. As such, it also plays an important role in the protection of phagocytic leukocytes against their own products. Disturbances in this system may therefore lead to serious problems in the host-defence against microorganisms.

Glutathione peroxidase

In 1970, Holmes *et al.* found a deficiency of glutathione peroxidase in the neutrophils of two unrelated female CGD patients. Since then, two more deficiencies of this enzyme have been found, one in the neutrophils of a brother of one of the female CGD patients (Malawista & Gifford 1975), and one in an unrelated male patient (Matsuda *et al.* 1976). In each case, the patients displayed severe recurrent infections with *Staphylococcus aureus*, and their neutrophils showed a strongly decreased respiratory burst during phagocytosis.

In only one of these patients was the NAD(P)H oxidase activity assayed: the NADH oxidase activity was unmeasurable, but the NADPH oxidase activity was equal to that in the control cells (Matsuda *et al.* 1976). In three of the four patients the G6PD activities in the leukocytes were measured and found to be normal (Holmes *et al.* 1970; Matsuda *et al.* 1976). Thus, this leaves only the glutathione peroxidase deficiency (20–35% of normal) as the cause of the neutrophil dysfunction. From the reactions shown in Fig. 1 it is not readily apparent, however, how a deficiency in glutathione peroxidase can lead to a diminished production of hydrogen peroxide. Two possible explanations have been put forward (Holmes *et al.* 1970).

The first theory is based on the observation by Strauss *et al.* (1969) that the glutathione reductase activity in phagocytosing leukocytes increases 2–3 times within 15 s after addition of particles to the cells, at a time when the NADPH oxidase activity is still at its resting level. From this observation Strauss and his colleagues drew the conclusion that the formation of H_2O_2 in the NADPH oxidase reaction might depend on a preceding reduction of glutathione. The concomitantly produced $NADP^+$ might then stimulate the HMP shunt pathway, and the NADPH generated would serve as the substrate for the NADPH oxidase. Finally, the increased hydrogen peroxide would be removed by the glutathione peroxidase reaction. They proposed the following order of reactions (2–5).

$$\text{GSSG} + \text{NADPH} + \text{H}^+ \xrightarrow{\text{glutathione reductase}} 2\text{GSH} + \text{NADP}^+ \tag{2}$$

$$2\text{NADP}^+ + \text{glucose-6-PO}_4 \xrightarrow{\text{HMP shunt}} 2\text{NADPH} + 2\text{H}^+ + \text{CO}_2 + \text{ribulose-5-PO}_4 \tag{3}$$

$$\text{NADPH} + \text{H}^+ + \text{O}_2 \xrightarrow{\text{NADPH oxidase}} \text{NADP}^+ + \text{H}_2\text{O}_2 \tag{4}$$

$$2\text{GSH} + \text{H}_2\text{O}_2 \xrightarrow{\text{glutathione peroxidase}} \text{GSSG} + 2\text{H}_2\text{O} \tag{5}$$

This scheme represents a cycle and a deficiency in any reaction (e.g. the glutathione peroxidase reaction) would cause it to stop. The result would be a decrease in hydrogen peroxide production, leading to a diminished bactericidal activity and CGD-like symptoms.

Several criticisms of this hypothesis are possible. First, the reported activation of glutathione reductase at the start of phagocytosis has never been confirmed by other workers (Rossi *et al.* 1972). Second, this set of reactions implies that *all* the hydrogen peroxide must react with reduced glutathione to keep the cycle going. This would leave no H_2O_2 available for bactericidal reactions, a conclusion which is in conflict with the overwhelming evidence to the contrary.

The second explanation put forward by Holmes *et al.* (1970) to explain the lack of respiratory burst in glutathione peroxidase-deficient leukocytes was inactivation of the hydrogen peroxide-generating system when the glutathione-dependent peroxide and (oxygen) radical removal is limited. This explanation does not seem very likely either, in view of the fact that glutathione can be oxidized to an appreciable extent also in the absence of any enzyme (Baehner *et al.* 1970; Noseworthy & Karnovsky 1972). However, in the next sections of this paper we shall present additional evidence for this theory.

Some puzzling features of the glutathione peroxidase deficiency syndrome remain, however. First, a relatively high rest activity of this enzyme is found in the leukocytes of each case, together with a normal activity in the red cells (measured only in the two patients by Holmes *et al.* 1970). Second, animal models with experimentally induced glutathione peroxidase deficiency (by reduced selenium intake) have yielded contradictory results about the effect of such a deficiency on the microbicidal capacity and the oxidative metabolism of phagocytic leukocytes (Serfass & Ganther 1975; Bassford *et al.* 1977; Bass *et al.* 1977). Thus, no strict correlation between the glutathione peroxidase activity and the functional leukocyte defect is apparent. Further studies with regard to the early stages of the respiratory burst are needed to establish whether the cells are damaged by an initially-normal generation of oxygen products or whether they do not generate these products at all owing to another mechanism.

Glutathione reductase

Deficiencies of glutathione reductase in erythrocytes are common and are practically always caused by an insufficient intake of riboflavin (Beutler 1969). We have described a family with three homozygous children deficient in glutathione reductase in the erythrocytes and in the leukocytes; this defect could not be restored by riboflavin *in vivo* or FAD *in vitro* (Loos *et al.* 1976). Although these individuals do not suffer from an increased incidence of infections, their neutrophils and monocytes show a peculiar respiratory burst when particles are ingested (Weening *et al.* 1977). Initially, the oxygen consumption of the phagocytic leukocytes increases in a normal fashion, as does the superoxide generation, the hydrogen peroxide production and the iodination (Fig. 2). After a few minutes, however, these reactions slow down and gradually stop in the glutathione reductase-deficient cells, except for the superoxide generation which continues in a normal fashion.

This abnormal metabolic response raises two questions. Why do the first-mentioned reactions stop and why does the release of O_2^- keep going? As in the glutathione peroxidase-deficiency, the defect in the respiratory burst can, in principle, be explained either by the reaction cycle proposed by Strauss *et al.* (1969) (see p. 244) or by oxidative damage due to insufficient detoxification of H_2O_2 and/or (oxygen) radicals. Two observations support the second explanation. First, after pre-incubation for 10 min in the presence of an H_2O_2-generating system, normal cells still showed about 70% of the original respiratory response to phagocytosis, whereas the glutathione reductase cells had completely lost this capacity. Therefore, the glutathione reductase-deficient leukocytes are more

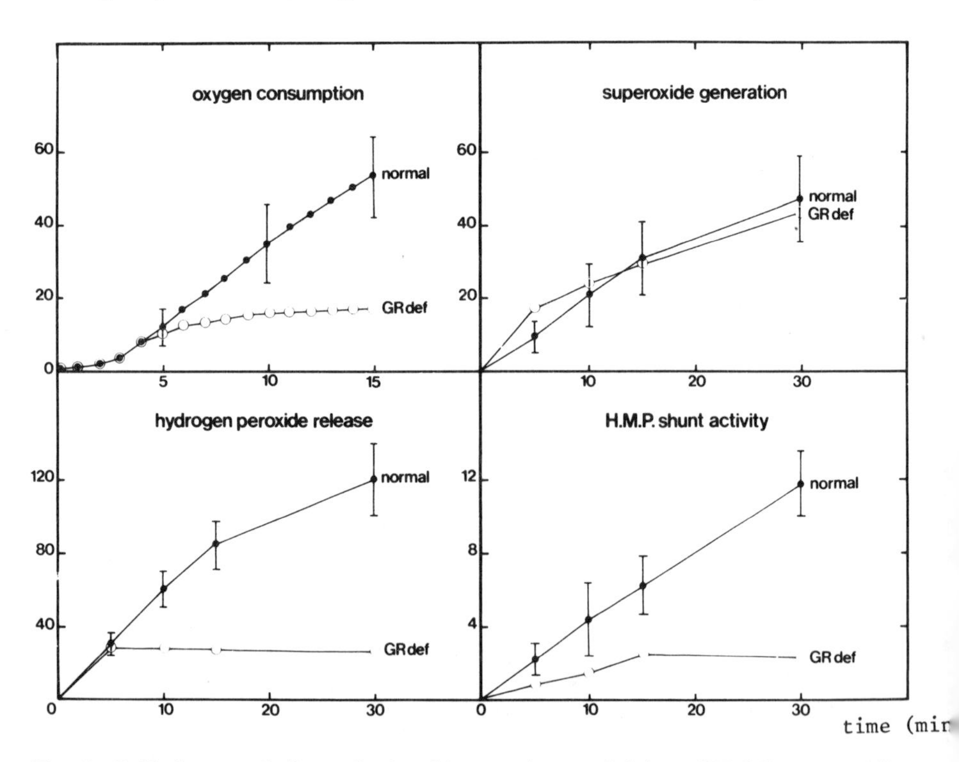

FIG. 2. Oxidative metabolism of glutathione reductase-deficient (GRdef) neutrophils during zymosan phagocytosis. All values are given in nmol/10^6 cells; differences between the parameters are caused by experimental differences. The normal values are given as mean ± S.D. ($n > 10$). (For experimental details see Roos *et al.* 1979*a*).

susceptible to oxidative damage than normal leukocytes are. Second, in the presence of a superoxide scavenger (cytochrome *c*), the cells showed a completely normal oxidative response to particles. Thus, cytochrome *c* must have protected the deficient cells against the damaging effects of its own O_2^- or dismutation products. Consequently, a normal cytochrome *c* reduction is found (Fig. 2), which also proves that a normal oxidative response to particle ingestion can be mounted in spite of an 80–90% decrease in glutathione reductase activity. We conclude, therefore, that the scheme proposed by Strauss *et al.* (1969) (see p. 244) is not operative in phagocytic leukocytes.

It also follows from these observations that the glutathione redox cycle is an important system for the protection against oxygen products and radicals: although the glutathione reductase-deficient cells contain a normal activity of catalase, superoxide dismutase and myeloperoxidase, these cells were quickly inactivated by oxidative damage. To investigate this idea further, we have measured the levels of reduced glutathione during phagocytosis (Roos *et al.* 1979*a*). Fig. 3 shows that the level of GSH decreases in normal cells about 20% during the first 10–15 min of phagocytosis, followed by a slow restoration of this level*. In contrast, the GSH in the glutathione reductase-deficient cells

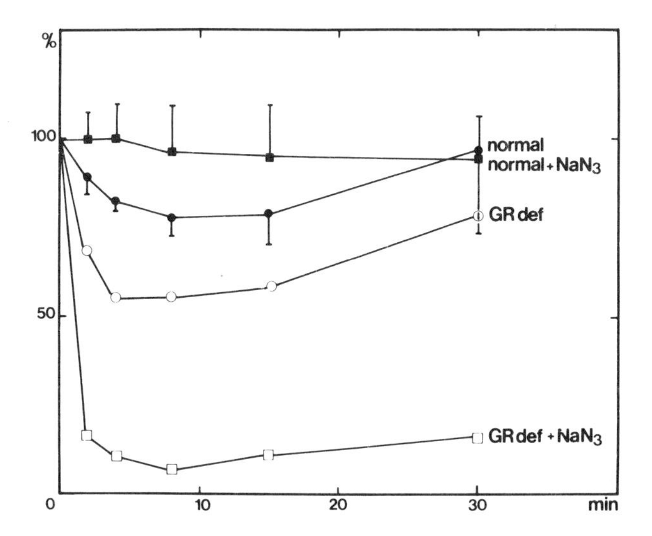

FIG. 3. Changes in neutrophil glutathione levels during zymosan phagocytosis. Normal cells and glutathione reductase-deficient (GRdef) cells were incubated with and without 2mM-NaN_3. The change in GSH levels is given as the percentage of the original level before addition of zymosan particles. Normal values are given as mean ± S.D. ($n = 5$). Original GSH level is 1.8 nmol/10^6 cells.

* This restoration has since been shown to be due to reduction of non-glutathione thiol compounds (Voetman *et al.* 1979).

decreases rapidly to 40–60% of the original level during phagocytosis. Moreover, in the presence of azide (to inhibit catalase and myeloperoxidase) the level of GSH immediately drops to about zero. Concomitantly, the inactivation of cell functions was also accelerated by azide.

We conclude from these observations that the glutathione redox system cooperates with other protective systems against oxidative damage in phagocytic leukocytes.

The lack of recurrent infections in the individuals with glutathione-reductase deficiency, in sharp contrast to the patients with glutathione-peroxidase deficiency, may possibly be explained by the short, normal generation of bactericidal products in the leukocytes of the former. Apparently, this reaction suffices for normal antibacterial activity. At high ratios of *Staphylococcus aureus* to glutathione reductase-deficient neutrophils, however, a killing deficiency *in vitro* has been found (Roos *et al.* 1979*a*). In glutathione peroxidase-deficient patients, the inactivation of the oxidative system may be so fast that insufficient oxygen products are generated for adequate antibacterial activity.

Glutathione synthetase

The last deficiency in the glutathione redox system known to affect the function of phagocytic leukocytes is one of glutathione synthetase, the enzyme which adds glycine to γ-glutamylcysteine to complete the synthesis of glutathione.

Spielberg *et al.* (1979) have examined the leukocytes of a boy with less than 4% of normal glutathione synthetase activity and less than 25% of normal reduced glutathione content. This patient has frequent episodes of acute otitis media, often in combination with neutropenia. His neutrophils show an impaired bacterial killing *in vitro*. The generation of oxygen products during phagocytosis appeared to be normal or even elevated. However, according to us, this may be due to the fact that only a small part of the H_2O_2 was measured, and only during short incubations. Therefore, this dysfunction may still be caused by oxidative damage of the cells.

It should be mentioned that not always will glutathione-synthetase deficiency lead to defects in phagocytic functions: in a patient with 5% of normal GSH levels in the erythrocytes, we found half-normal GSH levels in the neutrophils and no abnormality in the function or in the oxidative metabolism of these cells. Perhaps only in severe, generalized synthetase deficiency will the function of the phagocytes be affected. [*Note added in proof:* Recently this has also been described by Spielberg *et al.* (1978).]

References

ALLEN, R. C., STJERNHOLM, R. L. & STEELE, R. H. (1972) Evidence for the generation of an electronic excitation state(s) in human polymorphonuclear leukocytes and its participation in bactericidal activity. *Biochem. Biophys. Res. Commun. 47*, 679–684

BABIOR, B. M. & KIPNES, R. S. (1977) Superoxide-forming enzyme from human neutrophils: evidence for a flavin requirement. *Blood 50*, 517–524

BABIOR, B. M., KIPNES, R. S. & CURNUTTE, J. T. (1973) The production by leukocytes of superoxide, a potential bactericidal agent. *J. Clin. Invest. 52*, 741–744

BABIOR, B. M., CURNUTTE, J. T. & KIPNES, R. S. (1975) Pyridine nucleotide-dependent superoxide production by a cell-free system from human granulocytes. *J. Clin. Invest. 56*, 1035–1042

BABIOR, B. M., CURNUTTE, J. T. & MCMURRICH, B. J. (1976) The particulate superoxide-forming system from human neutrophils: properties of the system and further evidence supporting its participation in the respiratory burst. *J. Clin. Invest. 58*, 989–996

BAEHNER, R. L. (1975) Subcellular distribution of nitroblue tetrazolium reductase (NBT-R) in human polymorphonuclear leukocytes (PMN). *J. Lab. Clin. Med. 86*, 785–792

BAEHNER, R. L. & KARNOVSKY, M. L. (1968) Deficiency of reduced nicotinamide-adenine dinucleotide oxidase in chronic granulomatous disease. *Science (Wash. D.C.) 162*, 1277–1279

BAEHNER, R. L. & NATHAN, D. G. (1968) Quantitative nitroblue tetrazolium test in chronic granulomatous disease. *N. Engl. J. Med. 278*, 971–976

BAEHNER, R. L., GILMAN, N. & KARNOVSKY, M. L. (1970) Respiration and glucose oxidation in human and guinea pig leukocytes: comparative studies. *J. Clin. Invest. 42*, 692–700

BAEHNER, R. L., JOHNSTON, R. B. JR. & NATHAN, D. G. (1972) Comparative study of the metabolic and bactericidal characteristics of severely glucose-6-phosphate dehydrogenase-deficient polymorphonuclear leukocytes and leukocytes from children with chronic granulomatous disease. *J. Reticuloendothel. Soc. 12*, 150–169

BAEHNER, R. L., BOXER, L. A. & DAVIS, J. (1976) The biochemical basis of nitroblue tetrazolium reduction in normal human and chronic granulomatous disease polymorphonuclear leukocytes. *Blood 48*, 309–313

BALDRIDGE, C. W. & GERARD, R. W. (1933) The extra respiration of phagocytosis. *Am. J. Physiol. 103*, 235–236

BASS, D. A., DECHATELET, L. R., BURK, R. F., SHIRLEY, P. & SZEJDA, P. (1977) Polymorphonuclear leukocyte bactericidal activity and oxidative metabolism during glutathione peroxidase deficiency. *Infect. Immun. 18*, 78–84

BASSFORD, R. E., BARTUS, B., KAPLAN, S. S. & PLATT, D. (1977) Glutathione peroxidase and phagocytosis, in *Movement, Metabolism and Bactericidal Mechanisms of Phagocytes* (Rossi, F., Patriarca, P. & Romeo, D., eds.), pp. 265–275, Piccin Medical Books, Padua

BEUTLER, E. (1969) Effect of flavin compounds on glutathione reductase activity: *in vivo* and *in vitro* studies. *J. Clin. Invest. 48*, 1957–1966

BIGGAR, W. D., BURON, S. & HOLMES, B. (1976) Chronic granulomatous disease in an adult male: a proposed X-linked defect. *J. Pediatr. 88*, 63–70

BORS, W., LENGFELDER, E., SARAN, M., FUCHS, C. & MICHEL, C. (1976) Reactions of oxygen radical species with methional: a pulse-radiolysis study. *Biochem. Biophys. Res. Commun. 70*, 81–87

BRIGGS, R. T., DRATH, D. B., KARNOVSKY, M. L. & KARNOVSKY, M. J. (1975) Localization of NADH oxidase on the surface of human polymorphonuclear leukocytes by a new cytochemical method. *J. Cell Biol. 67*, 566–586

BRIGGS, R. T., KARNOVSKY, M. L. & KARNOVSKY, M. J. (1977) Hydrogen peroxide production in chronic granulomatous disease. *J. Clin. Invest. 59*, 1088–1098

CAGAN, R. H. & KARNOVSKY, M. L. (1964) Enzymatic basis of the respiratory stimulation during phagocytosis. *Nature (Lond.) 204*, 255–257

CHESON, B. D., CHRISTENSEN, R. L., SPERLING, R., KOHLER, B. E. & BABIOR, B. M. (1976)

The origin of the chemiluminescence of phagocytosing granulocytes. *J. Clin. Invest. 58*, 789–796

COOPER, M. R., DECHATELET, L. R., MCCALL, C. E., LA VIA, M. F., SPURR, C. L. & BAEHNER, R. L. (1972) Complete deficiency of leukocyte glucose-6-phosphate dehydrogenase with defective bactericidal activity. *J. Clin. Invest. 51*, 769–778

CORBEEL, L., WEENING, R. S., ROOS, D., STRADSBAEDER, S., GEBOES, K., STANDAERT, L., EGGERMONT, E., CASTEELS-VAN DAALE, M., DELMOTTE, B. & EECKELS, R. (1978) Familial deficiency of granulocyte bactericidal capacity associated with growth retardation. *Acta Pediatr. Belg. 31*, 15–25

CURNUTTE, J. T., KIPNES, R. S. & BABIOR, B. M. (1975) Defect in pyridine nucleotide-dependent superoxide production by a particulate fraction from the granulocytes of patients with chronic granulomatous disease. *N. Engl. J. Med. 293*, 628–632

CURNUTTE, J. T., KARNOVSKY, M. L. & BABIOR, B. M. (1976) Manganese-dependent NADPH oxidation by granulocyte particles: the role of superoxide and the nonphysiological nature of the manganese requirement. *J. Clin. Invest. 57*, 1059–1067

DECHATELET, L. R., MCPHAIL, L. C., MULLIKEN, D. & MCCALL, C. E. (1975) An isotopic assay for NADPH oxidase activity and some characteristics of the enzyme from human polymorphonuclear leukocytes. *J. Clin. Invest. 55*, 714–721

DEWALD, B., BAGGIOLINI, M., CURNUTTE, J. T. & BABIOR, B. M. (1979) Subcellular localization of the superoxide-forming enzyme in human neutrophils. *J. Clin. Invest., 63*, 21–29

DILWORTH, J. A. & MANDELL, G. L. (1977) Adults with chronic granulomatous disease of 'childhood'. *Am. J. Med. 63*, 233–243

DRATH, D. B. & KARNOVSKY, M. L. (1974) Bactericidal activity of metal-mediated peroxidase-ascorbate systems. *Infect. Immun. 10*, 1077–1083

ELSBACH, P. (1973) On the interaction between phagocytes and micro-organisms. *N. Engl. J. Med. 289*, 846–852

EVANS, W. H. & KARNOVSKY, M. L. (1961) A possible mechanism for the stimulation of some metabolic functions during phagocytosis. *J. Biol. Chem. 236*, PC 30–32

FONG, K. L., MCCAY, P. B., POYER, J. L., KEELE, B. B. & MISRA, H. (1973) Evidence that peroxidation of lysosomal membranes is initiated by hydroxyl free radicals produced during flavin enzyme activity. *J. Biol. Chem. 248*, 7792–7797

FORD, D. K., PRICE, G. E., CULLING, CH. F. & VASSAR, PH. (1962) Familial lipochrome pigmentation of histiocytes with hyperglobulinemia, pulmonary infiltration, splenomegaly, arthritis and susceptibility to infection. *Am. J. Med. 33*, 478–489

FRIDOVICH, I. (1974) Superoxide dismutases. *Adv. Enzymol. 41*, 35–97

GOLDSTEIN, I. M., ROOS, D., KAPLAN, H. B. & WEISSMANN, G. (1975) Complement and immunoglobulins stimulate superoxide production by human leukocytes independently of phagocytosis. *J. Clin. Invest. 56*, 1155–1163

GOLDSTEIN, I. M., CERQUIERA, M., LIND, S. & KAPLAN, H. B. (1977) Evidence that the superoxide-generating system of human leukocytes is associated with the cell surface. *J. Clin. Invest. 59*, 249–254

GRAY, G. R., STAMATOYANNOPOULOS, G., NAIMAN, S. C., KILMAN, M. R. & KLEBANOFF, S. J. (1973) Neutrophil dysfunction, chronic granulomatous disease, and non-spherocytic haemolytic anaemia caused by complete deficiency of glucose-6-phosphate dehydrogenase. *Lancet 2*, 530–534

HELD, A. M. & HURST, J. K. (1978) Ambiguity associated with use of singlet oxygen trapping agents in myeloperoxidase-catalyzed oxidations. *Biochem. Biophys. Res. Commun. 81*, 878–885

HOHN, D. C. & LEHRER, R. I. (1975) NADPH oxidase deficiency in X-linked chronic granulomatous disease. *J. Clin. Invest. 55*, 707–713

HOLMES, B., PAGE, A. R. & GOOD, R. A. (1967) Studies of the metabolic activity of leukocytes from patients with a genetic abnormality of phagocytic functions. *J. Clin. Invest. 46*, 1422–1432

HOLMES, B., PARK, B. H., MALAWISTA, S. E., QUIE, P. G., NELSON, D. L. & GOOD, R. A.

(1970) Chronic granulomatous disease in females: a deficiency of leukocyte glutathione peroxidase. *N. Engl. J. Med.* *283*, 217–221

HOMAN-MÜLLER, J. W. T., WEENING, R. S. & ROOS, D. (1975) Production of hydrogen peroxide by phagocytizing human granulocytes. *J. Lab. Clin. Med.* *85*, 198–207

IVERSON, D., DECHATELET, L. R., SPITZNAGEL, J. K. & WANG, P. (1977) Comparison of NADH and NADPH oxidase activities in granules isolated from human polymorphonuclear leukocytes with a fluorimetric assay. *J. Clin. Invest.* *59*, 282–290

IYER, G. Y. N. & QUASTEL, J. H. (1963) NADPH and NADH oxidation by guinea pig polymorphonuclear leucocytes. *Canad. J. Biochem. Physiol.* *41*, 427–434

IYER, G. Y., ISLAM, M. F. & QUASTEL, J. H. (1961) Biochemical aspects of phagocytosis. *Nature (Lond.)* *192*, 535–541

JENSEN, M. S. & BAINTON, D. F. (1973) Temporal changes in pH within the phagocytic vacuole of the polymorphonuclear neutrophilic leukocyte. *J. Cell. Biol.* *56*, 379–388

JOHNSTON, R. B. JR. & NEWMAN, S. L. (1977) Chronic granulomatous disease. *Pediatr. Clin. N. Am.* *24*, 365–376

JOHNSTON, R. B. JR., KEELE, B. B. JR., MISRA, H. P., LEHMEYER, J. E., WEBB, L. S., BAEHNER, R. L. & RAJAGOPOLAN, K. V. (1975) The role of superoxide anion generation in phagocytic bactericidal activity. Studies with normal and chronic granulomatous disease leukocytes. *J. Clin. Invest.* *55*, 1357–1372

KELLOGG, E. W. & FRIDOVICH, I. (1975) Superoxide, hydrogen peroxide, and singlet oxygen in lipid peroxidation by a xanthine oxidase system. *J. Biol. Chem.* *250*, 8812–8817

KLEBANOFF, S. J. (1975) Antimicrobial mechanisms in neutrophilic polymorphonuclear leukocytes. *Semin. Hematol.* *12*, 117–142

KLEBANOFF, S. J. & ROSEN, H. (1979) The role of myeloperoxidase in the microbicidal activity of polymorphonuclear leukocytes, in *This Volume*, pp. 263–284

KRINSKY, N. I. (1974) Singlet excited oxygen as a mediator of the antibacterial action of leukocytes. *Science (Wash. D.C.)* *186*, 363–365

LANDING, B. H. & SHIRKEY, H. S. (1957) A syndrome of recurrent infection and infiltration of viscera by pigmented lipid histiocytes. *Pediatrics* *20*, 431–438

LEHRER, R. I. (1975) The fungicidal mechanisms of human monocytes. I. Evidence for myeloperoxidase-linked and myeloperoxidase-independent candidacidal mechanisms. *J. Clin. Invest.* *55*, 338–346

LOOS, J. A., ROOS, D., WEENING, R. S. & HOUWERZIJL, J. (1976) Familial deficiency of glutathione reductase in human blood cells. *Blood* *48*, 53–62

MCPHAIL, L. C., DECHATELET, L. R., SHIRLEY, P. S., WILFERT, C., JOHNSTON, R. B. JR. & MCCALL, C. E. (1977) Deficiency of NADPH oxidase activity in chronic granulomatous disease. *J. Pediatr.* *90*, 213–217

MALAWISTA, S. E. & GIFFORD, R. H. (1975) Chronic granulomatous disease of childhood (CGD) with leukocyte glutathione peroxidase (LPG) deficiency in a brother and a sister: a likely autosomal recessive inheritance. *Clin. Res.* *23*, 416 A

MANDELL, G. L. & HOOK, E. W. (1969) Leukocyte bactericidal activity in chronic granulomatous disease: correlation of bacterial hydrogen peroxide production and susceptibility to intracellular killing. *J. Bacteriol.* *100*, 531–532

MARSH, W. L., ØYEN, R. & NICHOLS, M. E. (1976) Kx antigen, the McLeod phenotype and chronic granulomatous disease: further studies. *Vox Sang.* *31*, 356–362

MATSUDA, I., OKA, Y., TANIGUCHI, N., FURUYAMA, M., KODAMA, S., ARASHIMA, S. & MITSUYAMA, T. (1976) Leukocyte glutathione peroxidase deficiency in a male patient with chronic granulomatous disease. *J. Pediatr.* *88*, 581–583

MILLER, T. E. (1969) Killing and lysis of gram-negative bacteria through the synergistic effect of hydrogen peroxide, ascorbic acid, and lysozyme. *J. Bacteriol.* *98*, 949–955

NATHAN, D. G., BAEHNER, R. L. & WEAVER, D. K. (1969) Failure of nitroblue tetrazolium reduction in the phagocytic vacuoles of leukocytes in chronic granulomatous disease. *J. Clin. Invest.* *48*, 1895–1904

NELSON, R. D., HERRON, M. J., SCHMIDTKE, J. R. & SIMMONS, R. L. (1977) Chemiluminescence response of human leukocytes: influence of medium components on light production. *Infect. Immun. 17*, 513–520

NOSEWORTHY, J. JR. & KARNOVSKY, M. L. (1972) Role of peroxide in the stimulation of the hexose monophosphate shunt during phagocytosis by polymorphonuclear leukocytes. *Enzyme 13*, 110–131

PATRIARCA, P., CRAMER, R., DRI, P., FANT, L., BASFORD, R. E. & ROSSI, F. (1973) NADPH oxidizing activity in rabbit polymorphonuclear leukocytes: localization in azurophilic granules. *Biochem. Biophys. Res. Commun. 53*, 830–837

PAUL, B. B. & SBARRA, A. J. (1968) The role of the phagocyte in host-parasite interactions. XII. The direct quantitative estimation of H_2O_2 in phagocytizing cells. *Biochim. Biophys. Acta 156*, 168–178

PINCUS, S. H. & KLEBANOFF, S. J. (1971) Quantitative leukocyte iodination. *N. Engl. J. Med. 284*, 744–750

REED, P. W. (1969) Glutathione and the hexose monophosphate shunt in phagocytizing and hydrogen peroxide-treated rat leukocyte. *J. Biol. Chem. 244*, 2459–2464

REED, P. W. & TEPPERMAN, J. (1969) Phagocytosis-associated metabolism and enzymes in the rat polymorphonuclear leukocyte. *Am. J. Physiol. 216*, 223–230

ROBERTS, J. & QUASTEL, J. H. (1964) Oxidation of reduced triphosphopyridine nucleotide by guinea pig polymorphonuclear leucocytes. *Nature (Lond.) 202*, 85–86

RODEY, G. E., PARK, B. H., FORD, D. K., HOLMES-GRAY, B. & GOOD, R. A. (1970*a*) Defective bactericidal activity of peripheral blood leukocytes in lipochrome histiocytosis. *Am. J. Med. 49*, 322–327

RODEY, G. E., JACOBS, H. S., HOLMES, B., MCARTHUR, J. R. & GOOD, R. A. (1970*b*) Leucocyte G-6-PD levels and bactericidal activity. *Lancet 1*, 355–356

ROOS, D. & BALM, A. J. M. (1979) The oxidative metabolism of monocytes, in *Biochemistry of the RES*, vol. 2 (Sbarra, A. J. & Strauss, R. R., eds.), Plenum Press, New York, in press

ROOS, D., VAN SCHAIK, M. L. J., WEENING, R. S. & WEVER, R. (1977) Superoxide generation in relation to other oxidative reactions in human polymorphonuclear leukocytes, in *Superoxide and Superoxide Dismutases* (Michelson, A. M., McCord, J. M. & Fridovich, I., eds.), pp. 307–316, Academic Press, London & New York

ROOS, D., WEENING, R. S., VOETMAN, A. A., VAN SCHAIK, M. L. J., BOT, A. A. M., MEERHOF, L. J. & LOOS, J. A. (1979*a*) Protection of phagocytic leukocytes by endogenous glutathione. Studies in a family with glutathione reductase deficiency. *Blood*, in press

ROOS, D., WEENING, R. S. & LOOS, J. A. (1979*b*) The protective role of glutathione, in *Inborn Errors of Immunity and Phagocytosis*, MTP Press, Lancaster

ROOT, R. K. & METCALF, J. A. (1977) H_2O_2 release from human granulocytes during phagocytosis. Relationship to superoxide anion formation and cellular catabolism of H_2O_2: studies with normal and cytochalasin B-treated cells. *J. Clin. Invest. 60*, 1266–1279

ROOT, R. K., METCALF, J., OSHINO, N. & CHANCE, B. (1975) H_2O_2 release from human granulocytes during phagocytosis. I. Documentation, quantification, and some regulating factors. *J. Clin. Invest. 55*, 945–955

ROSEN, H. & KLEBANOFF, S. J. (1976) Chemiluminescence and superoxide production by myeloperoxidase-deficient leukocytes. *J. Clin. Invest. 58*, 50–60

ROSSI, F. & ZATTI, M. (1964) Biochemical aspects of phagocytosis in polymorphonuclear leucocytes. NADH and NADPH oxidation by the granules of resting and phagocytizing cells. *Experientia 20*, 21–23

ROSSI, F. & ZATTI, M. (1966) Effect of phagocytosis on the carbohydrate metabolism of polymorphonuclear leukocytes. *Biochim. Biophys. Acta 121*, 110–119

ROSSI, F., ROMEO, D. & PATRIARCA, P. (1972) Mechanism of phagocytosis-associated oxidative metabolism in polymorphonuclear leukocytes and macrophages. *J. Reticuloendothelial Soc. 12*, 127–149

SBARRA, A. J. & KARNOVSKY, M. L. (1959) The biochemical basis of phagocytosis. I. Metabolic changes during the ingestion of particles by polymorphonuclear leukocytes. *J. Biol. Chem. 234*, 1355–1362

SEGAL, A. W. (1979) Chronic granulomatous disease—biochemistry and oxygen metabolism, in *Inborn Errors of Immunity and Phagocytosis*, MTP Press, Lancaster

SEGAL, A. W. & PETERS, T. J. (1976) Characterisation of the enzyme defect in chronic granulomatous disease. *Lancet 1*, 1363–1365

SEGAL, A. W. & PETERS, T. J. (1977) Analytical subcellular fractionation of human granulocytes with special reference to the localization of enzymes involved in microbicidal mechanisms. *Clin. Sci. Mol. Med. 52*, 429–442

SELVARAJ, R. J. & SBARRA, A. J. (1967) The role of the phagocyte in host-parasite interactions. VII. Di- and triphosphopyridine nucleotide kinetics during phagocytosis. *Biochim. Biophys. Acta 141*, 243–249

SERFASS, R. E. & GANTHER, H. E. (1975) Defective microbicidal activity in glutathione peroxidase-deficient neutrophils of selenium-deficient rats. *Nature (Lond.) 255*, 640–641

SHOHET, S. B., PITT, J., BAEHNER, R. L. & POPLACK, D. G. (1974) Lipid peroxidation in the killing of phagocytized pneumococci. *Infect. Immun. 19*, 1321–1328

SLATER, T. F. (1979) Mechanisms of protection against the damage produced in biological systems by oxygen-derived radicals, in *This Volume*, pp. 143–159

SPIELBERG, S. P., GARRICK, M. D., CORASH, L. M., BUTLER, J. D., TIETZE, F., ROGERS, L. & SCHULMAN, J. D. (1978) Biochemical heterogeneity in glutathione synthetase deficiency. *J. Clin. Invest. 62*, 1417–1420

SPIELBERG, S. P., BOXER, L. A., OLIVER, J. M., ALLEN, J. M. & SCHULMAN, J. D. (1979) Oxidative damage to neutrophils in glutathione synthetase deficiency. *Br. J. Haematol.*, in press

STJERNHOLM, R. L. & MANAK, R. C. (1970) Carbohydrate metabolism in leukocytes. XIV. Regulation of pentose cycle activity and glycogen metabolism during phagocytosis. *J. Reticuloendothel. Soc. 8*, 550–560

STOSSEL, T. P., MASON, R. J. & SMITH, A. L. (1974) Lipid peroxidation by human blood phagocytes. *J. Clin. Invest. 54*, 638–645

STRAUSS, R. R., PAUL, B. B., JACOBS, A. A. & SBARRA, A. J. (1969) The role of the phagocyte in host-parasite interactions. XIX. Leukocytic glutathione reductase and its involvement in phagocytosis. *Arch. Biochem. Biophys. 135*, 265–271

TAKANAKA, K. & O'BRIEN, P. J. (1975) Mechanisms of H_2O_2 formation by leukocytes. Properties of the NAD(P)H oxidase activity of intact leukocytes. *Arch. Biochem. Biophys. 169*, 436–442

TAUBER, A. I. & BABIOR, B. M. (1977) Evidence for hydroxyl radical production by human neutrophils. *J. Clin. Invest. 60*, 374–379

VAN DER MEER, J. W. M., VAN ZWET, TH. L., VAN FURTH, R. & WEEMAES, C. M. R. (1975) New familial defect in microbicidal function of polymorphonuclear leucocytes. *Lancet 2*, 630–632

VOETMAN, A. A., LOOS, J. A. & ROOS, D. (1979) Changes in the level of glutathione and other sulfhydryl compounds in human polymorphonuclear leukocytes at rest and during phagocytosis, in press

WEBB, L. S., KEELE, B. B. JR. & JOHNSTON, R. B. JR. (1974) Inhibition of phagocytosis-associated chemiluminescence by superoxide dismutase. *Infect. Immun. 9*, 1051–1056

WEENING, R. S., ROOS, D. & LOOS, J. A. (1974) Oxygen consumption of phagocytizing cells in human leukocyte and granulocyte preparations: a comparative study. *J. Lab. Clin. Med. 83*, 570–576

WEENING, R. S., WEVER, R. & ROOS, D. (1975) Quantification of the production of superoxide radicals by phagocytizing human granulocytes. *J. Lab. Clin. Med. 84*, 245–252

WEENING, R. S., ROOS, D., WEEMAES, C. M. R., HOMAN-MÜLLER, J. W. T. & VAN SCHAIK

M. L. J. (1976) Defective initiation of the metabolic stimulation in phagocytizing granulocytes: a new congenital defect. *J. Lab. Clin. Med. 88*, 757–768
WEENING, R. S., ROOS, D., VAN SCHAIK, M. L. J., VOETMAN, A. A., DE BOER, M. & LOOS, J. A. (1977) The role of glutathione in the oxidative metabolism of phagocytic leukocytes. Studies in a family with glutathione reductase deficiency, in *Movement, Metabolism and Bactericidal Mechanisms of Phagocytes* (Rossi, F., Patriarca, P. L. & Romeo, D., eds.), pp. 277–283, Piccin Medical Books, Padua
WEISS, S. J., KING, G. W. & LOBUGLIO, A. F. (1977) Evidence for hydroxyl radical generation by human monocytes. *J. Clin. Invest. 60*, 370–373
WEISS, S. J., RUSTAGI, P. K. & LOBUGLIO, A. F. (1978) Human granulocyte generation of hydroxyl radicals. *J. Exp. Med. 147*, 316–323
WINDHORST, D. B., PAGE, A. R., HOLMES, B., QUIE, P. G. & GOOD, R. A. (1968) The pattern of genetic transmission of the leukocyte defect in fatal granulomatous disease of childhood. *J. Clin. Invest. 47*, 1026–1034

Discussion

Cohen: Why are cells deficient in glucose-6-phosphate dehydrogenase susceptible to damage? The answer is obvious for red cells: NADPH can only be made through that dehydrogenase and the second enzyme of the hexose-monophosphate shunt. Does not isocitrate dehydrogenase (EC 1.1.1.42) or other enzymes provide NADPH in other cell types?

Roos: In neutrophils the hexose-monophosphate shunt is the NADPH-delivery system for the oxidase and for the glutathione redox system. The shunt activity is stimulated many-fold during phagocytosis.

Cohen: Might there be some structural coupling between NADPH generation and NADPH oxidase in phagocytes?

Roos: Yes, this is one of the currently favoured theories.

Williams: If you are not postulating an extra enzyme which converts superoxide in these systems into H_2O_2, what brings about that conversion?

Roos: Probably, an enzyme system, located in the plasma membrane, reduces oxygen to superoxide which may react with the protons to form hydrogen peroxide without any enzyme.

Williams: We now need to know the rates and steady-state concentrations to substantiate that idea. Why should the system bother to go through O_2^- when perfectly good systems exist to convert O_2 into H_2O_2 directly without releasing O_2^-? If the cells did not release it, they would not have needed superoxide dismutase outside them.

Roos: Don't most enzymes that generate hydrogen peroxide do so, to a considerable degree, by way of superoxide?

Fridovich: No; some enzymes make substantial amounts of superoxide but most enzymes that generate hydrogen peroxide (e.g. all the flavin oxidases) do so without making O_2^-.

Roos: Yet the leukocyte enzyme is a flavin oxidase and produces O_2^-.

Fridovich: It is generally accepted that the pH inside the phagosome is lower than neutrality: about 4.5–5.0. At this pH the spontaneous dismutation (i.e. the reaction of $HO_2 + O_2^-$) must be considered; it is rapid, with a rate constant about 5×10^{-7} l mol^{-1} s^{-1}. So, the formation of H_2O_2 from O_2^- without SOD in phagosomes is not a problem. As to the question why?: *ex post facto* we can conjecture that with both O_2^- and peroxide there can be generated still more reactive intermediates which can attack the ingested microorganism.

Reiter: If superoxide is the active reagent, does one observe the same level of killing when one fills lipósomes with glucose oxidase, which does not generate superoxide?

Roos: Superoxide itself is not bactericidal in cell-free systems; the main bactericidal process is executed by hydrogen peroxide in combination with superoxide.

Reiter: Is galactose oxidase, for instance, which does produce superoxide, more bactericidal?

Fridovich: Galactose oxidase is inhibited by superoxide dismutase but it does not produce substantial amounts of free superoxide (Hamilton *et al.* 1978).

Cohen: In measuring the effects of azide on the concentration of glutathione (Fig. 3) you were indirectly measuring the steady-state concentration of peroxide within the cells. Since the competition between glutathione peroxidase and catalase depends on the concentration of H_2O_2 (that is, in erythrocytes, catalase becomes increasingly effective at concentrations of peroxide greater than 1 μmol/l; cf. Cohen & Hochstein 1963), the concentration of peroxide within the glutathione reductase-deficient cells must be greater than 1 μmol/l.

Roos: The glutathione reductase deficiency allows the accumulation of hydrogen peroxide.

Klebanoff: Did azide or cyanide increase oxygen consumption, as previously found (Klebanoff & Hamon 1972), and thus possibly increase the production of oxygen radicals in glutathione reductase-deficient cells, in addition to affecting their ultimate disposition?

Roos: Yes; the oxygen consumption is stimulated by azide as well as by cyanide. We interpret that to mean that catalase is inhibited and so no oxygen is formed from peroxide. This was found both in normal and in glutathione reductase-deficient cells. However, addition of these inhibitors does not increase the production of radicals, it only allows their accumulation. The inactivation of the glutathione reductase-deficient cells was about as fast in the presence as it was in the absence of azide or cyanide. We expected a faster inactivation in the presence of these inhibitors, because the level of reduced glutathione in these cells decreases faster in these conditions.

Flohé: Then why do you claim that the glutathione system has nothing to do with oxygen consumption? According to your earlier remarks (p. 248), inhibition of catalase in white blood cells should increase GSSG formation. Doesn't GSSG stimulate the pentose-phosphate shunt (Egglestone & Krebs 1974), which in turn supplies the substrate for NADPH oxygenase?

Roos: GSSG stimulates the shunt by reacting with NADPH. I conclude from the fact that the reduction of cytochrome *c* by superoxide is normal in neutrophils with 15% of normal glutathione reductase activity, that this enzyme is not important for the formation of superoxide in these cells.

Fridovich: Azide has a dramatic effect on the glutathione reductase-deficient cells from the patients you discussed (Fig. 3) but those cells do not show the respiratory burst, do they?

Roos: In fact, they do for a few minutes.

Fridovich: And enough peroxide is produced in those few minutes to damage the cells?

Roos: Yes; in that time the azide allows all the hydrogen peroxide to accumulate, then the cells are damaged and stop metabolizing oxygen.

Segal: What effect did azide alone have on concentrations of reduced glutathione in non-phagocytosing cells?

Roos: The concentrations stayed high, as in normal cells.

Segal: You propose that these cells stop consuming oxygen because they are damaged by their own products. Is there any evidence of damage, such as release of lactate dehydrogenase or uptake of Trypan blue?

Roos: No lactate dehydrogenase is released and there is no active uptake of Trypan blue. We observed no signs of cell lysis.

Segal: The drop in glutathione concentration may stop further phagocytosis of particles (Oliver *et al.* 1976).

Roos: No; phagocytosis of bacteria continues normally for 20 min when the process is saturated (Roos *et al.* 1979).

Segal: In that case the cells cannot be that sick.

Roos: The subjects (I should not call them patients) are not seriously ill. In contrast, glutathione peroxidase-deficient patients are as ill as CGD patients. Why, we don't know, but we favour the following explanation. The glutathione reductase-deficient cells show a brief normal period of production of superoxide and peroxide (Fig. 2) and may be able to cope with bacterial infections *in vivo. In vitro* the killing of bacteria by the glutathione reductase-deficient leukocytes was normal unless one gave these cells a large amount of bacteria. In that case normal cells still killed most of the bacteria in 15 min. Probably, the reaction to zymosan particles is exaggerated — with a large particle the cell has to work hard and releases more oxygen products than it does when it engulfs a bacterium. Consequently, the glutathione reductase-deficient cells may also have been inactivated sooner by the oxygen radicals during incubation with zymosan than with bacteria. The cells of glutathione reductase-deficient patients have normal glutathione peroxidase activity so they can degrade hydrogen peroxide while the concentration of reduced glutathione drops — that period may be long enough for protection. That possibility cannot apply to peroxidase-deficient cells. As a consequence, these cells may be inactivated almost immediately after the start of phagocytosis.

Segal: If phagocytosis of *bacteria* is normal in the glutathione reductase-deficient cells, that does not mean that oxygen consumption and production of hydrogen peroxide are not terminated prematurely by failure of the cell to phagocytose the zymosan that was used as a stimulus in these experiments. It would be important to compare the uptake of zymosan with oxygen consumption in cells from these patients and normal controls.

Roos: Phagocytosis is not necessary for metabolic stimulation (Goldstein *et al.* 1975).

Willson: Is oxygen consumption necessarily equated with processes involving O_2^- or can we say that processes involving O_2^- are needed to initiate oxygen consumption which then continues independently (with the aid of a reducing system)?

Roos: The latter is possible but there is no evidence for it in leukocytes.

Willson: Such a mechanism could perhaps explain the finding by Drs Segal & Allison that SOD inhibits the reduction of cytochrome *c* but does not affect oxygen consumption: only a little O_2^- is required to start some cycle that is independent of superoxide (e.g. lipid peroxidation); reduction of iron(III) to iron(II) or some other redox system may be needed to keep the cycle going.

Fridovich: Several groups (Curnutte & Babior 1974; Root & Metcalf 1977; Salin & McCord 1974; Weening *et al.* 1975; Drath & Karnovsky 1975) have eliminated the possibility that O_2^- is a trigger. They quantified O_2^- in terms of SOD-inhibitable reduction of cytochrome *c* and concluded that about half the total oxygen consumption can be accounted for by O_2^- detectable outside the cells (assuming that none of the SOD or cytochrome *c* crosses into the cell).

Segal: With regard to glutathione peroxidase deficiency, we have studied selenium-deficient cows which have 10% of the total normal enzyme activity; their oxygen consumption is normal in response to the addition of IgG-coated latex particles. It seems unlikely to me that in patients with glutathione reductase-deficiency the cytochrome *c* reduction continues for much longer than phagocytosis-induced oxygen consumption because the cytochrome *c* acts as a superoxide radical scavenger, protecting the cell from oxidative damage. Cytochrome *c* is excluded from the closed phagocytic vacuole; therefore, as we all believe, the product is formed in the vacuole.

Fridovich: Why do you say 'we all believe'? The data argue against that. Something in the membrane is activated and can generate the products of oxygen reduction (O_2^- and peroxide), directing them to the outside or, after invagination, to the cytosol. The lining of the cytosol is not the only site of action. Cells have been activated without phagocytosis, and show similar results (Roos *et al.* 1976).

Segal: That seems metabolically wasteful and doesn't make biological sense to me. Those data describe cytochrome *c* reduction. As I tried to outline, this evidence is not sufficient basis for such a major physiological principle.

Fridovich: To be able to criticize the idea, though, you should first re-examine the sensitivity of the method you used to collect evidence.

Winterhalter: The evidence presented at the Banyuls meeting (see Michelson *et al.* 1977) compels us to accept that O_2^- must be involved at the beginning of the killing effect. I am not saying that it is the reactive species. In particular, three experiments stand out: (*a*) with polylysyl-SOD (Salin & McCord 1977); (*b*) with superoxide dismutase bound to latex particles which are co-phagocytosed with the bacteria (Johnston & Lehmeyer 1977) and (*c*) those which showed the protection afforded to white cells, which are killed by their own superoxide and related species, by relatively high concentrations of superoxide dismutase and by much lower concentrations of the polylysyl-enzyme (Salin & McCord 1977).

Fridovich: Most of us would agree. Dr Segal, I would like you to do a calculation based on the following consideration. The molar extinction coefficient for reduced cytochrome *c* at 550 nm is well known. You should calculate the number of micromoles of cytochrome *c* reduced per minute in your system and then calculate how much change in oxygen consumption could be caused by adding cytochrome *c* to see whether your method is sensitive enough to detect that small effect.

Flohé: Dr Segal, in your experiments with white cells with 10% glutathione peroxidase activity, what were you looking for? Are the experiments comparable with those of Dr Roos in terms of time or were you looking only at very short periods? Dr Roos concluded that the peroxidase system is essential for the protection of the whole cell.

Segal: We were looking at the initial rate of oxygen consumption, which was normal. A few patients have been described who have chronic granulomatous disease and glutathione peroxidase deficiency; this combination does not make sense, because it should not cause the syndrome of chronic granulomatous disease.

Flohé: Why not?

Segal: Why should it block the initiation of oxygen consumption by neutrophils?

Flohé: It should not necessarily block the initiation, but, as Dr Roos pointed out, if the cells start to produce considerable quantities of O_2^- and hydrogen peroxide, they might be more prone to commit suicide, since one of the protective mechanisms is no longer operating.

Segal: But in chronic granulomatous disease there is no oxygen consumption by the cells from the outset and no reason for them to be damaged by this oxidative process.

Roos: Good kinetic studies of the metabolic rates in the cells of the glutathione peroxidase-deficient patients are needed.

Cohen: In glutathione peroxidase deficiency the capacity to generate GSSG is lacking. Might not oxidized glutathione or mixed disulphides play some critical role?

Roos: Burchill *et al.* (1978) have given some evidence that protein–glutathione disulphides may be involved in the disassembly of microtubules.

Flohé: At least in tissues other than leukocytes, an increase in GSSG directly or indirectly stimulates the pentose-phosphate shunt (Kinoshita & Masurat 1957; Jacob & Jandl 1966; Hochstein & Utley 1968; Bénard & de Groot 1969; Egglestone & Krebs 1974).

Williams: With regard to the H_2O_2 system, you described how H_2O_2 may be used in two ways: through catalase and through GSH (the path you emphasized). What is the functional requirement for the catalase pathway?

Roos: It might protect certain parts of the phagocytic cell. In some preliminary experiments on cells from patients who lack catalase activity a heavy stress with hydrogen peroxide damages the cells but the metabolic stimulation during phagocytosis is normal. Thus the H_2O_2 generated during phagocytosis can be detoxified by the glutathione system alone, but protection against larger amounts of H_2O_2 requires the combined action of catalase and the glutathione redox system.

Williams: Are the rate constants for glutathione peroxidase comparable with those of catalase?

Michelson: Yes.

Roos: At low concentrations of hydrogen peroxide it is more effective than catalase (Cohen & Hochstein 1963).

Flohé: This frequently heard remark needs some comment. At low concentrations of H_2O_2, catalase and GSH peroxidase react with H_2O_2 at comparable rates. The rate constants for GSH peroxidase may be slightly higher. Both reactions are first order over a wide range of concentrations. Only at high, probably unphysiological, concentrations is the peroxide turnover of catalase limited by denaturation of the enzyme, and in the case of GSH peroxidase, regeneration of reduced enzyme by GSH becomes rate-limiting. It is, therefore, hard to imagine how some mysterious 'crossover' in the efficacy of these two enzymes might be brought about at physiological peroxide concentrations, that is, at concentrations below micromolar.

Winterhalter: Acatalasaemic patients have a very unstable catalase enzyme. In long-lived cells, such as erythrocytes, the enzyme dies after a few days. But do acatalasaemic patients have normal catalase levels in their short-lived white cells?

Roos: The activity is decreased in the leukocytes also.

Winterhalter: Is their bactericidal capacity normal?

Roos: I don't know about the bactericidal capacity *in vitro* but the metabolic stimulation is normal and acatalasaemic patients do not suffer from life-threatening infections.

Cohen: A partial answer to Professor Williams' question is that cells from acatalasaemic ducks are resistant to hydrogen peroxide, provided it is added slowly (e.g. by a diffusion method) (Cohen & Hochstein 1963). When one looks at some of the characteristics of damage to erythrocytes (say, damage to haemoglobin), the cell is completely protected at the expense of glutathione. The glutathione is oxidized but then reduced as long as glucose is metabolized through the hexose shunt.

Williams: Does catalase have any role, then?

Fridovich: Yes! Acatalasaemic humans (occasionally) have serious defects around the gums where there are bacteria that make and secrete peroxide which damages the gum tissues because there is no catalase. It seems that, when the concentration of peroxide applied is high, catalase is very important.

Klebanoff: If catalase is removed but the cell functions normally, that does not necessarily mean that when catalase is present it is not the functional enzyme for degradation of hydrogen peroxide, since we are blessed with back-up systems of all kinds.

Segal: We conclude that catalase is important because the addition of hydrogen peroxide to intact neutrophils had no effect on the measurable concentration of SH groups in the absence of azide, however, in the presence of azide, the number of SH groups dropped with increasing hydrogen peroxide concentration (see Fig. 1) (unpublished results).

References

Bénard, B. & de Groot, L. J. (1969) The role of hydrogen peroxide and glutathione in glucose oxidation by the thyroid. *Biochim. Biophys. Acta 184*, 48–53

Burchill, B. R., Oliver, J. M., Pearson, C. B., Leinbach, E. D. & Berlin, R. D. (1978) Microtubule dynamics and glutathione metabolism in phagocytizing human polymorphonuclear leukocytes. *J. Cell Biol. 76*, 439–447

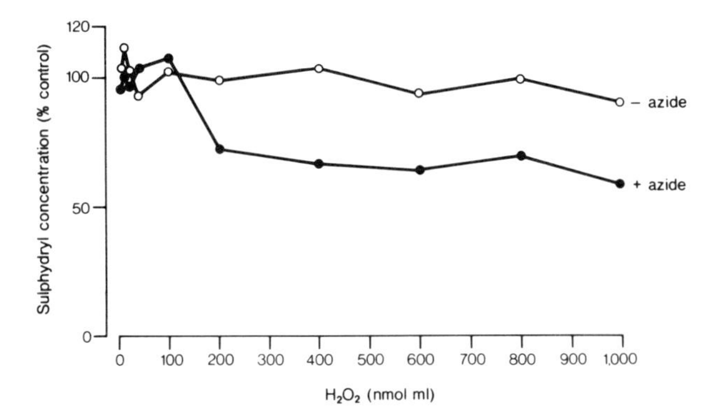

FIG. 1 (Segal). Addition of exogenous hydrogen peroxide to human neutrophils; effect on the concentration of soluble thiol groups.

COHEN, G. & HOCHSTEIN, P. (1963) Glutathione peroxidase: the primary agent for the elimination of hydrogen peroxide in erythrocytes. *Biochemistry 2*, 1420–1428

CURNUTTE, J. T. & BABIOR, B. M. (1974) Biological defense mechanisms — the effect of bacteria and serum on superoxide production by granulocytes. *J. Clin. Invest. 53*, 1662–1672

DRATH, O. B. & KARNOVSKY, M. L. (1975) Superoxide production by phagocytic leukocytes. *J. Exp. Med. 141*, 257–262

EGGLESTONE, L. V. & KREBS, H. A. (1974) Regulation of the pentose phosphate cycle. *Biochem. J. 183*, 425–435

GOLDSTEIN, I. M., ROOS, D., KAPLAN, H. B. & WEISSMAN, G. (1975) Complement and immunoglobulins stimulate superoxide production by human leukocytes independently of phagocytosis. *J. Clin. Invest. 56*, 1155–1163

HAMILTON, G. A., ADOLPH, P. K., DE JERSEY, J., DUBOUS, G. C., DYRKACZ, G. R. & LIBBY, R. D. (1978) Trivalent copper, superoxide and galactose oxidase. *J. Am. Chem. Soc. 100*, 1899–1912

HOCHSTEIN, P. & UTLEY, M. (1968) Hydrogen peroxide detoxification by glutathione peroxidase and catalase in rat liver homogenates. *Mol. Pharmacol. 4*, 574–579

JACOB, H. S. & JANDL, J. H. (1966) Effects of sulfhydryl inhibition on red blood cells. III. Glutathione in the regulation of the hexose monophosphate pathway. *J. Biol. Chem. 241*, 4243–4250

JOHNSTON, R. B. & LEHMEYER, J. E. (1977) The involvement of oxygen metabolites from phagocytic cells in bactericidal activity and inflammation, in Michelson *et al.* (1977), pp. 291–306

KINOSHITA, J. H. & MASURAT, T. (1957) Studies on glutathione in bovine lens. *Arch. Ophthalmol. (Paris) 57*, 266–274

KLEBANOFF, S. J. & HAMON, C. B. (1972) Role of myeloperoxidase-mediated antimicrobial systems in intact leukocytes. *J. Reticuloendothel. Soc. 12*, 170–196

MICHELSON, A. M., MCCORD, J. M. & FRIDOVICH, I. (eds.) (1977) *Superoxide and Superoxide Dismutase*, Academic Press, London and New York

OLIVER, J. M., ALBERTINI, D. F. & BERLIN, R. D. (1976) Effect of glutathione-oxidizing agents on microtubule assembly and microtubule-dependent surface properties of human neutrophils. *J. Cell Biol. 71*, 921–932

ROOS, D., GOLDSTEIN, I. M., KAPLAN, H. B. & WEISSMAN, G. (1976) Dissociation of phagocytosis, metabolic stimulation and lysosomal enzyme release in human leukocytes. *Agents Actions 6*, 256–259

ROOS, D., WEENING, R. S., VOETMAN, A. A., VAN SCHAIK, M. L. J., BOT, A. A. M., MEERHOF, L. J. & LOOS, J. A. (1979) Protection of phagocytic leukocytes by endogenous glutathione. Studies in a family with glutathione reductase deficiency. *Blood*, in press

ROOT, R. K. & METCALF, J. A. (1977) H_2O_2 release from human granulocytes during phagocytosis. Relationship to superoxide anion formation and cellular catabolism of H_2O_2: studies with normal and cytochalasin B-treated cells. *J. Clin. Invest. 60*, 1266–1279

SALIN, M. L. & MCCORD, J. M. (1974) Superoxide dismutases in polymorphonuclear leukocytes. *J. Clin. Invest. 54*, 1005–1009

SALIN, M. L. & MCCORD, J. M. (1977) Free radicals in leukocyte metabolism and inflammation, in Michelson *et al.* (1977), pp. 257–270

WEENING, R. S., WEVER, R. & ROOS, D. (1975) Quantitative aspects of the production of superoxide radicals by phagocytizing human granulocytes. *J. Lab. Clin. Med. 85*, 245–252

The role of myeloperoxidase in the microbicidal activity of polymorphonuclear leukocytes

S. J. KLEBANOFF and H. ROSEN

Department of Medicine, University of Washington School of Medicine, Seattle, Washington

Abstract Myeloperoxidase (MPO), H_2O_2 and a halide form a powerful antimicrobial system effective against bacteria, fungi, viruses and mammalian cells. After phagocytosis, MPO is released into the phagosome from adjacent granules where it interacts with H_2O_2 generated either by leukocytic or microbial metabolism and a halide such as chloride or iodide to form agents toxic to the ingested organisms. Evidence for H_2O_2 and MPO participation in the microbicidal activity of polymorphonuclear leukocytes (PMNs) has been obtained from patients with neutrophil dysfunction. In chronic granulomatous disease, PMNs have a microbicidal defect associated with the absence of the respiratory burst. The importance of H_2O_2 deficiency in the PMN dysfunction is emphasized by its reversal by H_2O_2. PMNs which lack MPO also have a major fungicidal and bactericidal defect. Bactericidal activity is particularly low during the early postphagocytic period, after which the organisms are killed. Although emphasizing the importance of MPO-mediated antimicrobial systems particularly during the early postphagocytic period, these findings also indicate the presence of MPO-independent systems which develop slowly but are ultimately effective. The MPO-independent antimicrobial systems may be oxygen-dependent or oxygen-independent. The acetaldehyde–xanthine oxidase system has been used as a model of the MPO-independent, oxygen-dependent antimicrobial systems of the PMN. A microbicidal effect by this system was observed which was inhibited by superoxide dismutase, catalase and scavengers of hydroxyl radicals ($OH^•$) and singlet oxygen ($^1O_2$). The microbicidal activity of acetaldehyde and xanthine oxidase is increased considerably by MPO and chloride. The formation of ethylene from methional or 2-oxo-4-methylthiobutyric acid by PMNs has been regarded as evidence for $OH^•$ formation. We have found ethylene formation to be largely dependent on MPO and evidence for the initiation of ethylene formation by $^1O_2$ has been obtained. Both the xanthine oxidase system and the MPO–H_2O_2–halide system convert diphenylfuran into *cis*-dibenzoylethylene, an effect which is compatible with, although not proof of, the formation of $^1O_2$ by these systems.

Phagocytosis by polymorphonuclear leukocytes (PMNs) is associated with a burst of oxidative metabolism (Sbarra & Karnovsky 1959); oxygen consumption is increased many-fold and much if not all the extra oxygen consumed is converted into hydrogen peroxide (H_2O_2) (Iyer *et al.* 1961) via a superoxide anion (O_2^-) intermediate (Babior *et al.* 1973). The highly-reactive products of oxygen reduction and excitation, namely, H_2O_2, O_2^-, hydroxyl radicals ($OH^\bullet$) and singlet molecular oxygen ($^1O_2$) have been implicated as microbicidal agents in the PMN, with the microbicidal activity of H_2O_2 being considerably increased by myeloperoxidase (MPO) and a halide (for reviews see Klebanoff 1975; DeChatelet 1975; Babior 1978; Klebanoff & Clark 1978). In this paper we shall consider these oxygen-dependent antimicrobial systems of the PMN.

THE MPO-MEDIATED ANTIMICROBIAL SYSTEM

MPO, H_2O_2 and a halide form a powerful antimicrobial system effective against a variety of microorganisms (Klebanoff 1975; Klebanoff & Clark 1978) (Fig. 1). MPO is present in extremely high concentrations in the lysosomal (azurophil, primary) granules of the PMN. After phagocytosis, the membranes of cytoplasmic granules fuse with the membrane of the phagocytic vacuole, the connecting membrane ruptures and the granule contents including MPO are discharged into the vacuolar space.

The H_2O_2 required for the MPO-mediated antimicrobial system can be supplied by either leukocytic or microbial metabolism. The PMN responds to phagocytosis (or more accurately to perturbation of the plasma membrane) with a marked increase in oxygen consumption, and it is now well established that much if not all this extra oxygen consumed is converted into H_2O_2. It is the prevailing view that an enzyme, activated by phagocytosis, catalyses the oxidation of reduced pyridine nucleotides, with the reduction of oxygen first to the superoxide anion and then, by dismutation, to H_2O_2. The precise location of this oxidase, or the primary nucleotide on which it acts (NADH or NADPH)

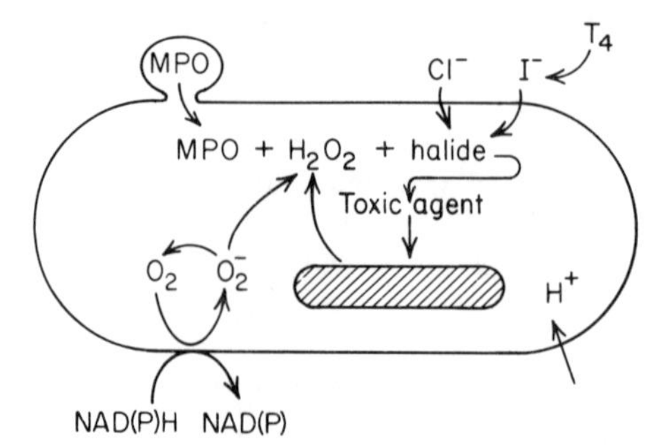

FIG. 1. The myeloperoxidase-mediated antimicrobial system: T_4, thyroid hormones (From Klebanoff & Clark 1978, reproduced by permission of North-Holland Publishing Co.)

is not known with certainty; however, there is considerable evidence to indicate that H_2O_2 can accumulate in the phagocytic vacuole as a result of its action.

Certain bacterial species, e.g., pneumococci, streptococci and lactobacilli, use flavoproteins for terminal oxidations with the production of H_2O_2. They also lack the haem enzyme catalase (although non-haem catalases and peroxidases may be present) and as a result H_2O_2 accumulates in the medium. The H_2O_2 formed may be autoinhibitory, particularly in the presence of MPO and a halide (Klebanoff 1968), and in mixed cultures H_2O_2-generating organisms can be used to kill non-H_2O_2-generating bacteria (Hamon & Klebanoff 1973), fungi (Hamon & Klebanoff 1973), viruses (Klebanoff & Belding 1974) or mammalian cells (Klebanoff & Smith 1970; Clark *et al.* 1975) by the peroxidase system.

Chloride is present in the PMN in concentrations considerably greater than those required in the isolated MPO-mediated antimicrobial system and is probably the predominant halide in the intact cell. Iodide is considerably more effective than chloride on a molar basis; however, its concentration in serum is small. The thyroid hormones are deiodinated by PMNs during phagocytosis (Klebanoff & Green 1973; Woeber & Ingbar 1973) and may serve as an additional source of iodide for the intact cell. Iodide absorbed as such or released from the thyroid hormones should contribute to the halide pool and thus to the toxicity, although this contribution may not be large.

The pH optimum of the MPO-mediated antimicrobial system is generally acid (pH 4.5–5.0) (Klebanoff 1967; McRipley & Sbarra 1967), although it varies with the H_2O_2 and halide concentrations (Sbarra *et al.* 1977). The pH within the phagocytic vacuole is also acid with the reported values ranging from 6.0–6.5 to 3.0 or below (Jensen & Bainton 1973; Jacques & Bainton 1978). Thus the pH within the phagocytic vacuole appears to be favourable for the operation of the MPO-mediated antimicrobial system.

The MPO-mediated antimicrobial system is inhibited by catalase, by an excess of H_2O_2 and by several low-molecular-weight reducing agents (Klebanoff 1967, 1968; McRipley & Sbarra 1967). Inhibitors may be introduced into the vacuole during degranulation, by diffusion from the cytosol or as a component of the ingested organism. With regard to the latter, staphylococcal strains rich in catalase are more virulent to mice and are more resistant to the microbicidal activity of PMNs *in vitro* than are strains with a low catalase content (Mandell 1975); this fact suggests that catalase is a virulence factor for some bacteria.

The considerable evidence for an important role for the MPO-mediated antimicrobial system in the intact cell has been reviewed recently (Klebanoff 1975; Klebanoff & Clark 1978) and may be summarized as follows:

(1) The components of the MPO system (MPO, H_2O_2, halide) are present in the PMN and are secreted into the phagocytic vacuole.

(2) The interaction of these components adjacent to the ingested organism is indicated by the iodination of microbial surface components, a reaction which requires MPO, H_2O_2 and iodide, and by the cytochemical demonstration of the interaction of MPO and H_2O_2 in the vacuolar space.

(3) H_2O_2 formed during the respiratory burst is required for optimum antimicrobial activity. The importance of the phagocytosis-induced respiratory burst is emphasized by the microbicidal defect seen when normal PMNs are exposed to an atmosphere of nitrogen (Mandell 1974) and by the major microbicidal defect found in leukocytes which lack the respiratory burst, i.e. in those from patients with chronic granulomatous disease (CGD) (Quie *et al.* 1967; Holmes *et al.* 1967). The defect in CGD appears to be due in large part to a deficiency of H_2O_2 since it is reversed by the introduction of H_2O_2 into the vacuole (Baehner *et al.* 1970; Johnston & Baehner 1970; Root 1974). Glucose oxidase, an enzyme which forms H_2O_2 apparently without the intermediacy of superoxide, can be used. Further, those organisms which generate H_2O_2, namely, pneumococci, streptococci and lactobacilli, are killed well by CGD leukocytes (Kaplan *et al.* 1968; Klebanoff & White 1969; Mandell & Hook 1969) and are rarely found in the lesions in this condition, whereas H_2O_2-negative mutants of the same organisms are killed poorly by CGD leukocytes (Holmes & Good 1972; Pitt & Bernheimer 1974).

(4) MPO is required for optimum microbicidal activity in the PMN. The evidence is as follows:

(*a*) Leukocytes which lack MPO have a marked microbicidal defect (Lehrer & Cline 1969; Lehrer *et al.* 1969; Klebanoff 1970). Patients with hereditary MPO deficiency have PMNs with a fungicidal and bactericidal defect second only to CGD in its severity. The bactericidal curve is characterized by a lag period after which the organisms die. This residual antimicrobial activity together with other host defence mechanisms combine to keep most patients with hereditary MPO deficiency in reasonably good health. Some, however, have had unusual infections (see Klebanoff & Clark 1978).

(*b*) Certain peroxidase inhibitors, for example, azide (Klebanoff 1970; Koch 1974), cyanide (Klebanoff 1970) and sulphonamides (Lehrer 1971), inhibit the microbicidal activity of normal PMNs. Since these agents do not affect the residual microbicidal activity of MPO-deficient PMNs, their effect on normal cells appears to be due to the inhibition of MPO.

MPO-INDEPENDENT ANTIMICROBIAL SYSTEMS

The residual antimicrobial activity of MPO-deficient leukocytes emphasizes the presence in these cells and presumably also in normal cells of antimicrobial systems which do not require MPO. These systems develop slowly but are ultimately effective, at least against bacteria. Several MPO-independent systems have been described; some require oxygen and others are effective in anaerobic conditions. Only the oxygen-dependent antimicrobial systems will be considered here.

The evidence for the presence of oxygen-dependent but MPO-independent antimicrobial systems in the PMN is as follows:

(1) Leukocytes which lack the respiratory burst, i.e. from patients with CGD, have a greater microbicidal defect than leukocytes from patients with hereditary MPO deficiency. This suggests the presence in MPO-deficient leukocytes of antimicrobial systems which are dependent on oxidative metabolism.

(2) The staphylocidal activity observed after a lag period in MPO-deficient PMNs is inhibited by hypoxia (Klebanoff & Hamon 1972).

(3) In contrast to CGD, the phagocytosis-induced respiratory burst is not depressed in hereditary MPO deficiency (Lehrer & Cline 1969); indeed, oxygen consumption (Klebanoff & Hamon 1972), superoxide production (Rosen & Klebanoff 1976), H_2O_2 generation (Klebanoff & Pincus 1971) and glucose C-1 oxidation (Klebanoff & Pincus 1971) are greater than that of normal cells.

(4) The highly reactive products of oxygen reduction and excitation (i.e., O_2^-, $OH^\bullet$, H_2O_2 and $^1O_2$) have antimicrobial activity in the absence of MPO.

Several investigators have described a microbicidal effect by systems which generate O_2^- and, by dismutation, H_2O_2 (see Klebanoff 1975; DeChatelet 1975; Babior 1978; Klebanoff & Clark 1978). In some instances, this antimicrobial activity is inhibited by catalase but not by superoxide dismutase (SOD), thus implicating H_2O_2 as the toxic agent. In other instances, catalase, SOD and $OH^\bullet$ scavengers are inhibitory and $OH^\bullet$ generated by the interaction of H_2O_2 and O_2^- according to the Haber-Weiss reaction (1) (Haber & Weiss 1934) has been

$$O_2^- + H_2O_2 \rightarrow O_2 + OH^- + OH^\bullet \tag{1}$$

proposed as the microbicidal agent. The Haber–Weiss reaction as depicted above has been questioned as an efficient mechanism for the generation of $OH^\bullet$ (McClune & Fee 1976; Halliwell 1976); however, inhibitory studies suggest that a strong oxidant is formed with the properties of $OH^\bullet$ by this or a closely related mechanism. In this regard, McCord & Day (1978) recently proposed that the formation of $OH^\bullet$ from O_2^- and H_2O_2 may depend on a metal ion as follows:

$$\begin{array}{l} O_2^- + M^{n+} \rightarrow O_2 + M^{(n-1)+} \\ \underline{M^{(n-1)+} + H_2O_2 \rightarrow M^{n+} + OH^- + OH^\bullet} \\ O_2^- + H_2O_2 \rightarrow O_2 + OH^- + OH^\bullet \end{array} \tag{2}$$

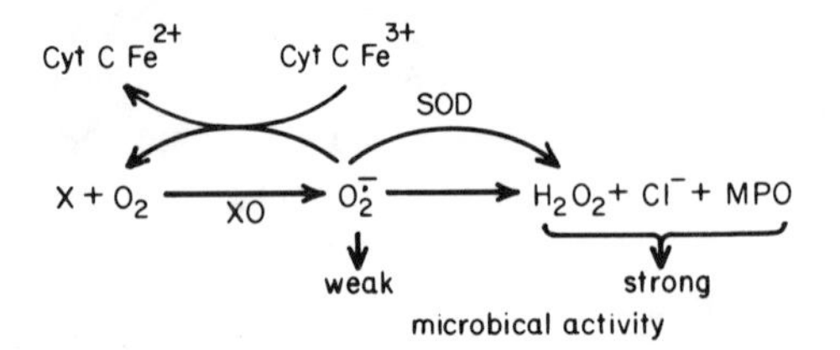

FIG. 2. Microbicidal activity of the xanthine (X)–xanthine oxidase (XO)–chloride–myeloperoxidase (MPO) system. (From Klebanoff & Clark 1978, reproduced by permission of North-Holland Publishing Co.)

The microbicidal activity of systems that generate O_2^- and (by dismutation) H_2O_2 is increased considerably by MPO and a halide. In initial studies (Klebanoff 1974), xanthine and xanthine oxidase were used as the source of O_2^- and H_2O_2 and little antimicrobial activity was observed when the xanthine oxidase system alone was used. The microbicidal activity was greatly increased by the addition of MPO and a halide; scavengers of O_2^- such as ferricytochrome *c* decreased the microbicidal effect of this supplemented system and the inhibition was reversed by SOD (Fig. 2).

In more recent studies, acetaldehyde was used instead of xanthine as substrate for xanthine oxidase since xanthine, and the product of its oxidation by xanthine oxidase, uric acid, appear to scavenge singlet oxygen (Kellogg & Fridovich 1977) and possibly other reactive intermediates. The microbicidal activity of the xanthine oxidase system was considerably increased (Table 1). Acetaldehyde (10^{-2} mol/l) and xanthine oxidase (4 mU) in phosphate buffer (pH 7.0) decreased the viable cell count from 4.5×10^6 to 5×10^3 (99.9%) in 60 min. The further addition of MPO (8 mU) and chloride (5×10^{-2} mol/l) to the acetaldehyde–xanthine oxidase system greatly increased the toxicity, with a bactericidal effect observed at an acetaldehyde concentration only 1% of that required

TABLE 1

Staphylocidal activity of xanthine oxidase systems

Acetaldehyde (mol/l)	*Viable cell count (10^6 organisms/ml)*		
	Acetaldehyde	*Acetaldehyde + xanthine oxidase*	*Acetaldehyde + xanthine oxidase + MPO and Cl^-*
10^{-2}	3.6[a]	0.005[b]	
10^{-3}	3.7	4.2	
10^{-4}		4.9	0.01[b]
10^{-5}		3.9	4.7

[a] Geometric mean of three experiments.
[b] Significance of difference from initial viable cell count (4.5×10^6/ml) < 0.05; all other values not significantly different.

TABLE 2

Inhibition of the staphylocidal activity of the acetaldehyde–xanthine oxidase system

Additions	*Viable cell count (10^6 organisms/ml)*
Control	4.5
Complete xanthine oxidase system	0.008
+ SOD (1 μg/ml)	3.1[a]
+ SOD (heated)	0.006
+ catalase (60 μg/ml)	4.1[a]
+ catalase (heated)	0.03
+ mannitol (10^{-2} mol/l)	0.5[a]
+ benzoate (10^{-2} mol/l)	4.1[a]
+ DABCO[b] (10^{-2} mol/l)	4.1[a]
+ histidine (10^{-3} mol/l)	3.4[a]

[a] Inhibitor compared to no inhibitor, $P < 0.05$.
[b] 1,4-Diazabicyclo[2,2,2]octane.

for the xanthine oxidase system alone. These findings suggest that the toxic products of the xanthine oxidase system are required in considerably higher concentration in the absence of MPO and chloride than in their presence, for an equivalent microbicidal effect.

The bactericidal activity of the acetaldehyde–xanthine oxidase system was inhibited by SOD, catalase, the $OH^{\bullet}$ scavengers mannitol and benzoate, and the $^1O_2$ quenchers, 1,4-diazabicyclo[2,2,2]octane (DABCO) and histidine (Table 2). Although the specificity of some of the inhibitors for a particular reactive species is not absolute, the findings suggest that the bactericidal activity of the acetaldehyde–xanthine oxidase system depends on O_2^-, H_2O_2 and the products of their interaction, $OH^{\bullet}$ and $^1O_2$. The formation of $OH^{\bullet}$ and $^1O_2$ by the PMN is considered below.

FORMATION OF HYDROXYL RADICALS

Since both O_2^- and H_2O_2 are formed by PMNs during phagocytosis, their interaction to generate $OH^{\bullet}$ might be expected. A method for the detection of $OH^{\bullet}$ by PMNs has recently been proposed based on the formation of ethylene from methional (Tauber & Babior 1977) or from 2-oxo-4-methylthiobutyric acid (Weiss *et al.* 1978). This conversion is initiated by electron abstraction from the sulphur atom of the thioether (Yang 1969). Although $OH^{\bullet}$ is particularly efficient in this regard, other one-electron oxidants also are effective (Bors *et al.* 1976). When ethylene formation is inhibited by SOD, catalase and $OH^{\bullet}$ scavengers and is stimulated by H_2O_2, $OH^{\bullet}$ generated by the interaction of O_2^- and H_2O_2 has been proposed as the oxidant (Beauchamp & Fridovich 1970). Ethylene formation by PMNs is strongly inhibited by SOD and, to a

TABLE 3

Ethylene formation from 2-oxo-4-methylthiobutyric acid by MPO-deficient PMNs[a]

Additions	*Ethylene formation (pmol [10^7 PMN]$^{-1}$ h^{-1})*
Normal PMNs	13 200
Normal PMNs + MPO	12 800
MPO-deficient PMNs	960
MPO-deficient PMNs + MPO	28 600
MPO-deficient PMNs + MPO (heated)	940

[a] See Klebanoff & Rosen (1978) for experimental details.

lesser degree, by catalase (see, however, Tauber & Babior 1977) and $OH^{\bullet}$ scavengers (Tauber & Babior 1977; Weiss *et al.* 1978).

We have confirmed the formation of ethylene from 2-oxo-4-methylthiobutyric acid by phagocytosing PMNs (Klebanoff & Rosen 1978; Table 3); our studies, however, indicate that this conversion largely depends on MPO. The evidence is as follows.

(1) Ethylene formation by leukocytes which lack MPO (i.e. from patients with hereditary MPO deficiency) is less than 10% of that of normal cells (Klebanoff & Rosen 1978; Table 3). The addition of highly purified MPO to MPO-deficient leukocytes increases ethylene formation to a level greater than that of similarly treated normal leukocytes and this effect of MPO is lost on heat-treatment (100 °C for 15 min). We have found the phagocytosis-induced respiratory burst to be greater in MPO-deficient than in normal PMNs (Klebanoff & Pincus 1971; Klebanoff & Hamon 1972; Rosen & Klebanoff 1976). The ethylene formation by MPO-deficient PMNs with added MPO thus may reflect the increased activity of the respiratory burst in these cells.

(2) Ethylene formation from 2-oxo-4-methylthiobutyric acid by normal PMNs is inhibited by the peroxidase inhibitors azide and cyanide (Weiss *et al.* 1978; Klebanoff & Rosen 1978; Table 4); these agents do not inhibit the residual ethylene formation by MPO-deficient PMNs (Table 4).

TABLE 4

Effect of azide and cyanide on ethylene formation from 2-oxo-4-methylthiobutyric acid by normal and MPO deficient leukocytes[a]

Inhibitor	*Ethylene formation (pmol [10^7 PMN]$^{-1}$ h^{-1})*	
	Normal	*MPO-deficient*
None	15 600	550
Azide (2×10^{-4} mol/l)	630	610
Cyanide (2×10^{-4} mol/l)	8 800	770

[a] See Klebanoff & Rosen (1978) for experimental details.

(3) Ethylene is formed from 2-oxo-4-methylthiobutyric acid by a model system consisting of MPO, H_2O_2, chloride and EDTA (Klebanoff & Rosen 1978; Table 5). Each component of the system is required for optimum activity although complete loss of activity was observed only when H_2O_2 was deleted. As with the intact PMN, ethylene formation is inhibited by azide, cyanide and SOD and, as expected from the H_2O_2 requirement, catalase was strongly inhibitory. The singlet oxygen quenchers, DABCO and histidine, also strongly inhibited ethylene formation by the MPO system, whereas the hydroxyl-radical scavengers mannitol and benzoate were only partially inhibitory ($< 50\%$) at concentrations 100 and 10 times that of the oxomethylthiobutyric acid, respectively. The inhibition of the microbicidal activity of PMNs by SOD, catalase and $OH^•$ scavengers (Johnston *et al.* 1975) has been cited as evidence for the formation of hydroxyl radicals and their use in the microbicidal activity of the cells. It is of interest, therefore, that ethylene formation by the MPO-dependent model system also is inhibited by SOD, catalase and, to a lesser degree, by $OH^•$ scavengers.

These findings suggest either that MPO is required for $OH^•$ formation by the PMN or that ethylene is formed largely by a mechanism independent of $OH^•$ radicals. The inhibitory effect of $^1O_2$ quenchers on ethylene formation by the MPO–H_2O_2–chloride–EDTA system (Table 5) is compatible with a $^1O_2$ mechanism. In this regard ethylene is formed from 2-oxo-4-methylthiobutyric acid by rose bengal in the presence of oxygen and light and this reaction is

TABLE 5

Ethylene formation from 2-oxo-4-methylthiobutyric acid by an MPO-dependent model system[a]

Additions	*Ethylene formation (pmol)*
MPO + H_2O_2 + chloride + EDTA	2230
− EDTA	800
− chloride	1430
− H_2O_2	10
− MPO	340
+ azide (10^{-3} mol/l)	350
+ cyanide (10^{-3} mol/l)	100
+ SOD (5 μg/ml)	200
+ SOD (heated)	2640
+ catalase (60 μg/ml)	130
+ catalase (heated)	2640
+ DABCO[b] (0.1 mol/l)	−20
+ histidine (10^{-2} mol/l)	510
+ mannitol (0.1 mol/l)	1170
+ benzoate (10^{-2} mol/l)	1330

[a] See Klebanoff & Rosen (1978) for experimental details.
[b] 1,4-Diazabicyclo[2,2,2]octane.

stimulated by ^2H_2O and inhibited by the $^1O_2$ quenchers azide, DABCO and histidine (Klebanoff & Rosen 1978). These are the properties expected of a $^1O_2$-mediated Type II dye-sensitized photooxidation (Foote 1976) and suggest that $^1O_2$ can initiate ethylene formation.

FORMATION OF SINGLET OXYGEN

Interest in the formation of $^1O_2$ by PMNs and its involvement in microbicidal activity stemmed from the finding that phagocytosis is associated with emission of light (Allen *et al.* 1972). Chemiluminescence indicates the formation of an electronically excited state but not the nature of the excited species. Spectral analysis has revealed broad peak activity rather than the specific spectrum of $^1O_2$ (Cheson *et al.* 1976; Andersen *et al.* 1977); however, this is not surprising as secondary excitation of particle or cell components induced by singlet oxygen or other products of the respiratory burst is expected. Additional indirect evidence for the generation of $^1O_2$ by PMNs is the finding that *Sarcina lutea*, which are rich in carotenoid pigments, are less readily destroyed by PMNs than are pigmentless white mutant organisms (Krinsky 1974); carotenoid pigments are efficient scavengers of $^1O_2$. Finally, evidence has been presented that is compatible with the formation of $^1O_2$ by systems believed to be operative in the PMN, namely, the MPO-mediated antimicrobial system and superoxide-dependent reactions.

Evidence for the formation of $^1O_2$ by the MPO–H_2O_2–halide system is as follows:

(1) The light emitted by phagocytosing PMNs depends in part on MPO (Rosen & Klebanoff 1976). This is indicated by (*a*) the decreased chemiluminescence of MPO-deficient leukocytes during the early post-phagocytic period, (*b*) the inhibition of the chemiluminescence of normal PMNs by the peroxidase inhibitor azide, an effect which is largely dependent on MPO since it is very much depressed when MPO-deficient PMNs are used, and (*c*) by the emission of light by a cell-free system consisting of MPO, H_2O_2 and a halide. It should be emphasized again that chemiluminescence, although compatible with, is not proof of $^1O_2$ formation.

(2) Diphenylfuran is converted into its $^1O_2$-product *cis*-dibenzoylethylene by the MPO system, particularly with bromide and chloride as the halide (Rosen & Klebanoff 1977). Agents which scavenge $^1O_2$, such as β-carotene, bilirubin, histidine and DABCO, inhibit this oxidative cleavage of diphenylfuran by the MPO system. The conversion is stimulated by the substitution of ^2H_2O for water, a procedure which prolongs the lifetime of $^1O_2$ and generally stimulates $^1O_2$-dependent reactions. As for chemiluminescence, these findings are subject to

other interpretations since diphenylfuran can be converted into *cis*-dibenzoylethylene by $^1O_2$-independent mechanisms and the specificity of the quenchers and ^2H_2O for $^1O_2$-dependent reactions is not absolute (Rosen & Klebanoff 1977; Held & Hurst 1978).

Superoxide-anion involvement in the light emission by PMNs is suggested by the inhibitory effect of SOD and by the emission of light by superoxide-generating systems. Amongst the proposed mechanisms for the formation of $^1O_2$ from superoxide are:

(1) spontaneous dismutation (Khan 1970, 1977)

$$HO_2^{\bullet} + O_2^- + H^+ \rightarrow {}^1O_2 + H_2O_2 \quad (3)$$

(2) interaction with $OH^{\bullet}$ (Arneson 1970)

$$O_2^- + OH^{\bullet} \rightarrow {}^1O_2 + OH^- \quad (4)$$

(3) interaction with H_2O_2 (Kellogg & Fridovich 1975)

$$O_2^- + H_2O_2 \rightarrow {}^1O_2 + OH^- + OH^{\bullet} \quad (5)$$

Our evidence supports the formation of $^1O_2$ by reaction (5). Diphenylfuran is converted into *cis*-dibenzoylethylene by acetaldehyde and xanthine oxidase and this conversion is inhibited by SOD, catalase and by the $^1O_2$ scavengers azide, DABCO and histidine (Table 6). The hydroxyl-radical scavengers mannitol and benzoate were without effect. Other investigators, however, have been unable to detect the formation of $^1O_2$ from O_2^- (Nilsson & Kearns 1974; King *et al.* 1975).

TABLE 6

Diphenylfuran oxidation by acetaldehyde–xanthine oxidase

Additions	cis-*Dibenzoylethylene (pmol)*
Acetaldehyde (4×10^{-2} mol/l)	0
Xanthine oxidase (8 mU/ml)	0
Acetaldehyde + xanthine oxidase	26
+ SOD (1 μg/ml)	6
+ catalase (60 μg/ml)	4
+ DABCO[a] (10^{-2} mol/l)	3
+ histidine (10^{-3} mol/l)	6
+ mannitol (10^{-2} mol/l)	26
+ benzoate (10^{-2} mol/l)	25

[^{3}H]Diphenylfuran (50 pmol) was incubated with the components indicated in phosphate buffer, pH 7.0, for 30 min and the *cis*-dibenzoylethylene formed was determined as previously described (Rosen & Klebanoff 1977).

[a] 1,4-Diazabicyclo[2,2,2]octane.

CONCLUSION

In summary, it is our view that most organisms ingested by normal PMNs are killed by an antimicrobial system consisting of MPO, H_2O_2 and a halide. This antimicrobial system predominates during the early post-phagocytic period and its broad specificity and high potency make it unlikely that many organisms will survive its action. Those that do are attacked by other antimicrobial systems that are both oxygen-dependent and oxygen-independent and these systems are particularly crucial when the MPO system is inhibited or absent. In the absence of MPO, both oxygen-dependent and oxygen-independent systems are effective, whereas in chronic granulomatous disease, where a defect in oxidative metabolism affects all oxygen-dependent systems including the MPO-dependent one, only oxygen-independent antimicrobial systems are effective. It is probable that various reactive species, namely, hypohalous acids generated by the MPO system, singlet oxygen generated both by the MPO system and from the superoxide anion, hydroxyl radicals, H_2O_2, organic peroxides, etc., contribute to the oxygen-dependent toxicity and that organisms vary in susceptibility to these agents owing to their content of protective enzymes (e.g. catalase, superoxide dismutase) and to the presence of exposed essential surface sites susceptible to attack.

ACKNOWLEDGEMENTS

The studies described here were supported by US Public Health Service Grants AI07763 and CA18354. We thank Kay Tisdel for her help in the preparation of this manuscript and Ann Waltersdorph and Joanne Fluvog for their technical assistance.

References

ALLEN, R. C., STJERNHOLM, R. L. & STEELE, R. H. (1972) Evidence for the generation of an electronic excitation state(s) in human polymorphonuclear leukocytes and its participation in bactericidal activity. *Biochem. Biophys. Res. Commun. 47*, 679–684

ANDERSEN, B. R., BRENDZEL, A. M. & LINT, T. F. (1977) Chemiluminescence spectra of human myeloperoxidase and polymorphonuclear leukocytes. *Infect. Immun. 17*, 62–66

ARNESON, R. M. (1970) Substrate-induced chemiluminescence of xanthine oxidase and aldehyde oxidase. *Arch. Biochem. Biophys. 136*, 352–360

BABIOR, B. M. (1978) Oxygen-dependent microbial killing by phagocytes. *N. Engl. J. Med. 298*, 659–668, 721–725

BABIOR, B. M., KIPNES, R. S. & CURNUTTE, J. T. (1973) Biological defense mechanisms. The production by leukocytes of superoxide, a potential bactericidal agent. *J. Clin. Invest. 52*, 741–744

BAEHNER, R. L., NATHAN, D. G. & KARNOVSKY, M. L. (1970) Correction of metabolic deficiencies in the leukocytes of patients with chronic granulomatous disease. *J. Clin. Invest. 49*, 865–870

BEAUCHAMP, C. & FRIDOVICH, I. (1970) Mechanism for the production of ethylene from methional. Generation of the hydroxyl radical by xanthine oxidase. *J. Biol. Chem. 245*, 4641–4646

BORS, W., LENGFELDER, E., SARAN, M., FUCHS, C. & MICHEL, C. (1976) Reactions of oxygen radical species with methional: a pulse radiolysis study. *Biochem. Biophys. Res. Commun. 70*, 81–87

CHESON, B. D., CHRISTENSEN, R. L., SPERLING, R., KOHLER, B. E. & BABIOR, B. M. (1976) The origin of the chemiluminescence of phagocytosing granulocytes. *J. Clin. Invest. 58*, 789–796

CLARK, R. A., KLEBANOFF, S. J., EINSTEIN, A. B. & FEFER, A. (1975) Peroxidase–H_2O_2–halide system: cytotoxic effect on mammalian tumor cells. *Blood 45*, 161–170

DECHATELET, L. R. (1975) Oxidative bactericidal mechanisms of polymorphonuclear leukocytes. *J. Infect. Dis. 131*, 295–303

FOOTE, C. S. (1976) Photosensitized oxidation and singlet oxygen: consequences in biological systems, in *Free Radicals in Biology*, vol. 2 (Pryor, W. A., ed.), pp. 85–133, Academic Press, New York

HABER, F. & WEISS, J. (1934) The catalytic decomposition of hydrogen peroxide by iron salts. *Proc. R. Soc. Lond. A 147*, 332–351

HALLIWELL, B. (1976) An attempt to demonstrate a reaction between superoxide and hydrogen peroxide. *FEBS (Fed. Eur. Biochem. Soc.) Lett. 72*, 8–10

HAMON, C. B. & KLEBANOFF, S. J. (1973) A peroxidase-mediated *Streptococcus mitis*-dependent antimicrobial system in saliva. *J. Exp. Med. 137*, 438–450

HELD, A. M. & HURST, J. K. (1978) Ambiguity associated with use of singlet oxygen trapping agents in myeloperoxidase-catalyzed oxidations. *Biochem. Biophys. Res. Commun. 81*, 878–885

HOLMES, B. & GOOD, R. A. (1972) Laboratory models of chronic granulomatous disease. *J. Reticuloendothel. Soc. 12*, 216–237

HOLMES, B., PAGE, A. R. & GOOD, R. A. (1967) Studies of the metabolic activity of leukocytes from patients with a genetic abnormality of phagocytic function. *J. Clin. Invest. 46*, 1422–1432

IYER, G. Y. N., ISLAM, D. M. F. & QUASTEL, J. H. (1961) Biochemical aspects of phagocytosis. *Nature (Lond.) 192*, 535–541

JACQUES, Y. V. & BAINTON, D. F. (1978) Changes in pH within the phagocytic vacuoles of human neutrophils and monocytes. *Lab. Invest. 39*, 179–185

JENSEN, M. S. & BAINTON, D. F. (1973) Temporal changes in pH within the phagocytic vacuole of the polymorphonuclear neutrophilic leukocyte. *J. Cell Biol. 56, 379*–388

JOHNSTON, R. B. JR. & BAEHNER, R. L. (1970) Improvement of leukocyte bactericidal activity in chronic granulomatous disease. *Blood 35*, 350–355

JOHNSTON, R. B. JR., KEELE, B. B. JR., MISRA, H. P., LEHMEYER, J. E., WEBB, L. S., BAEHNER, R. L. & RAJAGOPALAN, K. V. (1975) The role of superoxide anion generation in phagocytic bactericidal activity. Studies with normal and chronic granulomatous disease leukocytes. *J. Clin. Invest. 55*, 1357–1372

KAPLAN, E. L., LAXDAL, T. & QUIE, P. G. (1968) Studies of polymorphonuclear leukocytes from patients with chronic granulomatous disease of childhood: bactericidal capacity for streptococci. *Pediatrics 41*, 591–599

KELLOGG, E. W. TERT. & FRIDOVICH, I. (1975) Superoxide, hydrogen peroxide and singlet oxygen in lipid peroxidation by a xanthine oxidase system. *J. Biol. Chem. 250*, 8812–8817

KELLOGG, E. W. TERT. & FRIDOVICH, I. (1977) Liposome oxidation and erythrocyte lysis by enzymically generated superoxide and hydrogen peroxide. *J. Biol. Chem. 252*, 6721–6728

KHAN, A. U. (1970) Singlet molecular oxygen from superoxide anion and sensitized fluorescence of organic molecules. *Science (Wash. D.C.) 168*, 476–477

KHAN, A. U. (1977) Theory of electron transfer generation and quenching of singlet oxygen [$^1\Sigma_g^+$ and $^1\Delta_g$] by superoxide anion. The role of water in the dismutation of O_2^-. *J. Am. Chem. Soc. 99*, 370–371

KING, M. M., LAI, E. K. & MCKAY, P. B. (1975) Singlet oxygen production associated with enzyme-catalyzed lipid peroxidation in liver microsomes. *J. Biol. Chem. 250*, 6496–6502

KLEBANOFF, S. J. (1967) Iodination of bacteria: a bactericidal mechanism. *J. Exp. Med. 126*, 1063–1078

KLEBANOFF, S. J. (1968) Myeloperoxidase-halide-hydrogen peroxide antimicrobial system. *J. Bacteriol. 95*, 2131–2138

KLEBANOFF, S. J. (1970) Myeloperoxidase: contribution to the microbicidal activity of intact leukocytes. *Science (Wash. D.C.) 169*, 1095–1097

KLEBANOFF, S. J. (1974) Role of the superoxide anion in the myeloperoxidase-mediated antimicrobial system. *J. Biol. Chem. 249*, 3724–3728

KLEBANOFF, S. J. (1975) Antimicrobial mechanisms in neutrophilic polymorphonuclear leukocytes. *Semin. Hematol. 12*, 117–142

KLEBANOFF, S. J. & BELDING, M. E. (1974) Virucidal activity of H_2O_2-generating bacteria: requirement for peroxidase and a halide. *J. Infect. Dis. 129*, 345–348

KLEBANOFF, S. J. & CLARK, R. A. (1978) *The Neutrophil: Function and Clinical Disorders*, North-Holland, Amsterdam

KLEBANOFF, S. J. & GREEN, W. L. (1973) Degradation of thyroid hormones by phagocytosing human leukocytes. *J. Clin. Invest. 52*, 60–72

KLEBANOFF, S. J. & HAMON, C. B. (1972) Role of myeloperoxidase-mediated antimicrobial systems in intact leukocytes. *J. Reticuloendothel. Soc. 12*, 170–196

KLEBANOFF, S. J. & PINCUS, S. H. (1971) Hydrogen peroxide utilization in myeloperoxidase-deficient leukocytes: a possible microbicidal control mechanism. *J. Clin. Invest. 50*, 2226–2229

KLEBANOFF, S. J. & ROSEN, H. (1978) Ethylene formation by polymorphonuclear leukocytes: role of myeloperoxidase. *J. Exp. Med. 148*, 490–506

KLEBANOFF, S. J. & SMITH, D. C. (1970) Peroxidase-mediated antimicrobial activity of rat uterine fluid. *Gynecol. Invest. 1*, 21–30

KLEBANOFF, S. J. & WHITE, L. R. (1969) Iodination defect in the leukocytes of a patient with chronic granulomatous disease of childhood. *N. Engl. J. Med. 280*, 460–466

KOCH, C. (1974) Effect of sodium azide upon normal and pathological granulocyte function. *Acta Path. Microbiol. Scand. 82*, 136–142

KRINSKY, N. I. (1974) Singlet excited oxygen as a mediator of the antibacterial action of leukocytes. *Science (Wash. D.C.) 186*, 363–365

LEHRER, R. I. (1971) Inhibition by sulfonamides of the candidacidal activity of human neutrophils. *J. Clin. Invest. 50*, 2498–2505

LEHRER, R. I. & CLINE, M. J. (1969) Leukocyte myeloperoxidase deficiency and disseminated candidiasis: the role of myeloperoxidase in resistance to *Candida* infection. *J. Clin. Invest. 48*, 1478–1488

LEHRER, R. I., HANIFIN, J. & CLINE, M. J. (1969) Defective bactericidal activity in myeloperoxidase-deficient human neutrophils. *Nature (Lond.) 223*, 78–79

MANDELL, G. L. (1974) Bactericidal activity of aerobic and anaerobic polymorphonuclear neutrophils. *Infect. Immun. 9*, 337–341

MANDELL, G. L. (1975) Catalase, superoxide dismutase, and virulence of *Staphylococcus aureus. In vitro* and *in vivo* studies with emphasis on staphylococcal–leukocyte interaction. *J. Clin. Invest. 55*, 561–566

MANDELL, G. L. & HOOK, E. W. (1969) Leukocyte bactericidal activity in chronic granulomatous disease: correlation of bacterial hydrogen peroxide production and susceptibility to intracellular killing. *J. Bacteriol. 100*, 531–532

MCCLUNE, G. J. & FEE, J. A. (1976) Stopped flow spectrophotometric observation of superoxide dismutation in aqueous solution. *FEBS (Fed. Eur. Biochem. Soc.) Lett. 67*, 294–298

MCCORD, J. M. & DAY, E. D. JR. (1978) Superoxide-dependent production of hydroxyl radical catalyzed by iron-EDTA complex. *FEBS (Fed. Eur. Biochem. Soc.) Lett. 86*, 139–142

MCRIPLEY, R. J. & SBARRA, A. J. (1967) Role of the phagocyte in host–parasite interactions. XII. Hydrogen peroxide-myeloperoxidase bactericidal system in the phagocyte. *J. Bacteriol. 94*, 1425–1430

NILSSON, R. & KEARNS, D. R. (1974) Role of singlet oxygen in some chemiluminescence and enzyme oxidation reactions. *J. Phys. Chem. 78*, 1681–1683
PITT, J. & BERNHEIMER, H. P. (1974) Role of peroxide in phagocytic killing of pneumococci. *Infect. Immun. 9*, 48–52
QUIE, P. G., WHITE, J. G., HOLMES, B. & GOOD, R. A. (1967) *In vitro* bactericidal capacity of human polymorphonuclear leukocytes: diminished activity in chronic granulomatous disease in childhood. *J. Clin. Invest. 46*, 668–679
ROOT, R. K. (1974) Correction of the function of chronic granulomatous disease (CGD) granulocytes (PMN) with extracellular H_2O_2. *Clin. Res. 22*, 452A
ROSEN, H. & KLEBANOFF, S. J. (1976) Chemiluminescence and superoxide production by myeloperoxidase-deficient leukocytes. *J. Clin. Invest. 58*, 50–60
ROSEN, H. & KLEBANOFF, S. J. (1977) Formation of singlet oxygen by the myeloperoxidase-mediated antimicrobial system. *J. Biol. Chem. 252*, 4803–4810
SBARRA, A. J. & KARNOVSKY, M. L. (1959) The biochemical basis of phagocytosis. I. Metabolic changes during the ingestion of particles by polymorphonuclear leukocytes. *J. Biol. Chem. 234*, 1355–1362
SBARRA, A. J., SELVARAJ, R. J., PAUL, B. B., MITCHELL, G. W. JR. & LOUIS, F. (1977) Some newer insights of the peroxide mediated antimicrobial system, in *Movement, Metabolism and Bactericidal Mechanisms of Phagocytes* (Rossi, F., Patriarca, P. L. & Romeo, D., eds.), pp. 295–304, Piccin Medical Books, Padua, Italy
TAUBER, A. I. & BABIOR, B. M. (1977) Evidence for hydroxyl radical production by human neutrophils. *J. Clin. Invest. 60*, 374–379
WEISS, S. J., RUSTAGI, P. K. & LOBUGLIO, A. F. (1978) Human granulocyte generation of hydroxyl radical. *J. Exp. Med. 147*, 316–323
WOEBER, K. A. & INGBAR, S. H. (1973) Metabolism of L-thyroxine by phagocytosing human leukocytes. *J. Clin. Invest. 52*, 1796–1803
YANG, S. F. (1969) Further studies on ethylene formation from α-keto-γ-methylthiobutyric acid or β-methylthiopropionaldehyde by peroxidase in the presence of sulfite and oxygen. *J. Biol. Chem. 244*, 4360–4365

Discussion

Michelson: Dr Klebanoff, how specifically does 1mM-histidine (see Table 6) scavenge singlet oxygen? Would there not be fairly good scavenging of hydroxyl radicals? Furthermore, the concentrations of DABCO you used (10 mmol/l) are likely to quench almost any active oxygen species.

Klebanoff: I agree with your reservations; the inhibitor studies are suspect, as is the oxidation of diphenylfuran, as definitive proof of singlet oxygen formation. The data, however, are compatible with a singlet oxygen mechanism, although other interpretations are certainly possible.

Michelson: 'Compatible', but not proof. A ^2H_2O experiment would be critical.

Klebanoff: In the rose bengal experiment ethylene formation is stimulated by ^2H_2O (Klebanoff & Rosen 1978).

Michelson: Rose bengal produces other products besides singlet oxygen, even a small amount of O_2^-, for instance (Balny & Douzou 1974). No photochemical or other system produces one species of oxygen alone.

Fridovich: Foote (1977) recently expressed this by saying that those compounds that react rapidly with singlet oxygen are not specific for it and those compounds that are specific for it do not react rapidly. Dr Klebanoff's results are the best possible in the absence of a good reactive and specific trapping agent for singlet oxygen.

Klebanoff: Cholesterol appears to be a specific reagent for the detection of singlet oxygen (Kulig & Smith 1973), but it is not a particularly reactive scavenger.

Michelson: Cilento's group (Bechara *et al.* 1978) used dibromoanthracene-disulphonate and diphenylanthracenedisulphonate as triplet and singlet state, respectively, energy transfer compounds — not just for singlet oxygen.

Fridovich: In the MPO-dependent model system (Table 5) removal of EDTA had a large effect. What is the function of the EDTA?

Klebanoff: EDTA is definitely required for optimum activity. Addition of cobalt, zinc, manganese or copper largely inhibits the system (Klebanoff & Rosen 1978). As this inhibition can be overcome by addition of a stoichiometric amount of EDTA, one possibility is that EDTA removes trace-metal inhibitors.

Fridovich: . . . which inhibit the MPO itself?

Klebanoff: Yes — or at least, the MPO-catalysed formation of ethylene. Alternatively, EDTA itself may function as a component of the system. Let us suppose that the initial reaction is the interaction of MPO, H_2O_2 and Cl^- to form singlet oxygen which then abstracts an electron from methional or 2-oxo-4-methylthiobutyric acid with the production of superoxide. This could react with an iron(III)–EDTA complex to form the iron(II)–EDTA complex. The iron (II)–EDTA complex could, in turn, react with H_2O_2 to regenerate the iron(III)–complex and release hydroxyl radicals (McCord & Day 1978) which could contribute to the ethylene formation. This would account for the fact that ethylene formation is roughly halved by mannitol.

Fridovich: Xanthine oxidase also rapidly inactivates itself in ordinary buffers unless EDTA or another chelating agent is present, because trace metals can gain access to a reactive thiol group in the reduced enzyme but not in the resting oxidized enzyme. A critical test would be to see whether MPO — in an ordinary assay for MPO, not in an ethylene-production system — was self-inactivated during its reaction and if EDTA prolonged its life-time in that assay.

Klebanoff: MPO is rapidly inactivated by excessive H_2O_2 in the absence of an adequate amount of an appropriate electron donor.

Fridovich: Does EDTA prevent the inactivation of MPO by H_2O_2?

Klebanoff: I don't know.

Fridovich: Haber & Weiss (1934) studied the iron salt-catalysed decomposition of H_2O_2 and proposed many radical intermediates. Those radicals could inactivate the enzyme. EDTA might prevent that.

Goldstein: Although no piece of evidence has yet proved the participation of singlet oxygen in pathological systems, your findings appear compelling. We have looked at the photohaemolysis of protoporphyric red cells in which singlet oxygen is believed to be present (Lamolla *et al.* 1973) and yet in our studies the haemolysis of these red cells does not proceed by a singlet oxygen mechanism. It would be interesting to study this further.

What happens when ^2H_2O is added to intact white cells?

Klebanoff: Cell function is lost when the ^2H_2O concentration is high. At present, Dr Henry Rosen in our laboratory is trying to couple cholesterol to latex beads in an attempt to recover the specific singlet oxygen products after phagocytosis.

Goldstein: Direct study in bacteria is a better approach than using artificial cells. In a liposome one can demonstrate the cholesterol product from singlet oxygen but we have not been able to do this for intact red cells. Nor can we observe any effect with ^2H_2O with intact red cells when they undergo protoporphyrin photohaemolysis.

Winterhalter: Isn't the problem that cholesterol is outside but protoporphyrin is inside the red cells?

Goldstein: Whatever it is, these protoporphyrin-dependent haemolyses do not proceed *via* singlet oxygen.

Winterhalter: Hydrocholesterol generation can be demonstrated *in vitro* on irradiation of a system containing protoporphyrin and cholesterol.

Goldstein: I am not saying that singlet oxygen is not formed, but that the red cells are not being directly haemolysed by singlet oxygen.

Fridovich: Kellogg & I (1977) noted the effects on erythrocytes of the xanthine oxidase–acetaldehyde system. Our results were very similar to those described by Dr Klebanoff and we concluded that singlet oxygen was involved. We cannot be sure whether it was *singlet* but we know that the active species was not O_2^-, H_2O_2 or $OH^\bullet$ and it was made only in the presence of O_2^- and hydrogen peroxide.

Michelson: I should not be unhappy if myeloperoxidase can produce singlet oxygen. After all, horseradish peroxidase gives triplet acetone from isobutanal (Bechara *et al.* 1978). Can myeloperoxidase use O_2^- as a substrate instead of H_2O_2?

Klebanoff: It reacts with O_2^- to form an 'oxyperoxidase' which like oxyhaemoglobin has oxygen attached to the haem iron. Oxyperoxidase can, under certain circumstances, react with certain electron donors or electron acceptors (Yamazaki & Yokota 1973).

Reiter: Mycoplasmas (pleuropneumonia-like organisms) possess, in contrast to other bacteria, no cell wall (or outer membrane), require steroids for growth and contain cholesterol in their inner membranes. This might be a useful model for singlet oxygen experiments.

Williams: Does the reactivity of this MPO show the acid-dependence typical of other peroxidases? When these reactive species are generated in the system you described, the degree of bacterial killing will depend on the pH, because the rate of disproportionation of O_2^- controls the amount of O_2^- and that reaction depends on pH. If one were to adjust the pH in these systems (e.g. by buffers), the different steps in this system (which are differentially pH-dependent) could be sorted out.

Klebanoff: The pH dependence of peroxidase reactions is a function (to some degree) of the substance being oxidized. The MPO–H_2O_2–halide system, under most circumstances, has a pH optimum in the acid range. Bactericidal activity, however, can be detected at neutral or slightly alkaline pHs. Below pH 4–5, the enzyme is inactivated.

Williams: The linear dependence which you describe is just the oxidation of halide by MPO (which is almost linear with hydrogen ion concentration). Manipulation of the pH would be useful to see what is happening to O_2^- since its steady-state is pH-dependent. I am suspicious of O_2^- being an attacking agent *per se*. It is always put in reaction schemes but, instead, is either singlet oxygen or $OH^•$ the attacking reagent? Acidity favours formation of H_2O_2 and makes the peroxidase as reactive as possible.

Klebanoff: The prevailing view is that O_2^- is effective, either through formation of hydroxyl radicals or singlet oxygen. H_2O_2 would be a very reactive product in the presence of the other components of the peroxidase system.

Williams: In other words, O_2^- itself is not the attacking species?

Fridovich: Dr Bielski's data (pp. 43–48) illustrate the slow rates of reaction of O_2^- with many amino acids; the organic chemistry of O_2^- is only now beginning to be investigated (Lee-Ruff 1977). Certainly, O_2^- reduces cytochrome *c* and NBT and oxidizes catecholamines and catechols, and so on, but it is not nearly as reactive as $^1O_2$ or $OH^•$. The reports of protection by SOD or catalase imply the action of O_2^- and peroxide together in producing more reactive species.

Willson: Radiation chemists have long known that hydroxyl radicals react with chloride but that the reaction depends strongly on pH, the rate only being rapid in acidic solution. If hydroxyl radicals are formed, would they be free in the bulk of the medium or could they react with chloride in an area of high proton concentration? The latter reaction would create a more specific radical (e.g. Cl_2^-) which would react with, say, methional, methionine, or some of the other more sensitive amino acids much more readily than mannitol or iodide. Hydroxyl radicals react rapidly with iodide to give a species which could have bactericidal properties, too. Have you considered these radical species?

Klebanoff: The myeloperoxidase–halide–H_2O_2 system does not generate hydroxyl radicals — at least, we have no evidence that it does. The xanthine oxidase system does, and we use it as a model for the myeloperoxidase-independent antimicrobial systems of the neutrophil. Its antimicrobial activity does not appear to be potentiated further by the addition of a halide.

Hill: How do you propose that hydrogen peroxide reacts with halide ions to give singlet oxygen? Is a reactive halide radical formed?

Klebanoff: Chloride is oxidized by the peroxidase system to hypochlorous acid. A classical reaction for the generation of $^1\Delta_gO_2$ is the interaction between hypochlorite and H_2O_2, particularly at alkaline pH, raising the possibility that hypochlorite formed by the peroxidase system reacts with excessive H_2O_2 to form singlet oxygen.

Hill: Isn't hypochlorite itself bactericidal?

Klebanoff: Yes, it is a powerful germicide. Active chloride is continuously generated in the phagocytic vacuole but whether this is effective itself or through the generation of oxygen products or both is not, as yet, settled.

Hill: Has hypochlorite been checked against organic indicators?

Klebanoff: Yes. Hypochlorous acid does convert diphenylfuran into *cis*-dibenzoylethylene particularly in the presence of chloride (Rosen & Klebanoff 1977). This may be due to the decomposition of hypochlorous acid to form $^1O_2$; however, a direct interaction with diphenylfuran without a $^1O_2$ intermediate has been proposed (Held & Hurst 1978). Of the halides, bromide was the most effective in the conversion of diphenylfuran into *cis*-dibenzoylethylene by the peroxidase system, with activity occurring at concentrations above 0.1 mmol/l (Rosen & Klebanoff 1977). Chloride was also highly effective but a higher concentration (>10 mmol/l) was required. Little conversion was observed with iodide or thiocyanate ions.

Reiter: Blood contains so much thiocyanate that the ion might diffuse into leukocytes.

Klebanoff: Thiocyanate has a dual effect on the peroxidase-mediated antimicrobial system. Thiocyanate can serve as a required component, as in saliva or milk, whereas when a halide (iodide, bromide, chloride) is added thiocyanate inhibits the microbicidal effect at relatively high concentrations. We have found the myeloperoxidase–H_2O_2–thiocyanate system to be bactericidal to *E. coli* at thiocyanate concentrations ranging from 10^{-5} to 10^{-3} mol/l. However, the bactericidal effect was lost on the increase in thiocyanate concentration to 10^{-2} mol/l and this concentration of thiocyanate also inhibited the bactericidal effect when iodide, bromide or chloride was added (Klebanoff & Clark 1978).

Reiter: Catalase does not reverse the bactericidal activity of the lactoperoxidase–thiocyanate–hydrogen peroxide system but you find that the activity of the myeloperoxidase–chloride–H_2O_2 system is reversed by catalase. The lactoperoxidase system is however reversed by cysteine, Na_2S, $Na_2S_2O_4$, NADH etc.

Klebanoff: Even with H_2O_2?

Reiter: Yes. Catalase reverses the lactoperoxidase system only at quite unphysiological levels (over 100 U/ml). Otherwise the system would never be active in milk or other secretions which contain both lactoperoxidase and catalase.

Williams: Is there any membrane that stops passage of thiocyanate? This anion is used in a common way of measuring membrane potential; thiocyanate passes rapidly through (most) membranes. [*See also next paper.*]

FERTILIZATION OF SEA URCHIN EGGS

Klebanoff: I should like to outline some studies by Drs C. A. Foerder, E. M. Eddy, B. M. Shapiro and myself on the fertilization of sea urchin eggs which we find is associated with a series of changes which have striking similarities to the changes induced by phagocytosis in the neutrophil. It has been known since Warburg's studies at the turn of the century that fertilization of sea urchin eggs is associated with a burst of oxygen consumption. We find that, as in the neutrophil, much of this oxygen consumed is converted into H_2O_2, as measured by formate oxidation in the presence of catalase or by scopoletin oxidation in the presence of horseradish peroxidase (Foerder *et al.* 1978). The sea urchin egg contains a peroxidase in cortical granules and a massive exocytosis occurs after fertilization with the release of this enzyme (Foerder & Shapiro 1977). Shortly after phagocytosis, the glycoprotein coat of the egg is raised and converted into a rigid structure owing to the formation of cross-links between tyrosine residues and this hardening reaction appears to be catalysed by the ovoperoxidase and to require H_2O_2 (Foerder & Shapiro 1977; Foerder *et al.* 1978). The ovoperoxidase is detected on the fertilization membrane by cytochemical techniques and the hardening reaction is inhibited by peroxidase inhibitors. As in phagocytosing PMNs, fertilization of sea urchin eggs (or parthenogenetic activation by the ionophore A23187) is associated with the emission of light (Foerder *et al.* 1978), with the conversion of oestradiol into an alcohol-precipitable form (oestradiol binding), with the degradation of the thyroid hormones and with the conversion of iodide into a trichloroacetic acid-precipitable form (idionation) (unpublished data). The bound iodine and oestradiol can be visualized autoradiographically on the fertilization membrane. These striking similarities between the egg and the neutrophil raises the possibility that peroxidase release and H_2O_2 generation may be phenomena common to a number of diverse cells with differing consequences. Of particular pertinence in view of the spermicidal activity of peroxidase, H_2O_2 and a halide (Smith & Klebanoff 1970) is the possibility that peroxidase and H_2O_2 released by the egg contribute to the block to polyspermy by killing adjacent sperm.

Michelson: Is this specific to sea-urchin eggs? You could make a fortune by giving cattle inhibitors of peroxidase to increase twinning!

Klebanoff: The studies have been performed only with sea urchin eggs.

Cohen: The coupling of the two tyrosines reminded me of two other biological phenomena: the hardening of the insect cuticle and the formation of lignin in plants and wood. Both of these processes are presumed to involve conversion of tyrosine into polyphenolic intermediates by the action of peroxidases and subsequent polymerization. Why do you specify ring-to-ring coupling of tyrosines in sea urchin eggs?

Klebanoff: The membrane was isolated, dissected and the dityrosine linkage demonstrated (Foerder & Shapiro 1977).

Williams: I suspect that the active enzymes are tyrosinases rather than peroxidases. The insect has two glands, one of which stores tyrosine-like compounds and the other which stores the enzymes for their oxidation, tyrosinases. The tyrosinases catalyse the reaction which produces the insect cuticle. The plants often use oxidases not peroxidases. Many oxidases and tyrosinases contain copper.

Fridovich: However, Gross *et al.* (1977) also implicated the peroxidases of plants with lignin formation.

Chvapil: In mammalian systems high content of zinc in the prostatic fluid and seminal plasma controls and prevents the oxygen burst in spermatozoa. During passage of the ejaculate through the genital tract zinc is continuously diluted and resorbed. When the spermatozoa meet the egg, the inhibitory effect of Zn^{2+} on the energy outburst is lost. Spermatozoa then behave as you described for sea urchin eggs. It was suggested that the role of Zn^{2+} is to hold the energy systems in spermatozoa in check until the time that fertilization is to take place.

Fridovich: Do you suggest that the spermatozoa also show a respiratory burst?

Chvapil: The inhibitory effect of Zn^{2+} on various functions of spermatozoa is reversible. By complexing Zn^{2+} for instance by histidine or cysteine buffer, the cells are reactivated; addition of up to 5μM-zinc stops these cells again.

Fridovich: Is the respiratory burst resistant to cyanide?

Chvapil: I don't know. The study by Skulachev *et al.* (1967) indicates, however, that Zn^{2+} ions could be used effectively to inhibit electron transport in mitochondria.

Fridovich: Is the fertilization of the sea-urchin eggs insensitive to cyanide?

Klebanoff: Fertilization, as judged by elevation of the fertilization membrane is unaffected by 1mM-cyanide; however, hardening of the fertilization membrane is inhibited (Foerder & Shapiro 1977). The fertilization-induced respiratory burst is largely, although not entirely, cyanide-insensitive (Perry & Epel 1977; D. Lockshon & B. M. Shapiro, unpublished work, 1978).

References

BALNY, C. & DOUZOU, P. (1974) Production of superoxide ions by photosensitization of dyes. *Biochem. Biophys. Res. Commun. 56*, 386–392

BECHARA, E. J. H., OLIVEIRA, O. M. M. F., DE BAPTISTA DURAN, N. R. C. & CILENTO, G. (1978) Peroxidase catalyzed generation of triplet acetone, in *Abstracts of the International Conference on Chemi and Bioenergized Processes*, p. 32, Guarujà, Brazil

FOERDER, C. A. & SHAPIRO, B. M. (1977) Release of ovoperoxidase from sea urchin eggs hardens the fertilization membrane with tyrosine crosslinks. *Proc. Natl. Acad. Sci. U.S.A. 74*, 4214–4218

FOERDER, C. A., KLEBANOFF, S. J. & SHAPIRO, B. M. (1978) Hydrogen peroxide production, chemiluminescence and the respiratory burst of fertilization: interrelated events in early sea urchin development. *Proc. Natl. Acad. Sci. U.S.A. 75*, 3183–3187

FOOTE, C. S. (1977) Personal communication

GROSS, G. G., JANSE, C. & ELSTNER, E. F. (1977) Involvement of malate, monophenols, and the superoxide radical in hydrogen peroxide formation by isolated cell walls from horseradish (*Armoracia lapathifolia* Gilib.). *Planta (Berl.) 136*, 271–276

HABER, F. & WEISS, J. (1934) The catalytic decomposition of hydrogen peroxide by iron salts. *Proc. R. Soc. Lond. A 147*, 332–351

HELD, A. M. & HURST, J. K. (1978) Ambiguity associated with use of singlet oxygen trapping agents in myeloperoxidase-catalyzed oxidations. *Biochem. Biophys. Res. Commun. 81*, 878–885

KELLOGG, E. W. TERT. & FRIDOVICH, I. (1977) Liposome oxidation and erythrocyte lysis by enzymically-generated superoxide and hydrogen peroxide. *J. Biol. Chem. 252*, 6721–6728

KLEBANOFF, S. J. & CLARK, R. A. (1978) *The Neutrophil: Function and Clinical Disorders*, North-Holland, Amsterdam

KLEBANOFF, S. J. & ROSEN, H. (1978) Ethylene formation by polymorphonuclear leukocytes: role of myeloperoxidase. *J. Exp. Med. 148*, 490–506

KULIG, M. J. & SMITH, L. L. (1973) Sterol metabolism. XXV. Cholesterol oxidation by singlet molecular oxygen. *J. Org. Chem. 38*, 3639–3642

LAMOLLA, A. A., YAMANE, T. & TROZZOLO, A. M. (1973) Cholesterol hydroperoxide formation in red cell membranes and photohemolysis in erythropoietic protoporphyria. *Science (Wash. D.C.) 179*, 1131–1133

LEE-RUFF, E. (1977) The organic chemistry of superoxide. *Chem. Soc. Rev. 6*, 195–214

MCCORD, J. M. & DAY, E. D. JR. (1978) Superoxide-dependent production of hydroxyl radical catalyzed by iron-EDTA complex. *FEBS (Fed. Eur. Biochem. Soc.) Lett. 86*, 139–142

PERRY, G. & EPEL, D. (1977) Calcium stimulation of a lipoxygenase activity accounts for the respiratory burst at fertilization of the sea urchin egg. *J. Cell. Biol. 75*, 40a

ROSEN, H. & KLEBANOFF, S. J. (1977) Formation of singlet oxygen by myeloperoxidase-mediated antimicrobial system. *J. Biol. Chem. 252*, 4803–4810

SKULACHEV, V. P., CHISTYAKOV, V. V., JESAITIS, A. A. & SMIRNOVA, E. G. (1967) Inhibition of the respiratory chain by zinc ions. *Biochem. Biophys. Res. Commun. 26*, 1–6

SMITH, D. C. & KLEBANOFF, S. J. (1970) A uterine fluid-mediated sperm-inhibitory system. *Biol. Reprod. 3*, 229–235

YAMAZAKI, I. & YOKOTA, K. (1973) Oxidation states of peroxidase. *Mol. Cell. Biochem. 2*, 39–52

The lactoperoxidase–thiocyanate–hydrogen peroxide antibacterium system

BRUNO REITER

National Institute for Research in Dairying, Reading

Abstract Lactoperoxidase present in various secretions oxidizes thiocyanate (SCN^-) in the presence of H_2O_2 to an unstable oxidation product—hypothiocyanite ($OSCN^-$), which is bactericidal for enteric pathogens including multiple antibiotic resistant strains of *E. coli.* The system damages the inner membrane causing leakage and cessation of uptake of nutrient, leading eventually to death of the organisms and lysis. The possible involvement of O_2^- and $^1O_2$ is discussed.

Lactoperoxidase and SCN^- occur at physiological levels in milk and other biological secretions (for review see Reiter & Oram 1967). The lactoperoxidase–thiocyanate–hydrogen peroxide system (LPS) is now considered to be one of the factors in milk which protect the neonate against bacterial infections in the gastrointestinal tract (for review see Reiter 1978*a*). Since we are now investigating the LPS *in vivo* as an antibacterial agent instead of the broad spectrum antibiotics, we need to know whether this system generates free radicals in analogy to the myeloperoxidase–chloride–H_2O_2 system (MPOS), which is generally known as the Klebanoff system.

ANTIBACTERIAL ACTIVITY OF THE LPS

The complete LPS was first demonstrated in milk where it temporarily inhibited group N streptococci (Reiter *et al.* 1963) and killed group A streptococci (Reiter *et al.* 1964). Other serogroups such as group B streptococci were also inhibited; this serogroup, known to cause bovine mastitis, has now been shown in some cases to colonize the female genitourinary tract and cause lethal meningitis in newborn babies (Ross 1978).

More recently it has been found that the LPS kills pathogenic Gram-negative organisms such as *Escherichia coli*, *Salmonella typhimurium* and *Pseudomonas aeruginosa* (Reiter *et al.* 1976) (Fig. 1). We have now tested a large range of

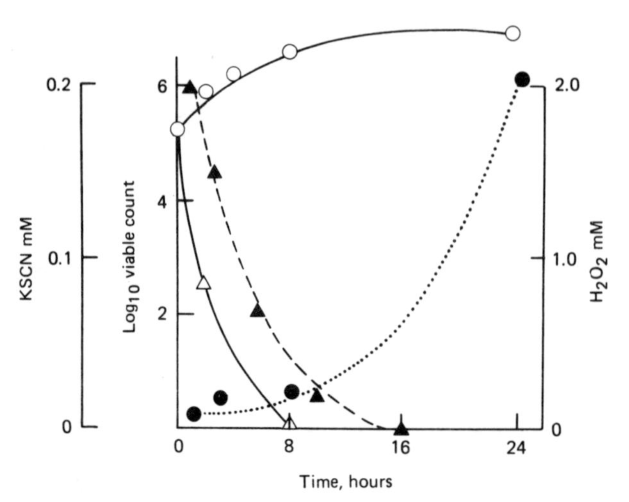

FIG. 1. Bactericidal effect of the lactoperoxidase–SCN^-–H_2O_2 system against *E. coli*; H_2O_2 production and oxidation of SCN^-. The synthetic medium contained: LP, 1.5 u/ml; SCN^-, 0.225 mmol/l; glucose oxidase, 0.1%; glucose 0.1%: ○——○, control, viable count without LPS; △——△ decrease of viable count in presence of LPS; ▲---▲ oxidation of SCN^-; ●····● H_2O_2 level during and after oxidation of SCN^-. For experimental details see Reiter *et al.* (1976).

serotypes of *E. coli* of human, bovine and porcine origin and found that all strains, including multiple antibiotic resistant strains were killed by the LPS (Reiter 1978*b*).

All milks so far tested (human, bovine, porcine) contain physiologically-active levels of LP. The SCN^- is derived basically from the metabolism of *S*-containing amino acids and detoxification of cyanide but its concentration depends largely on the glucoside level of the food (Wood 1975). The third component of the system, H_2O_2, either is generated by the organisms, as in the case of the streptococci, or must be supplied exogenously as with catalase-positive organisms, such as *E. coli*.

THE INTERMEDIARY OXIDATION PRODUCT OF SCN^-

Sörbo & Ljünggren (1958) first showed that myeloperoxidase and H_2O_2 oxidized SCN^- to a labile and short-lived intermediary oxidation product. We confirmed this for LP (Reiter *et al.* 1964) and suggested that sulphur dicyanide, $(SCN)_2$, was the active inhibitory compound. The end products of the oxidation of SCN^-, namely CO_2, NH_3 and SO_4^{2-}, were inactive (Oram & Reiter 1966*a,b*). However, it is now considered that the active intermediary oxidation product is $OSCN^-$ which can also be derived non-enzymically by hydrolysis from $(SCN)_2$ (Hoogendorn *et al.* 1977; Aune & Thomas 1977). Reducing agents, such as cysteine and $Na_2S_2O_4$, and anaerobiosis reverse the inhibitory activity of $OSCN^-$. LP also oxidizes iodide to iodine but, since the bactericidal activity of the lactoperoxidase, H_2O_2 and I^- system requires unphysiological levels of I^-, we abandoned further work on this system (Reiter *et al.* 1964).

THE MODE OF ACTION OF THE LPS

The LPS was shown to inhibit temporarily the multiplication of sensitive streptococci but strains of the same species can be resistant. It reduces the O_2 uptake of the organisms and inhibits their lactic acid production. Glycolytic enzymes containing essential thiol groups are oxidized and the activity of the LPS could be measured by its oxidation of NADH or NADPH (Fig. 2) (Reiter *et al.* 1964). Resistant strains were found to possess a 'reversal factor' which reversed the inhibition of glycolysis of sensitive strains and also catalysed the oxidation of $NADH_2$ in the presence of the intermediary oxidation products; it was purified and termed $NADH_2$-oxidizing enzyme (Oram & Reiter 1966*a*,*b*).

More recent studies with *E. coli* have shown that the LPS inhibits the uptake of amino acids (Table 1) and purines, and the synthesis of protein, DNA and RNA (Marshall & Reiter 1976), and eventually leads to lysis of the organism (Fig. 3). However, this seems to be only a secondary effect because within minutes of exposure to the LPS, K^+ and amino acids leak into the medium (Fig. 4) indicating damage to the inner membrane. This damage appears also

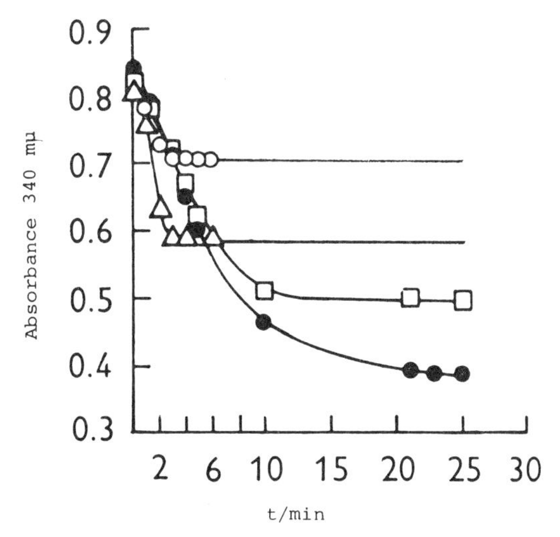

FIG. 2. Oxidation of NADH by the LPS: the reaction mixture consisted of: 300μM-potassium phosphate, pH 7.4; 0.4μM-NADH; 0.4μM-NaSCN; increasing H_2O_2 from 0.3–1.5 μmol/l. For experimental details see Reiter *et al.* (1964).

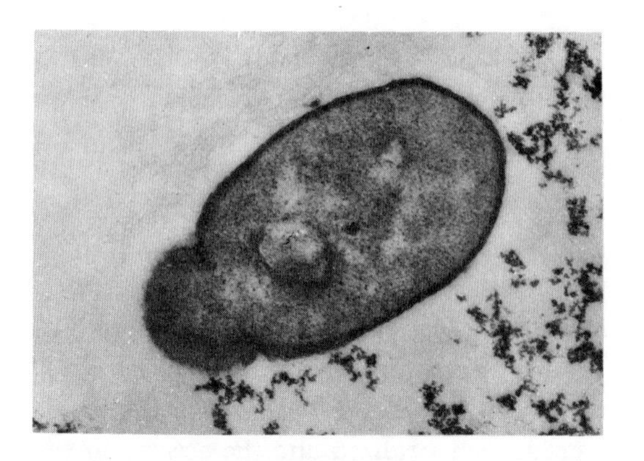

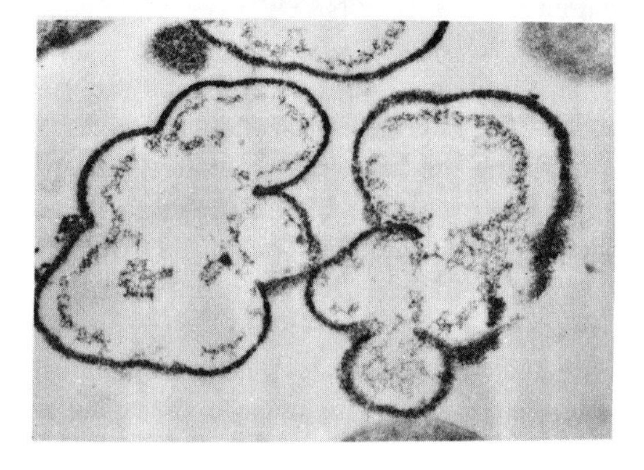

FIG. 3. Electron micrographs showing the effect of the LP-system on *E. coli*: (*A*) appearance of organisms after exposure to LPS for 90 min—polar damage; (*B*) ghosts of *E. coli* after four hours. For details see Reiter (1976).

TABLE 1

Uptake of [^{14}C]leucine, [^{14}C]glutamate,* [^{14}C]lysine* and [^{3}H]glucose by *E. coli* in the presence and absence of the LP-system (LPS) and the xanthine oxidase/xanthine system (XOS).

System present		*Labelled substrate*	*Uptake (measured as d.p.m.)*		
LPS	*XOS*		*after 2 min of treatment*	*after 10 min of treatment*	*after 30 min of treatment*
−	−	Leucine	1520	3335	8520
+	−	Leucine	115	150	250
−	−	Glucose	990	1915	2525
+	−	Glucose	175	250	175
−	−	Leucine	280	1180	4390
−	+	Leucine	220	555	940

Data are from Marshall & Reiter (1976). Methods were those of Beckerdite *et al.* (1974).
* Data not shown in Table.

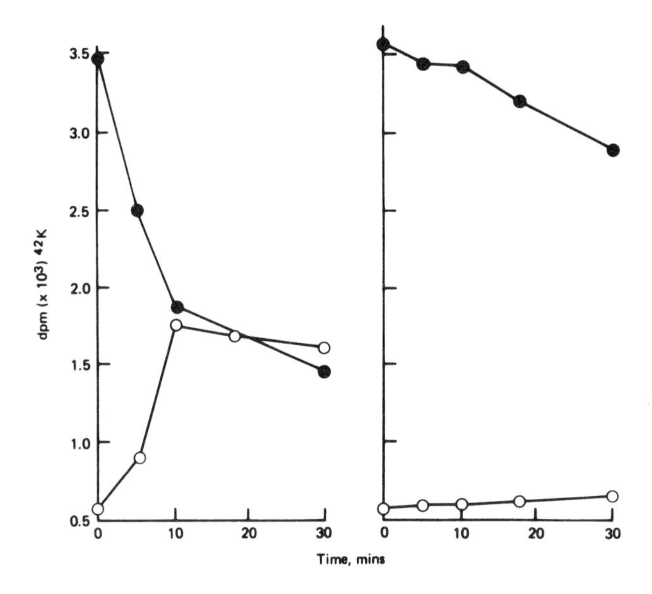

FIG. 4. Leakage of ^{42}K from *E. coli* on treatment with the LP-system: ○, ^{42}K in medium; ●, ^{42}K in organisms. The method used was that of Beckerdite *et al.* (1974). Early and discrete changes in permeability of *E. coli* and certain other Gram-negative bacteria during killing by granulocytes.

to inhibit the transport of glucose. The net uptake of glucose is immediately inhibited (Table 1) but phosphorylation by the phosphoenolpyruvate phosphotransferase is transiently enhanced. The glucose 6-phosphate, however, remains outside (Marshall 1978). Mickelson (1977) has also reported that the LPS inhibits the glucose uptake of *Streptococcus agalactiae* (Group B streptococci). Since the LPS appears to cut off the energy supply, it is not surprising that *E. coli* becomes non-motile (B. Reiter, C. B. Coles, A. Turvey & B. E. Brooker, unpublished work, 1978) and explains our earlier results on the inhibition of spermatozoal movement into and in cervical mucus (Reiter & Gibbons 1964).

It is still debated whether motility represents a virulence factor with enteropathogenic bacteria (see Guentzel & Barry 1975; Schrunk & Verwey 1976; Jones *et al.* 1976). However, there is less doubt that some pathogens attach themselves to the intestinal epithelium and produce enterotoxins which cause disease. We have observed that *E. coli* possessing the well known K_{88} antigen, which is specific for the attachment to porcine brush-border cells (reviewed by Jones 1975), fails to attach *in vitro* after exposure to the LPS. It is not clear yet whether this is caused by a change in hydrophobicity or surface charge (B. Reiter, C. B. Coles, A. Turvey & B. E. Brooker, unpublished work, 1978) (Fig. 5).

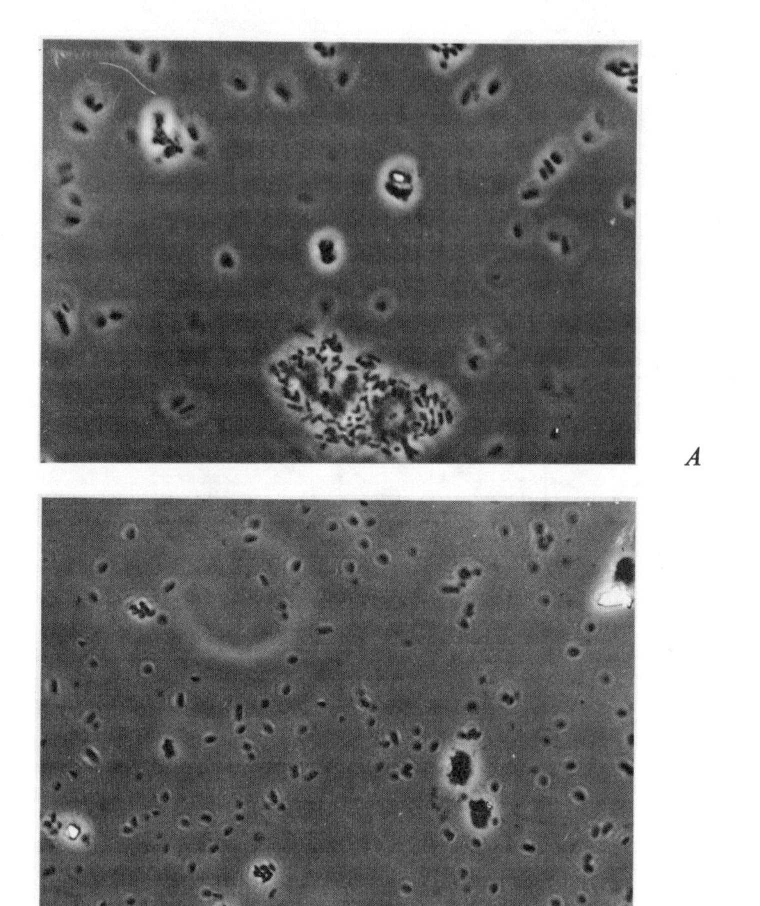

FIG. 5. Attachment of *E. coli* possessing K_{88} antigen to porcine brush-border cells in absence and presence of the LP-system: (*A*) attachment in the absence of the LPS; (*B*) attachment in the presence of the LPS. (For method, see Sellwood *et al.* 1975.)

FREE RADICALS

The involvement of free radicals always seemed an attractive hypothesis for the antibacterial effect of the LPS but remained difficult to prove. Yamazaki *et al.* (1960) detected free radicals during the horseradish–hydroquinone–H_2O_2 reaction using continuous flow e.s.r. spectroscopy. At about that time (1964) we found that quinones added to milk became bactericidal but we failed to detect any e.s.r. signal with the LPS. Similarly, an attempt to detect lipid peroxidation through the production of malonaldehyde failed (B. Reiter & V. M. E. Marshall, unpublished results, 1978).

A specific free radical $(SCN_2)^-$ can be generated by the reaction of hydroxyl radicals with SCN^- (Willson 1977). If this radical is active, it must be so by a mechanism that does not involve binding of the SCN^- ions because studies with $S^{14}CN^-$ show that SCN^- does not bind to the organisms during the LP reaction (Marshall 1978).

It may be possible, however, that the LPS reacts indirectly. Dolin (1971) was first to show that many organisms generate H_2O_2 from the oxidation of NADH catalysed by an NADH oxidase but since then it is accepted that O_2^- is first generated. Cellular damage in these species may normally be prevented by the presence of catalase and superoxide dismutase. We have previously shown that NADH (and NADPH) (Reiter *et al.* 1964) is oxidized by the LPS (Fig. 2) and it could be considered that the 'defence mechanisms' are overcome so leading indirectly to the damage of the organisms. However, our experiments with xanthine oxidase make this unlikely. Xanthine oxidase–xanthine leads to the formation of O_2^-. We found, however, that increased amounts of superoxide dismutase in a strain of *E. coli* grown in trypticase soy-yeast extract (low glucose) (Hassan & Fridovich 1977; see also pp. 77–85), made them more resistant to the bactericidal effect of xanthine oxidase–xanthine. In contrast, greater amounts of superoxide dismutase did not affect the bactericidal activity of the LPS. The 'damage' caused by the xanthine oxidase–xanthine system appears to be similar to that caused by the LPS because we found that the amino acid uptake is equally inhibited by the former system (Table 1) (B. Reiter, S. Philips & V. M. E. Marshall, unpublished results, 1978).

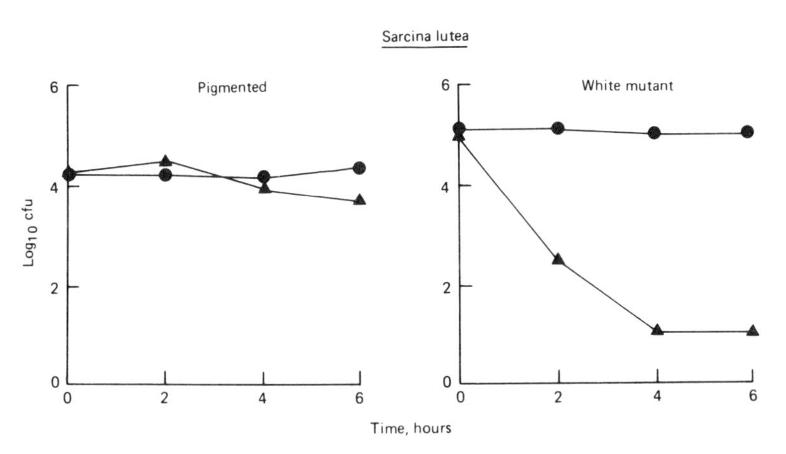

FIG. 6. Resistance of pigmented carotenoid-containing strains of *Sarcina lutea* to the bactericidal activity of the LPS and susceptibility of white mutants: ●, control; ▲, treated with lactoperoxidase. Mutants were produced according to Krinsky (1974); details of the methods are given in Reiter *et al.* (1976).

The involvement of $^1O_2$ seemed possible when we found that a pigmented strain of *Sarcina lutea* was resistant to the LPS although its white mutants were progressively susceptible. According to Krinsky (1974) a colourless mutant of *S. lutea* was much more readily killed by polymorphonuclear leukocytes than the pigmented organisms were because the carotenoid pigment (of the membrane) was considered to act as a quencher (Fig. 6) for $^1O_2$ generated by the myeloperoxidase–Cl^-–H_2O_2 system of the polymorphonuclear leukocytes. Since lactoperoxidase–H_2O_2 oxidizes Br^- to OBr^- and myeloperoxidase–H_2O_2 oxidizes Cl^- to OCl^-, both of which have been shown to generate $^1O_2$ (Piatt *et al.* 1977; Rosen & Klebanoff 1977), we attempted to test whether $OSCN^-$ also generated $^1O_2$ using the conversion of 2,5-diphenylfuran into *cis*-dibenzoylethylene (the specific product with singlet oxygen), but without success. Professor Klebanoff kindly repeated for us the lactoperoxidase reaction but detected only very low levels of *cis*-dibenzoylethylene compared with those found in the myeloperoxidase–Cl^- reaction (see p. 272). Unfortunately, there seems to be some ambiguity about the specificity of β-carotene (and other agents) as a quenching agent for $^1O_2$. Held & Hurst (1978) oxidized 2,5-diphenylfuran with OCl^- without intermediary formation of $^1O_2$, a finding that questions the validity of the concept of involvement of $^1O_2$ altogether.

Milk contains, in addition to lactoperoxidase and catalase, xanthine oxidase associated with the fat globule membrane. Indeed, milk fat is an excellent source for obtaining xanthine oxidase in pure form. That enzyme (0.008 U/ml) with xanthine (0.3 μg/ml) gave good bactericidal activity towards *E. coli* in a synthetic medium but failed to kill the organisms in milk. Although superoxide dismutase is present in milk (Hill 1975) we do not know yet whether the activity is high enough to neutralize the xanthine–xanthine oxidase system. We have also used this system for the generation of H_2O_2 to activate the LPS both in synthetic media and milk; in that combination the LPS becomes as activated as in the case of providing H_2O_2 directly or through glucose oxidase–glucose.

In conclusion, it does not seem to be firmly established whether the myeloperoxidase–Cl^- system depends on free-radical formation but in the case of the lactoperoxidase–SCN^- system the evidence is even scantier but may warrant further work. However, the formation of O_2^- by xanthine oxidase–xanthine is interesting because it gives us a tool with which to study the effect of O_2^- on cell membranes which can be compared with other systems such as the LPS and MPOS. Whatever the outcome, the analogy between the intracellular bactericidal system of polymorphonuclear leukocytes and that of milk is so close — besides complements, immunoglobulins, e.g. lactoferrin, lysozyme peroxidase — that I have felt justified in stating: '... if we accept that leukocytes are the primary defence against invading organisms we could regard the 'milk' colostrum as liquid leukocytes because so many of the antimicrobial factors are common to both' (Reiter 1978*b*).

References

AUNE, T. M. & THOMAS, E. L. (1977) Accumulation of hypothiocyanite ion during peroxidase-catalyzed oxidation of thiocyanate ion. *Eur. J. Biochem. 80*, 209–214

BECKERDITE, S., MOONEY, C., WEISS, J., FRANSON, R. & ELSBECH, P. (1974) *J. Exp. Med. 140*, 398–408

DOLIN, M. J. (1971) The DPNH-oxidizing enzymes of *Streptococcus faecalis. Arch. Biochem. Biophys. 55*, 415–435

GUENTZEL, M. N. & BARRY, L. J. (1975) Motility as a virulence factor for *Vibrio cholerae. Infect. Immun. 11*, 890–897

HASSAN, H. M. & FRIDOVICH, I. (1977) Regulation of superoxide dismutase synthesis in *Escherichia coli*: Glucose effect. *J. Bacteriol. 132*, 505–510

HELD, A. M. & HURST, J. K. (1978) Ambiguity associated with use of singlet oxygen trapping agents in myeloperoxidase-catalyzed oxidations. *Biochem. Biophys. Res. Commun. 81*, 878–885

HILL, R. D. (1975) Superoxide dismutase in bovine milk. *Austr. J. Dairy Technol. 30*, 26–28

HOOGENDORN, H., PIESSENS, J. P., SCHOLTES, W. & STODDARD, L. A. (1977) Hypothiocyanite ion: the inhibitor formed by the system lactoperoxidase-thiocyanate-hydrogen peroxide. I. Identification of the inhibiting compound. *Caries Res. 11*, 77–84

JONES, G. W. (1975) Adhesive properties of enteropathogenic bacteria, in *Microbiology 1975* (Schlessinger, D., ed.), pp. 137–142, American Society of Microbiology, Washington D.C.

JONES, G. W., ABRAMS, G. D. & FRETER, R. (1976) Adhesive properties of *Vibrio cholerae.* Adhesion to isolated rabbit brush border membranes and haemagglutinating activity. *Infect. Immun. 14*, 232

KRINSKY, N. I. (1974) Singlet excited oxygen as a mediator of the antibacterial action of leukocytes. *Science (Wash. D.C.) 186*, 363–365

MARSHALL, V. M. E. (1978) In vitro *and* in vivo *Studies on the Effect of the Lactoperoxidase-Thiocyanate-Hydrogen Peroxide System on* Escherichia coli, Ph.D. thesis, Reading University

MARSHALL, V. M. E. & REITER, B. (1976) The effect of the lactoperoxidase-thiocyanate-hydrogen peroxide system on the metabolism of *E. coli. Proc. Soc. Gen. Microbiol. 3*, 189

MICKELSON, M. N. (1977) Glucose transport in *Streptococcus agalactiae* and its inhibition by lactoperoxidase-thiocyanate-H_2O_2. *J. Bacteriol. 132*, 541–548

ORAM, J. D. & REITER, B. (1966*a*) The inhibition of streptococci by lactoperoxidase, thiocyanate and hydrogen peroxide. The effect of the inhibitory system on susceptible and resistant strains of group N streptococci. *Biochem. J. 100*, 373–381

ORAM, J. D. & REITER, B. (1966*b*) The inhibition of streptococci by lactoperoxidase, thiocyanate and hydrogen peroxide. The oxidation of thiocyanate and the nature of the inhibitory compound. *Biochem. J. 100*, 382–388

PIATT, J. F., CHEEMA, A. S. & O'BRIEN, P. J. (1977) Peroxidase catalyzed singlet oxygen formation from hydrogen peroxide. *FEBS (Fed. Eur. Biochem. Soc.) Lett. 74*, 251–253

REITER, B. (1976). *Symp. Soc. Appl. Act. 5*, 31–59

REITER, B. (1978*a*) Antimicrobial systems in milk. *J. Dairy Res. 45*, 131–147

REITER, B. (1978*b*) Review of nonspecific antimicrobial factors in colostrum, in *The Role of the Colostrum in Relation to Immunity and Survival in the Newborn Ruminant and Pig. Ann. Rech. Vét. 9*, 205–224

REITER, B. & GIBBONS, R. A. (1964) Some further aspects of the lactoperoxidase-thiocyanate-hydrogen peroxide inhibitory system with special reference to the behavior of spermatozoa in cervical mucus. *Annu. Rep. Natl. Inst. Res. Dairying*, p. 87

REITER, B. & ORAM, J. D. (1967) Bacterial inhibitors in milk and other biological fluids. *Nature (Lond.) 216*, 328–330

REITER, B., PICKERING, A., ORAM, J. D. & POPE, G. S. (1963) Peroxidase-thiocyanate inhibition of streptococci in raw milk. *J. Gen. Microbiol. 33*, xii

REITER, B., PICKERING, A. & ORAM, J. D. (1964) An inhibitory system — lactoperoxidase/thiocyanate/hydrogen peroxide — in raw milk, in *Microbial Inhibitors in Food* (Molin, N., ed.), pp. 297–305, Almqvist & Wiksell, Stockholm

REITER, B., MARSHALL, V. M. E., BJÖRCK, L. & ROSÉN, C.-G. (1976) The non-specific bactericidal activity of the lactoperoxidase-thiocyanate-hydrogen peroxide system of milk against *E. coli* and some Gram-negative pathogens. *Infect. Immun. 13*, 800–807

ROSEN, H. & KLEBANOFF, S. J. (1977) Formation of singlet oxygen by the myeloperoxidase mediated antimicrobial system. *J. Biol. Chem. 252*, 4803–4810

ROSS, P. W. (1978) The ecology of Group B streptococci, in *Streptococci* (Skinner, F. A. & Quesnel, L. B., eds.), Academic Press, London

SCHRUNK, G. D. & VERWEY, W. F. (1976) Distribution of cholera organisms in experimental *Vibrio cholerae*. Infections: proposed mechanisms of pathogenesis and antibacterial immunity. *Infect. Immun. 13*, 194–203

SELLWOOD, R., GIBBONS, R. A., JONES, G. W. & RUTTER, J. M. (1975) *J. Med. Microbiol. 8*, 405–411

SÖRBO, B. H. & LJÜNGGREN, J. G. (1958) The catalytic effect of peroxidase on the reaction between hydrogen peroxide and certain sulfur compounds. *Acta Chem. Scand. 12*, 415–459

WILLSON, R. L. (1977) 'Free' radicals and electron transfer in biology and medicine. *Chem. Ind. (5 March)*, 183–193

WOOD, J. L. (1975) Biochemistry, in *Chemistry and Biochemistry of Thiocyanic acid and its Derivatives* (Newman, A. H., ed.), pp. 145–221, Academic Press, London

YAMAZAKI, I., MASON, H. S. & PIETTE, L. (1960) Identification, by electron paramagnetic resonance spectroscopy of free radicals generated from substrate by peroxidase. *J. Biol. Chem. 235*, 2444–2449

The pulmonary and extrapulmonary effects of ozone

BERNARD D. GOLDSTEIN

**Department of Biochemistry, Brunel University, Uxbridge, Middlesex*

Abstract The toxicity of ozone is solely due to its action as an oxidant. It is an extremely reactive gas which rapidly forms intermediate oxidizing derivatives after inhalation. High concentrations cause death from pulmonary oedema. Both pulmonary and extrapulmonary toxicity have been observed at lower concentrations of ozone, including those currently present in urban air. Pulmonary cellular and subcellular membranes appear to be particularly susceptible. A primary mechanism of this effect is the oxidative decomposition of polyunsaturated fatty acids, which has been demonstrated in rodent lungs after inhalation of ozone. Supporting evidence includes the potentiation of ozone toxicity by vitamin E deficiency and an increased use of this antioxidant vitamin during repetitive exposure to ozone. Other membrane effects include oxidation of thiol groups and, perhaps, of tryptophan. Microsomal alterations include a loss of lung cytochrome P450 which may also be related to lipid peroxidation. Extrapulmonary toxicity is not directly due to ozone but may represent an effect due to lipid peroxide decomposition products, particularly malonaldehyde. This three-carbon dialdehyde has been shown to alter cell membrane fluidity and to have mutagenic properties; the latter perhaps due to cross-linkage of DNA to histone.

Ozone (O_3) is not considered an oxygen free radical in the usual chemical sense but it fits into the general context of this symposium by being both highly reactive and a pure oxidant. The latter has its advantages, in that studies of ozone toxicity, particularly *in vivo*, need not be concerned with the formation of a pure species, nor with possible perturbations due to those compounds generating the desired active entity. Ozone can also be studied as a purely harmful compound rather than, as with O_2^-, one that is both active in normal metabolic processes and has the capability of causing cell damage. Unravelling the molecular basis

* *Permanent address:* Departments of Environmental Medicine and Medicine, New York University Medical Center, New York

of ozone toxicity has implications for human health and for socioeconomic decisions with impacts worth literally thousands of millions of dollars. I plan to review briefly the overall toxicology of ozone and discuss the molecular processes of ozone toxicity. In particular I shall focus on the extrapulmonary manifestations of ozone toxicity which may provide a model for the way lipid peroxidation of cell membranes produces deleterious effects at distant non-membrane sites.

The subject of ozone toxicology has been reviewed several times in recent years (Menzel 1970; Cross *et al.* 1976; Goldstein 1977), most extensively by a committee of the US National Academy of Sciences (1977).

Ozone is naturally present in the lower atmosphere at levels of up to perhaps 0.04 parts per million (ppm) during daylight hours. It is also an unwanted by-product of modern society. The major source of ozone in urban air is an indirect sunlight-dependent process starting from oxides of nitrogen and hydrocarbons. This photochemical process generates several gas-phase free radicals and singlet oxygen, but the lifetime of these active species is believed to be too short for them to reach the lung during inhalation. Ozone is somewhat more stable and, although reasonably soluble in water, is sufficiently insoluble to allow it to penetrate deeply into the lung before impinging on an airway surface. In solution at physiological pH it should react rapidly within the mucus layer covering the airway or at the mucosal surface. It appears unlikely that ozone itself penetrates any more deeply. However, rate constants for the reaction of ozone in this milieu are not available.

In urban areas ozone concentrations tend to peak for an hour or two during late morning, owing to the photochemical process acting on morning rush hour automobile emissions. Precursors released from urban areas can be the source of elevated concentrations of ozone for many hundreds of miles downwind. The highest ozone concentrations, approaching 1 ppm, have been recorded in Southern California. Cities such as New York and London will generally have peak levels in the range of 0.15–0.30 ppm. As will be evident from the studies cited below, inhalation of such concentrations of ozone is clearly capable of producing biochemical changes in the lung and in extrapulmonary organs of laboratory animals. Another concentration deserves to be cited: in many species acute ozone exposure causes death from pulmonary oedema at an LD_{50} of 25 to 40 ppm h (concentration of ozone × time of exposure) although there is a wide difference in susceptibility of different species and even of strains of the same species (Goldstein *et al.* 1973). The point to be emphasized is that, of all agents currently of concern to environmental medicine, ozone perhaps has the narrowest range between the presumed acute lethal concentration and the known human exposure.

One of the more intriguing aspects of ozone toxicity is the phenomenon of tolerance: exposure to sublethal ozone concentrations confers protection against subsequent exposure to what would otherwise be lethal levels. This process has been related, in a broad sense, both to an apparent human adaptation to ozone exposure observed in pulmonary function studies (Hackney *et al.* 1977), and to biochemical findings of an increase in the levels of various enzymes and intermediates capable of protecting against oxidant stress in the lungs of animals exposed to sublethal ozone concentrations.

MOLECULAR EFFECTS

A role for free radicals in the effects of ozone was initially suggested by investigators who noted various radiomimetic characteristics in ozone toxicity (see Stokinger 1965 for review of earlier studies). These included effects similar to those of X-irradiation in such diverse systems as plants and human red cells, as well as the protection against ozone afforded by sulphur-containing radioprotective compounds (Fetner 1958; Brinkman *et al.* 1964; Fairchild *et al.* 1959). Hydroxyl radicals and sulphur-based radicals were among those suggested to be responsible for the tissue effects of ozone.

In 1967 we hypothesized that the ozone-induced degradation of unsaturated fatty acids was responsible for free-radical formation and in subsequent studies we demonstrated free-radical signals during the ozonization of linoleic acid (Goldstein & Balchum 1967; Goldstein *et al.* 1968*a*). More recent studies, most notably by Menzel and his colleagues (Menzel 1970; Menzel *et al.* 1972; Roehm *et al.* 1971*a,b*), Tappel and his colleagues (Chow & Tappel 1972, 1973; Chow *et al.* 1974; Fletcher & Tappel 1973) and others (Teige *et al.* 1974; Pryor *et al.* 1976), have greatly extended these initial observations and have provided a strong scientific framework for understanding the role of the oxidative decomposition of unsaturated fatty acids in ozone toxicity.

Studies of the direct ozonolysis of unsaturated fatty acids have generally supported the chemical mechanism originally suggested by Criegee (1957) (Fig. 1). This differs somewhat from the pathway of lipid peroxidation, which is discussed elsewhere in this volume, although the resulting decomposition

$$R{-}CH{=}CH{-}R' + O_3 \longrightarrow R{-}CH\overset{O-O-O}{\text{———}}CH{-}R' \longrightarrow$$

$$\underset{\text{ZWITTERION}}{R{-}\overset{+}{C}H{-}O{-}O^-} + \underset{\text{ALDEHYDE}}{R'{-}CH{=}O} \longrightarrow \underset{\text{OZONIDE}}{R{-}CH\overset{O-O}{\underset{O}{\diamond}}CH{-}R'} \longrightarrow$$

FIG. 1. The ozonolysis of olefinic bonds.

products are similar. The biological implications of the ozonolysis and the peroxidation of lipids are also presumably similar, but whether they are identical is unknown. Furthermore, it is not clear whether the lung unsaturated fatty acids are destroyed by the direct attack of ozone on the fatty acid molecule, by the initiation of lipid peroxidation by an ozone-induced free radical, or by a combination of these processes. For instance, the zwitterion shown in Fig. 1 could form peroxides which then might catalyse the peroxidation of other unsaturated fatty acid molecules (Roehm *et al.* 1971*b*). It could be argued that the much greater effectiveness of phenolic antioxidants against the autocatalytic lipid peroxidation process implies that the protection afforded by vitamin E *in vivo* indirectly indicates the occurrence of ozone-induced lipid peroxidation. On the other hand, Menzel *et al.* (1975*a,b*) have shown that at least some of the effects of ozone inhalation can be duplicated by using preformed fatty acid ozonides. (I shall use the term lipid peroxidation in the ensuing discussion partly for convenience but mainly to conform to current literature usage in discussing the oxidative decomposition of unsaturated fatty acids.)

The evidence that ozone exposure leads to lung lipid peroxidation *in vivo* includes direct identification of this process after various exposure regimens, including 0.4 ppm for four hours and 0.7 ppm for 5–7 days (Goldstein *et al.* 1969; Menzel *et al.* 1972; Chow & Tappel 1972; Fletcher & Tappel 1973). There is also a relatively extensive body of literature on the role of vitamin E in protecting against ozone toxicity even at concentrations as low as 0.1 ppm ozone (Mustafa 1975). We have demonstrated that vitamin E-deficient animals are more susceptible to lethal levels of ozone and, perhaps of greater interest, that daily exposure to sublethal levels of ozone results in an increased use of this antioxidant vitamin (Goldstein *et al.* 1970), a finding confirmed by Menzel *et al.* (1972).

I should emphasize that, although subnormal levels of vitamin E clearly potentiate ozone toxicity, it is still debatable whether supranormal concentrations of vitamin E are any more protective than normal concentrations. The possible efficacy of prescribing antioxidants for populations exposed to oxidant air pollution was discussed at a symposium on this subject held by the US National Institute of Environmental Health Sciences (1977). We recently had an opportunity to test my expressed scepticism about the usefulness of vitamin E for the general public (Goldstein 1976*a*). The results of our study were inconclusive (Hamburger *et al.* 1979). It was based on the observation, described in more detail below, that inhalation of ozone specifically decreases the ability of rat alveolar macrophages and circulating red cells to be agglutinated by the lectin concanavalin A (Goldstein *et al.* 1977; Hamburger & Goldstein 1979). Human red cells exposed *in vitro* to ozone responded similarly. We were fortunate to be able to study this finding in humans experimentally exposed to ozone

— in a continuing study of the effect of supranormal concentrations of vitamin E on the pulmonary physiological response of human volunteers to controlled ozone exposure being performed in Southern California under the direction of Dr Jack Hackney and Dr Ramon Buckley. The volunteers receive vitamin E or placebo for a few months and then undergo daily sessions in an exposure chamber in which they inhale either purified air or ozone (0.5 ppm). Blood samples taken at the cessation of each exposure session were sent to our laboratory in New York for analysis. We observed a tendency toward a lesser degree of red-cell agglutinability by concanavalin A after ozone inhalation than after exposure to clean air in those individuals with the lower serum concentrations of vitamin E (Hamburger *et al.* 1979). The results, however, were not statistically significant; *P* values for an effect of vitamin E ranging from 0.07–0.16 depending on the statistical approach. It is hoped that the pulmonary physiology studies will provide a more clear-cut answer.

The superoxide anion radical has been implicated in ozone toxicity, on the basis of a dose-related increase in rat lung superoxide dismutase activity after continuous exposure to 0.2–0.8 ppm ozone for one week (Mustafa *et al.* 1975). However, there is no readily-apparent chemical mechanism by which ozone can be transmuted into O_2^- and (see later) the increase in superoxide dismutase activity could be secondary to an altered distribution of the cells making up the lung homogenate. To study this further we made use of the superoxide-dismutase inhibitor diethyldithiocarbamate (DDC). Heikkila *et al.* (1976) demonstrated that intraperitoneal injection of DDC results in a dose-dependent inhibition of mouse brain, liver and blood superoxide dismutase. They suggested that DDC was a useful probe in that it should potentiate the effect of an agent which acts by producing superoxide anion radical. To our surprise, intraperitoneal injection of DDC (1.2 g/kg) resulted in a statistically-significant decrease in the mean survival of 20 mice exposed to ozone (21.3 ± 0.7 ppm h; cf. 26.4 ± 1.2 ppm h in buffer-injected control mice) (Goldstein *et al.* 1979). Injection of DDC also potentiated the lethal effects of paraquat, a compound previously reported to produce superoxide anion radical (Bus *et al.* 1974). However, further study revealed that injection of this amount of DDC produced not only a decrease in mouse lung and liver superoxide dismutase activity but also a loss of glutathione peroxidase activity (Fig. 2). Addition of DDC to liver homogenates produced similar effects. The loss of glutathione peroxidase activity appears to be related to the endogenous formation of superoxide anion radical in that it could be prevented by both anaerobic conditions and subsequently added superoxide dismutase, and could be potentiated by addition of dihydroxyfumarate, a compound which produces O_2^- (Goldstein *et al.* 1979). Although these findings may be pertinent to the interrelation of enzymic processes de-

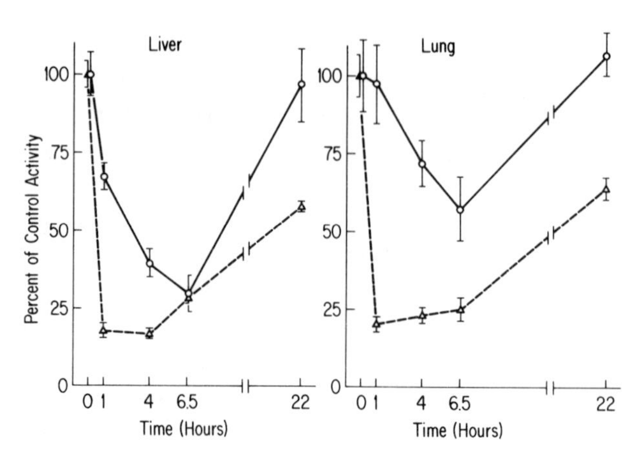

FIG. 2. Effect of intraperitoneal injection of diethyldithiocarbamate (1.2 g/kg) on mouse lung and liver superoxide dismutase (----) and glutathione peroxidase (——). Each point represents the mean ± S.E. for 6–12 mice. Control glutathione peroxidase activity in liver was 216 ± 22 and in lung was 78 ± 9 nmol NADPH oxidized $(min)^{-1}$ $(mg\ protein)^{-1}$. Control superoxide dismutase levels were 639 ± 27 μg/g liver and 168 ± 11 μg/g lung.

fending against oxidant stress, they unfortunately indicate that DDC has limited value as an *in vivo* probe for superoxide anion radical reputed to be produced by exogenous agents.

The role of glutathione peroxidase and related enzymes in protecting against pulmonary ozone toxicity has been thoroughly studied in rats by Chow *et al.* (1972, 1973, 1974, 1976). They have noted a dose-dependent increase in lung glutathione peroxidase, glutathione reductase, and glucose-6-phosphate dehydrogenase after exposure of rats to 0.2–0.8 ppm ozone for one week, with no apparent threshold for this effect. The increases in enzyme activity were inversely proportional to the dietary concentrations of vitamin E and directly proportional to the degree of lipid peroxidation in the lung. Elevated glutathione peroxidase activity has been interpreted as being secondary to lung lipid peroxidation and as being the basis for tolerance to ozone and, perhaps, of human adaptation to photochemical air pollution. This is an important point, particularly in relation to establishing permitted human exposure levels. The crucial question is whether these effects at low levels of ozone represent true toxicity or non-injurious adaptive responses. If the increase in the glutathione peroxidase system reflects a response to cell membrane lipid peroxidation, it could be argued that this would only protect intracellular constituents against such lipid peroxides and would not prevent the presumably-toxic damage to the cell membrane. Alternatively, the glutathione peroxidase response may reflect intracellular ozone-derived hydrogen peroxide without necessarily indicating cell damage. Other explanations are also conceivable. The major point is that better understanding of the basic interrelation between active oxygen species and tissue response would have a substantial impact on determining safe exposure levels for ozone.

Ozone will also react directly with several cellular reducing agents. Major attention has focused on the possible role of thiol groups in ozone toxicity. *In vitro* reduced glutathione reacts with ozone to give sulphoxides and sulphones, in addition to oxidized glutathione (Menzel 1971; Mudd *et al.* 1969). However, *in vivo* the reaction appears to give solely mixed disulphides, according to DeLucia *et al.* (1972, 1975) who observed that acute exposure to relatively high ozone levels (> 1.5 ppm) will decrease the number of non-protein thiol groups in rat lung. Longer term inhalation of ozone (0.8 ppm for one week) resulted in an increase in the number of non-protein thiol groups in rat and monkey lungs. Lung levels of ascorbic acid have also been found to be decreased after acute high level ozone exposure and increased after chronic ozone exposure. Although reduced pyridine nucleotides are readily oxidized by ozone *in vitro* (Mudd *et al.* 1974; Menzel 1971), this has not been observed *in vivo* (Nasr *et al.* 1971).

In addition to cysteine other amino acids are oxidized by ozone; the products in part depend on whether ozonolysis is done in aqueous or non-aqueous media (Mudd *et al.* 1969; Previero *et al.* 1963). We have been particularly interested in cell membrane tryptophan, having demonstrated that a loss in tryptophan-dependent fluorescence of native protein is the most sensitive available indicator of the red cell membrane effects of ozone *in vitro* (Goldstein & McDonagh 1975). Of note is the much greater loss in native protein fluorescence than in chemically-identifiable tryptophan. For instance, a 50% decrease in fluorescence from native protein in the red cell membrane is accompanied by the oxidation of perhaps 10% of total tryptophan molecules. This seems to be a general phenomenon of oxidative free-radical states in that we have observed a similar discrepancy in red cell membranes subjected to ultraviolet light, X-irradiation, protoporphyrin photohaemolysis, and after incubation with phenylhydrazine and related compounds. A much greater loss in native protein fluorescence than in tryptophan was also noted by Hopkins & Spikes (1970) as a result of the methylene-blue-sensitized photooxidation of lysozyme. The mechanism by which the loss in native protein fluorescence represents a magnification of tryptophan oxidation is unclear. It appears to be due to a non-dialysable quenching compound formed *in situ* — perhaps *N*-formylkynurenine or, in membranes, Schiff base cross-links with malonaldehyde.

Almost nothing is known about the effects of ozone on carbohydrate with the exception of an intimation of depolymerization of lung hyaluronic acid (Buell *et al.* 1965) and a loss in red cell membrane neuraminic acid *in vitro* (Goldstein *et al.* 1974). This is unfortunate in as much as a mucopolysaccharide is likely to be the first molecule in the lower respiratory tract reached by ozone.

There is also little known about the biochemical basis for physiological observations suggesting that chronic ozone exposure alters lung collagen. One study reports an increase in carbonyl compounds in rabbit lung elastin and collagen after acute inhalation of 1–5 ppm ozone (Buell *et al.* 1965). We have recently confirmed the observations of Davidkova *et al.* (1975) that malonaldehyde, a lipid peroxide decomposition product, can cross-link collagen. It is conceivable that repetitive ozone-induced lung lipid peroxidation may be the basis for altered lung elasticity.

Although there is also relatively little information about the direct reaction of ozone with nucleic acids, there is ample evidence of altered expression of genetic information after exposure to ozone. This includes bacterial mutagenesis (Hamelin & Chung 1975), chromosomal abnormalities in tissue culture (Sachsenmaier *et al.* 1965), increased incidence of pulmonary tumours in mice (Werthamer *et al.* 1970), as well as teratogenic effects in mice (Veninga 1967). An increased number of chromosome breaks have been observed in the circulating lymphocytes of hamsters exposed to 0.2 ppm ozone (Zelac *et al.* 1971). This finding led to the observation of an increased incidence of chromosomal aberrations in the lymphocytes of humans experimentally exposed to 0.5 ppm ozone for 6–10 h (Merz *et al.* 1975), although an attempt to confirm the human findings has been unsuccessful (McKenzie *et al.* 1977). As discussed in more detail below, the mechanism by which ozone produces chromosomal effects may reflect an interaction of lipid peroxide decomposition products with nuclear material rather than a direct reaction of ozone.

PULMONARY TOXICITY

Acute exposure to ozone results in pulmonary oedema which, at higher ozone levels, is the cause of death. Controlled exposure of humans to ozone has clearly demonstrated impairment of lung physiological function at acute exposure levels of 0.37 ppm, and perhaps lower (Hackney *et al.* 1977). Some evidence suggests additive or synergistic interactions between ozone and other air pollutants on the mechanical function of the lung (Hazucha & Bates 1975) and on cellular effects (Goldstein 1976*b*). The acute effects of ozone reported in epidemiological studies include impairment of running ability in high-school athletes (Wayne *et al.* 1967) and potentiation of asthma attacks (Schoettlin & Landau 1961).

The long-term effects of ozone in man are not known. In animals there is some evidence of decreased lung elasticity (Bartlett *et al.* 1974). Of possible pertinence to ozone-induced lung lipid peroxidation is the observation by Schlipköter & Bruch (1973) of increased levels of lipofuchsin pigment in the lungs of mice exposed to 0.86 ppm ozone for 11 months. Obviously, far more information is needed concerning the chronic effects of this relatively recent air pollutant.

In considering the pulmonary toxicology of ozone we need to keep in mind the particular difficulties involved in studying lung biochemistry. Many of these experiments, including our own, have been done using whole lung homogenates. Unfortunately the lung is a heterogeneous organ with a multiplicity of cell types. A significant biochemical effect in one cell type may be obscured, or observed biochemical alterations may reflect an unimportant adaptive response of cell types not involved in toxic tissue effects. To further complicate matters, acute ozone exposure leads to an influx of both blood cells and of oedema fluid which might account for altered levels of certain lung enzymes and intermediates (Cross *et al.* 1977). Chronic ozone exposure also changes the relative percentage of certain lung cell types: in particular Type I cells are replaced by Type II lung cells. It is, therefore, difficult to interpret whether observed biochemical changes are directly due to ozone or simply reflect altered mixtures of cells with inherently different levels of biochemical intermediates.

Cellular membranes have been suggested as the primary sites of ozone toxicity on the basis of studies on red cells and model membranes, as well as on bacteria and plants (Goldstein & Balchum 1967; Perchorowicz & Ting 1974; Scott & Lesher 1963). This assumption also rests on the observation of lung lipid peroxidation (see before). There is a paucity of studies evaluating defined pulmonary-membrane fractions after ozone exposure. Although many of the effects described below could be secondary to lipid peroxidation of organelle membranes, it should be emphasized that there are no conclusive data to confirm or refute this assumption.

Morphological alterations of lung mitochondria, usually swelling and degeneration, have been frequently noted as a consequence of ozone exposure. Rat lung mitochondrial function after inhalation of ozone has been studied extensively by Mustafa and his colleagues (see Mustafa 1977 for review). Acute exposure to ozone (2 ppm or more) produces an initial decrease in succinate-dependent mitochondrial oxygen consumption followed by an increase in this activity, which persists for at least three weeks (Mustafa *et al.* 1973; Mustafa 1974). Continuous exposure to lower levels of ozone also results in an increase in rat lung mitochondrial oxygen consumption. In one experiment rats fed diets containing concentrations of vitamin E either similar to those in a usual human diet or six times higher were exposed to 0.1 or 0.2 ppm ozone for seven days. A statistically-significant increase in lung mitochondrial oxygen consumption was noted in both groups of rats exposed to 0.2 ppm ozone, but only in the rats receiving the lower vitamin E level when exposed to 0.1 ppm ozone (Mustafa 1975).

Indirect evidence that microsomal processes play a role in the expression of ozone toxicity was obtained in studies demonstrating that an inducing regimen of phenobarbital potentiated ozone toxicity in rats, but pretreatment with allylisopropylacetamide, a compound which destroys cytochrome P450, protected against ozone toxicity (Goldstein & Balchum 1974). We have further demonstrated a decrease in lung microsomal cytochrome P450 concentration after a single inhalation by rabbits of 1 ppm ozone for 90 min. The nadir of this effect was not reached until three days after ozone exposure. *In vitro* studies demonstrated that the ozone-induced diminution in lung cytochrome P450 was accompanied by lipid peroxidation (Goldstein *et al.* 1975). Palmer *et al.* (1971, 1972) have shown that inhalation of 0.75 ppm ozone for three hours decreases the activity of benzopyrene hydroxylase, a mixed-function oxidase, in both rabbit tracheobronchial mucosa and Syrian hamster lung. Prolongation of pentobarbital sleeping time in mice after exposure to ozone has also been reported (Gardner *et al.* 1974).

These observations suggesting an ozone-induced alteration in lung microsomal function are possibly pertinent to the phenomenon of tolerance. A similar, but more thoroughly studied, situation exists for carbon tetrachloride. The protective effects of sublethal concentrations of carbon tetrachloride have been related to the observation that microsomal cytochromes act to potentiate CCl_4-induced hepatic lipid peroxidation during which the cytochromes are themselves destroyed. Accordingly, the degree of susceptibility to CCl_4 is believed to be directly related to hepatic cytochrome P450 levels, and, by producing a decrease in cytochrome P450 levels, prior sublethal administration of CCl_4 therefore protects against a subsequent lethal dose (Glende 1972). The data cited above could be interpreted as suggesting that tolerance to ozone is, at least in part, related to a similar interaction between ozone-induced lipid peroxidation and lung microsomal cytochromes. This is obviously conjecture, particularly as there are several other potential mechanisms for tolerance to ozone.

The interaction of ozone with lysosomes has been explored, in part owing to hypotheses relating ozone toxicity to the release of lysosomal enzymes (Dillard *et al.* 1972). The relatively extensive studies in this area can arbitrarily be summarized as follows. Acute ozone exposure decreases the activities of lysosomal enzymes within alveolar macrophages (Hurst *et al.* 1970; Holzman *et al.* 1968). This appears to represent both release of enzymes from the cell and inactivation of enzyme protein by ozone. In contrast, an increase in lysosomal enzyme activity is noted in whole lung homogenates after continuous exposure to 0.8 ppm ozone, but not 0.5 or 0.2 ppm, for eight days (Chow *et al.* 1974). This is unaffected by the level of tocopherol in the diet (Dillard *et al.* 1972). Sapse *et al.* (1968) reported that humans sensitive to the eye-irritating effects of photochemical air pollution have a reduced amount of lysozyme in their tears.

Alveolar macrophages have also been extensively studied owing to their apparent involvement in one of the most consistently observed adverse effects of low level ozone exposure: the potentiation of bacterial infections in animals (Miller & Ehrlich 1958; Coffin & Gardner 1972). Inhalation of as little as 0.08 ppm ozone for three hours produces a statistically-significant increase in the mortality of mice subsequently exposed to an aerosol of streptococci (Coffin & Blommer 1970). This effect appears to be due to an inability to kill inhaled bacteria (Coffin & Gardner 1972; E. Goldstein *et al.* 1971), a process which is usually ascribed to the pulmonary alveolar macrophage. The biochemical mechanism of this low-level ozone effect is unknown. Of interest is the reported loss of alveolar macrophage phagocytic ability after incubation with linoleic acid hydroperoxide (Khandwala & Gee 1973). An interference in the generation of bactericidal intermediates by the alveolar macrophage is also possible but has not been studied.

We recently observed a decrease in the ability of alveolar macrophages to be agglutinated by concanavalin A after exposure of rats to 0.5 ppm ozone for two hours (Goldstein *et al.* 1977). This finding was also observed in circulating red blood cells in rats and alveolar macrophages in rabbits and appeared to be specific for ozone, in that exposure to the irritant gas nitrogen dioxide produced the opposite effect (Goldstein *et al.* 1977). Ozone exposure produced no change in the measurable number or affinity of concanavalin A-binding sites. This suggests that the effect was related to the membrane response to this lectin, perhaps owing to an ozone-induced alteration in cell membrane fluidity.

EXTRAPULMONARY TOXICITY

Extrapulmonary effects of ozone in many organ systems have been reported (see US National Academy of Science 1977 for review). Our own studies have focused on red cell effects including demonstration of a decrease in red cell membrane acetylcholinesterase activity in mice (Goldstein *et al.* 1968*b*), which has since been observed in humans exposed to 0.5 ppm ozone (Buckley *et al.* 1975), and our recent observation of an ozone-induced alteration in the response of circulating red cells in rats to concanavalin A (Hamburger & Goldstein 1979).

An intriguing aspect of these extrapulmonary effects of ozone is the lack of obvious explanation for their occurrence. Ozone is clearly too reactive to penetrate beyond the respiratory tract mucosal surface cells. We have obtained suggestive evidence that a direct oxidizing derivative of ozone reaches the red cell, but only at near lethal ozone levels. This we did by using the aminotriazole technique of Cohen & Hochstein (1964) to test for increased red-cell hydrogen peroxide in rats and mice inhaling ozone, an effect that was only observed at levels higher than 5 ppm (Goldstein 1973).

On the basis of a small amount of evidence and a large amount of speculation, we have developed the hypothesis that the extrapulmonary effects of ozone are mediated, at least in part, by carbonyl compounds derived from ozone-induced lung lipid peroxidation. This concept may be pertinent to the question of how the cell membrane toxicity of active species of oxygen can result in secondary products which are damaging to intracellular constituents. This subject has recently been discussed by Pietronigro *et al.* (1977) who reported functional damage to DNA as a consequence of lipid peroxidation. The following discussion relates to lipid peroxidation in an intact cell, as well as to the extrapulmonary effects of ozone.

A basic assumption of the hypothesis is that the cell membrane target most likely to give rise to products capable of damaging intracellular components are unsaturated fatty acids. Other cell membrane components susceptible to oxidation, such as thiol groups and the indole group of tryptophan, appear less likely to have such distant effects. Defence mechanisms which are pertinent to a discussion of the distant effects of cell membrane lipid peroxidation include tocopherol, which clearly can limit the effects of lipid peroxidation, apparently mainly by preventing chain propagation. Intracellularly there exist several compounds and pathways which can prevent damage due to active species of oxygen, including specific enzymes such as catalase, superoxide dismutase and several peroxidases, of which the glutathione peroxidase system has received the most attention. These defensive mechanisms, along with the difference in polarity between the membrane and intracellular milieus, must be kept in mind when considering possible mediators of lipid peroxidation-initiated intracellular effects.

The agents derived from lipid peroxidation can be classified into three main categories:

Rapidly active derivatives
- free radicals
- singlet oxygen
- hydrogen peroxide

Lipid oxides
- hydroperoxides
- ozonides
- epoxides

End products
- carbonyls (malonaldehyde)
- alkanes
- alkenes

Much attention has been given to the possible effects of small radicals and active species of oxygen as the deleterious products of lipid peroxidation. There is little reason to question the idea that such derivatives produce further membrane damage. However, in considering these agents we should remember that some are so reactive or short-lived (e.g. $OH^{\bullet}$, singlet O_2) that the only way they could reach the intracellular milieu is through a long chain process. Although propagated processes can be demonstrated in relatively homogeneous test-tube systems, it appears unlikely that such chain reactions initiated at a hydrophobic membrane site could result in a similar rapidly-reacting-radical producing damage to an important intracellular molecule. This scepticism is based in part on the relative heterogeneity of the intact cell membrane making it difficult to propagate such processes, the presence within the membrane of chain terminators, and the apparent efficiency of intracellular radical scavenging processes should a radical actually reach that far. Active agents with somewhat longer half-lives (e.g. O_2^-, H_2O_2) are presumably also susceptible to various intracellular enzymes and free-radical trapping compounds.

Two additional points should be made about this argument against a direct role for lipid peroxidation-initiated free radicals in intracellular damage. Note that I am not suggesting that such active species cannot exist within cells. However, it can be argued that, when present intracellularly, such species are generally formed in a tightly coupled controlled biological process and that this differs greatly from the situation in which there is a random release of an active radical from the cell membrane into the intracellular milieu. There may be more than sufficient intracellular defences against the latter process. The second point that should be stressed is that I am discussing a sublethal cellular process. It is conceivable that a rapid and extensive peroxidation of the cell membrane could lead to sufficient radical flux to overwhelm intracellular protective processes. However, the resultant damage to intracellular targets would represent overkill in that there would already be sufficient membrane damage to cause cell death. This should be kept in mind when interpreting many *in vitro* models of intracellular damage due to lipid peroxidation, including our own, which often involve levels of initiation that would presumably result in drastic consequences to the cell membrane were they to happen *in vivo*.

A second major group of potentially-toxic agents derived from the oxidative degradation of polyunsaturated fatty acids consists of lipid molecules to which oxygen has been added. These include fatty acid hydroperoxides, endoperoxides, and ozonides. Of these, lipid hydroperoxides have been most thoroughly evaluated; they are toxic compounds which produce presumably-pathogenetic biochemical alterations in several cellular models as well as overt toxicity after administration (Cortesi & Privett 1972). Unsaturated fatty acid hydroperoxides in solution are reduced to the corresponding alcohols by glutathione peroxidase (Christophersen 1968). Ozonides have also been demonstrated to be toxic both *in vitro* and after *in vivo* administration (Menzel *et al.* 1975*a*; Cortesi & Privett 1972). As discussed above it has been suggested that the glutathione peroxidase system of the lung has a major role in the prevention of damage due to ozone-induced lipid ozonides or peroxides. It is, however, somewhat difficult to envisage relatively bulky and hydrophobic lipid peroxides or ozonides getting out of the cell membrane and into the intracellular milieu. McCay *et al.* (1976) noted that the addition of glutathione peroxidase to peroxidizing liver microsomes or mitochondria failed to produce any lipid alcohols. Glutathione peroxidase appears to act by preventing active species from reaching the membrane, thereby inhibiting initiation, rather than by detoxifying lipid peroxides once they are formed. This presumably reflects, at least in part, the inability of lipid peroxides to reach intracellular glutathione peroxidase.

The above argument could be summarized as stating that, in as much as free radicals and active states of oxygen are either too short-lived or are susceptible to cellular defence mechanisms, and as lipid hydroperoxides and related compounds are too bulky and hydrophobic, the only possible remaining mediators of intracellular toxicity due to membrane oxidation are the end products. This is an oversimplification; reasoning by exclusion is not a very satisfactory approach. There is however evidence that such end products, and particularly carbonyl compounds, are potentially toxic. Short-chain carbonyls are relatively soluble in water and, therefore, able to penetrate into aqueous intracellular compartments. There also is no obvious generalized detoxification mechanism for these compounds, although more information on this subject is needed.

The most-thoroughly studied carbonyl is malonaldehyde, a three-carbon dicarbonyl whose reaction with 2-thiobarbituric acid is the basis for the classic test of lipid peroxidation. Malonaldehyde cross-links the amino groups of protein, lipid, and nucleic acids through the formation of Schiff bases (Brooks & Klamerth 1968; Chio & Tappel 1969). The fluorescence of the resultant aminoiminopropene derivative is the basis for a more recent assay of lipid peroxidation (Fletcher *et al.* 1973). In our own studies we have used a slightly modified version of this assay to demonstrate *in vivo* red cell lipid peroxidation in humans receiving the oxidizing haemolytic drug diaminodiphenyl sulphone (dapsone) (Goldstein & McDonagh 1976) as well as in vitamin E-deficient neonates (B. D. Goldstein, unpublished data). This fluorescence is emitted primarily by older circulating red cells, thereby suggesting that oxidant drug-induced lipid peroxidation of the cell membrane is a gradual and cumulative process. Related studies have revealed greater fluorescence and a lower haematocrit in vitamin E-deficient rats compared to vitamin E-replete rats treated with the same amount of phenylhydrazine. This may in part be related to an increase in viscosity caused by cross-linking of the membrane by malonaldehyde (Goldstein *et al.* 1976). Cross-linking by malonaldehyde may also in part be responsible for the binding of haemoglobin to the cell membrane observed in oxidant haemolytic states (B. D. Goldstein, unpublished data). Incubation of rat or human red cells with malonaldehyde resulted in a marked decrease in agglutination by concanavalin A. Similar results were observed in red cells obtained from rats treated with acetylphenylhydrazine, in which there was fluorescent evidence of malonaldehyde cross-linking *in vivo*. Our finding that the red cells of rats inhaling ozone also had a decrease, although to a lesser extent, in lectin agglutinability might reflect a role for such carbonyl compounds in extrapulmonary ozone toxicity but of course could be coincidental.

As discussed above, it has been reported that ozone inhalation produces chromosomal abnormalities in circulating lymphocytes and, on the basis of *in vitro* studies, that cell membrane lipid peroxidation results in damage to the cell nucleus. Our studies in this area have been mainly concerned with evaluating possible mechanisms of interaction of malonaldehyde with nuclear constituents. We have not yet directly tested the hypothesis that malonaldehyde derived from ozone-induced cell membrane lipid peroxidation reaches the nucleus *in vivo*. Malonaldehyde is mutagenic to certain histidine-requiring auxotrophs of *S. typhimurium* (Mukai & Goldstein 1976). The usual ultraviolet-sensitive strains were unaffected by malonaldehyde. The pattern of strain sensitivity suggested that malonaldehyde acts by cross-linking DNA, interfering with post-replication repair processes, and producing frame-shift mutants. This is consistent with observations that malonaldehyde can cross-link DNA *in vitro* (Brooks & Klamerth 1968; Reiss *et al.* 1972) and can initiate mouse skin carcinogenesis (Shamberger *et al.* 1974).

We have also evaluated the cross-linking of DNA to histone by malonaldehyde, reasoning that in mammalian systems lysine-rich histones may be particularly susceptible to the formation of Schiff bases. Calf thymus DNA and histone could readily be cross-linked *in vitro*, and this reaction can be inhibited when the amine groups of histone are succinylated, when the arginine-rich basic protein protamine is substituted for histone, or when the material is incubated in salt concentrations which dissociated DNA from histone. Malonaldehyde also produced a dose-dependent inhibition in the growth of both CHV 79 cells and human fibroblasts in tissue culture, the latter occurring after a single addition of concentrations as low as 1 μmol/l to the medium. Evidence of DNA–protein cross-links was obtained in CHV 79 cells incubated with 0.1mM-malonaldehyde (S. J. Hamburger & B. D. Goldstein, unpublished data).

Also perhaps relevant to this area is a series of recent studies in which we have demonstrated that mice fed a diet rich in unsaturated fatty acids have a much greater incidence of skin tumours after initiation with dimethylbenzanthracene or β-propiolactone than mice fed an equivalent saturated fat diet (W. Troll & B. D. Goldstein, unpublished data). It is hoped that this system may provide an *in vivo* model for the interaction of carbonyl compounds derived from lipid peroxidation, of which malonaldehyde is only one, with nuclear constituents, and be of value for the study of the extrapulmonary effects of ozone as well as for carcinogenic agents. One obvious long-term goal of such studies is to address the possible biochemical concomitants of the chronic effects of ozone, particularly in relation to the unproven hypothesis linking active species of oxygen, lipid peroxidation, and ageing.

ACKNOWLEDGEMENTS

I thank my many colleagues who have participated in these studies; most recently Suzanne Hamburger, Marie Amoruso, Frank Mukai and Walter Troll. The work was supported by NIH grants ES-00673, ES-00260 and HL-18163, and by an NIH Senior International Fellowship (F06 TW 00158) from the Fogarty International Center.

References

BARTLETT, D. JR., FAULKNER, C. S. & COOK, K. (1974) Effect of chronic ozone exposure on lung elasticity in young rats. *J. Appl. Physiol. 37*, 92–96

BRINKMAN, R., LAMBERTS, H. B. & VENINGA, T. S. (1964) Radiomimetic toxicity of ozonised air. *Lancet 1*, 133–136

BROOKS, B. R. & KLAMERTH, O. L. (1968) Interaction of DNA with bifunctional aldehydes. *Eur. J. Biochem. 5*, 178–182

BUCKLEY, R. D., HACKNEY, J. D., CLARK, K. & POSIN, C. (1975) Ozone and human blood. *Arch. Environ. Health 30*, 40–43

BUELL, G. C., TOKIWA, Y. & MUELLER, P. K. (1965) Potential crosslinking agents in lung tissue. Formation and isolation after *in vivo* exposure to ozone. *Arch. Environ. Health 10*, 213–219

BUS, J. S., AUST, S. D. & GIBSON, J. E. (1974) Superoxide and singlet oxygen-catalyzed lipid peroxidation as a possible mechanism for paraquat (methyl viologen) toxicity. *Biochem. Biophys. Res. Commun. 58*, 749–755

CHIO, K. S. & TAPPEL, A. L. (1969) A synthesis and characterization of the fluorescent products derived from malonaldehyde and amino acids. *Biochemistry 8*, 2821–2826

CHOW, C. K. & TAPPEL, A. L. (1972) An enzymatic protective mechanism against lipid peroxidation damage to lungs of ozone-exposed rats. *Lipids 7*, 518–524

CHOW, C. K. & TAPPEL, A. L. (1973) Activities of pentose shunt and glycolytic enzymes in lungs of ozone-exposed rats. *Arch. Environ. Health 26*, 205–206

CHOW, C. K., DILLARD, C. J. & TAPPEL, A. L. (1974) Glutathione peroxidase system and lysozyme in rats exposed to ozone or nitrogen dioxide. *Environ. Res. 7*, 311–319

CHOW, C. K., HUSSAIN, M. Z., CROSS, C. E., DUNGWORTH, D. L. & MUSTAFA, M. G. (1976) Effect of low levels of ozone on rat lungs. *Exp. Mol. Pathol. 25*, 182–188

CHRISTOPHERSON, B. D. (1968) Formation of monohydroxy polyenic fatty acids from lipid peroxides by a glutathione peroxidase. *Biochem. Biophys. Acta 164*, 35–36

COFFIN, D. L. & BLOMMER, E. J. (1970) Alterations of the pathogenic role of streptococci group C in mice conferred by previous exposure to ozone, in *Aerobiology (Proc. Third Int. Symp.)* (Silver, I. H., ed.), Academic Press, New York

COFFIN, D. L. & GARDNER, D. E. (1972) Interaction of biological agents and chemical air pollutants. *Ann. Occup. Hyg. 15*, 219–234

COHEN, G. & HOCHSTEIN, P. (1964) Generation of hydrogen peroxide in erythrocytes by hemolytic agents. *Biochemistry 3*, 895–900

CORTESI, R. & PRIVETT, O. S. (1972) Toxicity of fatty ozonides and peroxides. *Lipids 7*, 715–721

CRIEGEE, R. (1957) The course of ozonization of unsaturated compounds. *Rec. Chem. Progr. 18*, 111–120

CROSS, C. E., DELUCIA, A. J., REEDY, A. K., HUSSAIN, M. Z., CHOW, C. K. & MUSTAFA, M. G. (1976) Ozone interactions with lung tissue. *Am. J. Med. 60*, 929–935

CROSS, C. E., DELUCIA, A. J., REEDY, A. K., HUSSAIN, M. Z., CHOW, C. K. & MUSTAFA, M. G. (1977) Biochemical measurements in lung homogenates as indicators of damage and repair elicited by toxic agents and air pollutants: artifacts caused by trapped blood. *Clin. Res. 27*, 163A

DAVIDKOVA, E., SVADLENKA, I. & DEYL, Z. (1975) Interactions of malonaldehyde with collagen. IV. Localisation of malonaldehyde binding site in collagen. *Z. Lebensm.-Unters.-Forsch. 158*, 279–283

DELUCIA, A. J., HOQUE, P. M., MUSTAFA, M. G. & CROSS, C. F. (1972) Ozone interaction with rodent lung: effect on sulfhydryls and sulfhydryl-containing enzyme activities. *J. Lab. Clin. Med. 80*, 559–566

DELUCIA, A. J., MUSTAFA, M. G., HUSSAIN, M. Z. & CROSS, C. E. (1975) Ozone interaction with rodent lung. III. Oxidation of reduced glutathione and formation of mixed disulfides between protein and nonprotein sulfhydryls. *J. Clin. Invest. 55*, 794–802

DILLARD, C. J., URRIBARRI, N., REDDY, K., FLETCHER, B., TAYLOR, S., DE LUMEN, B., LANGBERG, S. & TAPPEL, A. L. (1972) Increased lysosomal enzymes in lungs of ozone-exposed rats. *Arch. Environ. Health 25*, 426–431

FAIRCHILD, E. J., MURPHY, S. D. & STOKINGER, H. E. (1959) Protection by sulfur compounds against the air pollutants ozone and nitrogen dioxide. *Science (Wash. D.C.) 130*, 861–862

FETNER, R. H. (1958) Chromosome breakage in *Vicia faba* by ozone. *Nature (Lond.) 181*, 504–505

FLETCHER, B. L. & TAPPEL, A. L. (1973) Protective effects of dietary alphatocopherol in rats exposed to toxic levels of ozone and nitrogen dioxide. *Environ. Res. 6*, 165–175

FLETCHER, B. L., DILLARD, C. J. & TAPPEL, A. L. (1973) Measurement of fluorescent lipid peroxidation products in biological systems and tissues. *Anal. Biochem. 52*, 1–9

GARDNER, D. E., ILLING, J. W., MILLER, F. J. & COFFIN, D. L. (1974) The effect of ozone on pentobarbital sleeping time in mice. *Res. Commun. Chem. Path. Pharmacol. 9*, 689–699

GLENDE, E. A. JR. (1972) Carbon tetrachloride-induced protection against carbon tetrachloride toxicity. The role of the liver microsomal drug-metabolizing system. *Biochem. Pharmacol. 21*, 1697–1702

GOLDSTEIN, B. D. (1973) Hydrogen peroxide in erythrocytes. Detection in rats and mice inhaling ozone. *Arch. Environ. Health 26*, 279–280

GOLDSTEIN, B. D. (1976*a*) Lipid peroxidation in animals and man. *Environ. Health Perspect. 16*, 185A

GOLDSTEIN, B. D. (1976*b*) Combined exposure to ozone and nitrogen dioxide. *Environ. Health Perspect. 13*, 107–110

GOLDSTEIN, B. D. (1977) Cellular effects of ozone. *Rev. Environ. Health 2*, 177–202

GOLDSTEIN, B. D. & BALCHUM, O. J. (1967) Effect of ozone on lipid peroxidation in the red blood cell. *Proc. Soc. Exp. Biol. Med. 126*, 356–358

GOLDSTEIN, B. D. & BALCHUM, O. J. (1974) Modification of the response of rats to lethal levels of ozone by enzyme-inducing agents. *Toxicol. Appl. Pharmacol. 27*, 330–335

GOLDSTEIN, B. D. & McDONAGH, E. M. (1975) Effect of ozone on cell membrane protein fluorescence. 1. *In vitro* studies utilizing the red cell membrane. *Environ. Res. 9*, 179–186

GOLDSTEIN, B. D. & McDONAGH, E. M. (1976) Spectrofluorescent detection of *in vivo* red cell lipid peroxidation in patients treated with diaminodiphenyl sulfone. *J. Clin. Invest. 57*, 1302–1307

GOLDSTEIN, B. D., BALCHUM, O. J., DEMOPOULOS, H. B. & DUKE, P. S. (1968*a*) Electron paramagnetic resonance spectroscopy. *Arch. Environ. Health 17*, 46–49

GOLDSTEIN, B. D., PEARSON, B., LODI, C., BUCKLEY, R. D. & BALCHUM, O. J. (1968*b*) The effect of ozone on mouse blood *in vivo*. *Arch. Environ. Health 16*, 648–650

GOLDSTEIN, B. D., LODI, C., COLLINSON, C. & BALCHUM, O. J. (1969) Ozone and lipid peroxidation. *Arch. Environ. Health 18*, 631–635

GOLDSTEIN, B. D., BUCKLEY, R. D., GARDENAS, R. & BALCHUM, O. J. (1970) Ozone and vitamin E. *Science (Wash. D.C.) 169*, 605–606

GOLDSTEIN, B. D., LAI, L. Y., ROSS, S. R. & CUZZI-SPADA, R. (1973) Susceptibility of inbred mouse strains to ozone. *Arch. Environ. Health 27*, 412–413

GOLDSTEIN, B. D., LAI, L. Y. & CUZZI-SPADA, R. (1974) Potentiation of complement-dependent membrane damage by ozone. *Arch. Environ. Health 28*, 40–41

GOLDSTEIN, B. D., SOLOMON, S., PASTERNACK, B. S. & BICKERS, D. R. (1975) Decrease in rabbit lung microsomal cytochrome p-450 levels following ozone exposure. *Res. Commun. Chem. Pathol. Pharmacol. 10*, 759–762

GOLDSTEIN, B. D., FALK, G. W., BENJAMIN, L. J. & McDONAUGH, E. M. (1976) Alteration in the chloroform quenching of red cell membrane native protein fluorescence following incubation with malonaldehyde and other crosslinking agents. *Blood Cells 2*, 535–540

GOLDSTEIN, B. D., HAMBURGER, S. J., FALK, G. W. & AMORUSO, M. A. (1977) Effect of ozone and nitrogen dioxide on the agglutination of rat alveolar macrophages by concanavalin A. *Life Sci. 21*, 1637–1644

GOLDSTEIN, B. D., ROZEN, M. G., QUINTAVALLA, J. C. & AMORUSO, M. A. (1979) Decrease in mouse lung and liver glutathione peroxidase activity and potentiation of the lethal effects of ozone and paraquat by the superoxide dismutase inhibitor diethyldithiocarbamate. *Biochem. Pharmacol. 28*, 27–30

GOLDSTEIN, E., TYLER, W. S., HOEPRICH, P. D. & EAGLE, C. (1971) Adverse influence of ozone on pulmonary bactericidal activity of lung. *Nature (Lond.) 279*, 262–263

HACKNEY, J. D., LINN, W. S., KARUZA, S. K., BUCKLEY, R. D., LAW, D. C., BATES, D. V., HAZUCHA, M., PENGELLY, L. D. & SILVERMAN, F. (1977) Effects of ozone exposure in Canadians and Southern Californians. Evidence for adaptation? *Arch. Environ. Health 32*, 110–116

HAMBURGER, S. J. & GOLDSTEIN, B. D. (1979) Effect of ozone on the agglutination of erythrocytes by concanavalin A. I. Studies in rats. *Environ. Res.*, in press

HAMBURGER, S. J., GOLDSTEIN, B. D., BUCKLEY, R. D., HACKNEY, J. D. & AMORUSO, M. A. (1979) Effect of ozone on the agglutination of erythrocytes by concanavalin A. II. Study of human subjects receiving supplemental vitamin E. *Environ. Res.*, in press

HAMELIN, C. & CHUNG, Y. S. (1975) Characterization of mucoid mutants of *Escherichia coli* K-12 isolated after exposure to ozone. *J. Bacteriol. 122*, 19–24

HAZUCHA, M. & BATES, D. V. (1975) Combined effect of ozone and sulphur dioxide on human pulmonary function. *Nature (Lond.) 257*, 50–51

HEIKKILA, R. G., CABBAT, F. S. & COHEN, G. (1976) *In vivo* inhibition of superoxide dismutase in mice by diethyldithiocarbamate. *J. Biol. Chem. 251*, 2182–2185

HOLZMAN, R. S., GARDNER, D. E. & COFFIN, D. L. (1968) *In vivo* inactivation of lysozyme by ozone. *J. Bacteriol. 96*, 1562–1566

HOPKINS, T. R. & SPIKES, J. D. (1970). Conformational changes of lysozyme during photodynamic inactivation. *Photochem. Photobiol. 12*, 175–184

HURST, D. J., GARDNER, D. E. & COFFIN, D. L. (1970) Effect of ozone on acid hydrolysis of the pulmonary alveolar macrophage. *J. Reticuloendothel. Soc. 8*, 288–300

KHANDWALA, A. & GEE, J. B. L. (1973) Linoleic acid hydroperoxide: impaired bacterial uptake by alveolar macrophages, a mechanism of oxidant lung injury. *Science (Wash. D.C.) 182*, 1364–1365
MCCAY, P. B., GIBSON, D. D., FONG, K. & HORNBROOK, K. R. (1976) Effect of glutathione peroxidase activity on lipid peroxidation in biological membranes. *Biochem. Biophys. Acta 431*, 459–468
MCKENZIE, W. H., KNELSON, J. H., RUMMO, N. J. & HOUSE, D. E. (1977) Cytogenetic effects of inhaled ozone in man. *Mutat. Res. 48*, 95–102
MENZEL, D. B. (1970) Toxicity of ozone, oxygen, and radiation. *Annu. Rev. Pharmacol. 10*, 379–394
MENZEL, D. B. (1971) Oxidation of biologically active reducing substances by ozone. *Arch. Environ. Health 23*, 148–153
MENZEL, D. B., ROEH, J. N. & LEE, S. D. (1972) Vitamin E: the biological and environmental antioxidant. *Agric. Food Chem. 20*, 481–486
MENZEL, D. B., SLAUGHTER, R. J., BRYAN, A. M. & JAUREGUI, H. O. (1975*a*) Heinz bodies formed in erythrocytes by fatty acid ozonides and ozone. *Arch. Environ. Health 30*, 296–301
MENZEL, D. B., SLAUGHTER, R. J., BRYANT, A. M. & JAUREGUI, H. O. (1975*b*) Prevention of ozonide-induced Heinz bodies in human erythrocytes by vitamin E. *Arch. Environ. Health 30*, 234–236
MERZ, T., BENDER, H. A., KERR, H. D. & KULLE, T. J. (1975) Observations of aberrations in chromosomes of lymphocytes from human subjects exposed to ozone at a concentration of 0.5 ppm for 6 and 10 hours. *Mutat. Res. 31*, 299–302
MILLER, S. & EHRLICH, R. (1958) Susceptibility to respiratory infections of animals exposed to ozone. Susceptibility to *Klebsiella pneumoniae*. *J. Infect. Dis. 103*, 145–149
MUDD, J. B., LEAVITT, R., ONGUN, A. & MCMANUS, T. T. (1969) Reaction of ozone with amino acids and proteins. *Atmos. Environ. 3*, 669–681
MUDD, J. B., LEH, F. & MCMANUS, T. T. (1974) Reaction of ozone with nicotinamide and its derivatives. *Arch. Biochem. Biophys. 161*, 408–419
MUKAI, F. H. & GOLDSTEIN, B. D. (1976) Mutagenicity of malonaldehyde, a decomposition product of peroxidized polyunsaturated fatty acids. *Science (Wash. D.C.) 191*, 868
MUSTAFA, M. G. (1974) Augmentation of mitochondrial oxidative metabolism in lung tissue during recovery of animals from acute ozone exposure. *Arch. Biochem. Biophys. 165*, 537–538
MUSTAFA, M. G. (1975) Influence of dietary vitamin E on lung cellular sensitivity to ozone in rats. *Nutr. Rep. Int. 11*, 473–476
MUSTAFA, M. G. (1977) Biochemical effects of environmental oxidant pollutants in animal lungs, in *Biological Effects of Environmental Pollutants*, vol. 1 (Lee, S. D., ed.), Ann Arbor Science Publishers, Ann Arbor, Michigan
MUSTAFA, M. G., DELUCIA, A. J., YORK, G. K., ARTH., C. & CROSS, C. E. (1973) Ozone interaction with rodent lung. II. Effects on oxygen consumption of mitochondria. *J. Lab. Clin. Med. 82*, 357–365
MUSTAFA, M. G., MACRES, S. M., TARKINGTON, B. K., CHOW, D. K. & HUSSEIN, M. Z. (1975) Lung superoxide dismutase (SOD): stimulation by low-level ozone exposure. *Clin. Res. 23*, 138A
NASR, A. N. M., DINMAN, B. D. & BERNSTEIN, I. A. (1971) Nonparticulate air pollutants. II. Effect of ozone inhalation on nadide phosphate levels in tracheal mucosa. *Arch. Environ. Health 22*, 545–550
PALMER, M. S., SWANSON, D. H. & COFFIN, D. L. (1971) Effect of ozone on benzpyrene hydroxylase activity in the Syrian golden hamster. *Cancer Res. 31*, 730–733
PALMER, M. S., EXLEY, R. W. & COFFIN, D. L. (1972) Influence of pollutant gases on benzpyrene hydroxylase activity. *Arch. Environ. Health 25*, 439–442
PERCHOROWICZ, J. T. & TING, I. P. (1974) Ozone effects on plant cell permeability. *Am. J. Bot. 61*, 787–793

PIETRONIGRO, D. D., JONES, W. B. G., KALTY, K. & DEMOPOULOS, H. B. (1977) Interaction of DNA and liposomes as a model for membrane-mediated DNA damage. *Nature (Lond.) 267*, 78

PREVIERO, A., SCOFFONE, E., PAVETTA, P. & BENASSI, C. A. (1963) Comportamento degli ammino acidi di fronte all ozono. *Gazz. Chim. Ital. 93*, 841–848

PRYOR, W. A., STANLEY, J. P., BLAIR, E. & CULLEN, B. G. (1976) Autooxidation of polyunsaturated fatty acids. Part I. Effect of ozone on the autooxidation of neat methyl linoleate and methyl linolenate. *Arch. Environ. Health 31*, 201–210

REISS, U., TAPPEL, A. L. & CHIO, K. S. (1972) DNA-malonaldehyde reaction formation of fluorescent products. *Biochem. Biophys. Res. Commun. 48*, 921–926

ROEHM, J. N., HADLEY, J. G. & MENZEL, D. B. (1971*a*) Antioxidants vs. lung disease. *Arch. Intern. Med. 128*, 88–93

ROEHM, J. N., HADLEY, J. G. & MENZEL, D. B. (1971*b*) Oxidation of unsaturated fatty acids by ozone and nitrogen dioxide. A common mechanism of action. *Arch. Environ. Health 23*, 142–148

SACHSENMAIER, W., SIEBS, W. & TAN, T. A. (1965) Wirkung von Ozon auf Mauseascitestumorzellen und auf Hühnerfibroblasten in der Gewebekultur. *Z. Krebsforsch. 67*, 113–126

SAPSE, A. T., BONAVIDA, B., STONE, W. JR. & SCERCARZ, E. E. (1968) Human tear lysozyme. III. Preliminary study on lysozyme levels in subjects with smog eye irritation. *Am. J. Ophthalmol. 66*, 76–80

SCHLIPKÖTER, H. W. & BRUCH, J. (1973) Funktionelle und morphologische Veränderung bei Ozonexposition. *Zentralbl. Bakteriol. Parasitenkd. Infektionskr. Hyg. Erste Abt. Orig. Reihe B. Hyg. Betriebshyg. Präv. Med. 156*, 486–499

SCHOETTLIN, C. E. & LANDAU, E. (1961) Air pollution and asthmatic attacks in the Los Angeles area. *Public Health Rep. 76*, 545–548

SCOTT, D. B. M. & LESHER, E. C. (1963) Effect of ozone on survival and permeability of *Escherichia coli. J. Bacteriol. 85*, 567–576

SHAMBERGER, R. J., ANDREONE, T. L. & WILLIS, C. E. (1974) Antioxidants and cancer IV. Initiating activity of malonaldehyde as a carcinogen. *J. Natl. Cancer Inst. 53*, 1771–1773

STOKINGER, H. E. (1965) Ozone toxicology. A review of research and industrial experience: 1954–1964. *Arch. Environ. Health 10*, 719–731

TEIGE, B., MCMANUS, T. T. & MUDD, J. B. (1974) Reaction of ozone with phosphatidylcholine liposomes and the lytic effect of products on red blood cells. *Chem. Phys. Lipids 12*, 153–171

US NATIONAL ACADEMY OF SCIENCES COMMITTEE ON MEDICAL AND BIOLOGICAL EFFECTS OF ENVIRONMENTAL POLLUTANTS (1977) *Ozone and other Photochemical Oxidants*, Washington, D.C.

US NATIONAL INSTITUTE OF ENVIRONMENTAL HEALTH SCIENCES (1977) *Recent Developments in Toxicity of Environmental Oxidants* (1976) (Conference Abstracts). *Environ. Health Persp. 16*, 177–186

VENINGA, T. S. (1967) Toxicity of ozone in comparison with ionizing radiation. *Strahlentherapie 134*, 469–477

WAYNE, W. S., WEHRIE, P. F. & CARROLL, R. E. (1967) Oxidant air pollution and athletic performance. *J. Am. Med. Assoc. 199*, 901–904

WERTHAMER, S., SCHWARZ, L. H. & SOSKIND, L. (1970) Bronchial epithelial alterations and pulmonary neoplasia induced by ozone. *Pathol. Microbiol. 35*, 224–230

ZELAC, R. E., CROMROY, H. L., BOLCH, W. E., DANAVANT, B. G. & BEVIS, H. A. (1971) Inhaled ozone as a mutagen. I. Chromosome aberrations induced in Chinese hamster lymphocytes. *Environ. Res. 4*, 262–282

Discussion

Cohen: When you administered diethyldithiocarbamate (DDC) did you look at the nervous system or the adrenal medulla and observe the fall in glutathione peroxidase?

Goldstein: No. We looked at lung and liver. The effect is much greater in liver. We presume that the decrease in glutathione peroxidase activity is related to microsomes. When we spin out the liver microsomes before incubating the supernatant with DDC *in vitro* we observed less of a decrease in glutathione peroxidase activity. We suspect (but have no proof) that we are observing steady-state production of superoxide anion radical by liver microsomes which initiated the inactivation of glutathione peroxidase.

Cohen: So, is there a change in glutathione peroxidase activity in liver after the *in vivo* administration of DDC?

Goldstein: In vivo, yes.

Hill: Did you look at any protein not involved in reactions with radicals?

Goldstein: Only catalase — there was no effect.

Roos: What is known about the activity of the different enzymes in the type I and type II pneumocytes?

Goldstein: The trouble with certain reports is that the cells cultured as pneumocytes turn out on subsequent investigation to have been fibroblasts. One has to look at the literature very closely. There is a report of more phosphatidylserine and phosphatidylethanolamine in the membrane of type I cells. This finding implies that they are more peroxidizable (as they presumably have more unsaturated fatty acids). Some of us believe that lipid peroxidation could explain the phenomenon of tolerance.

Roos: Could the changes in superoxide dismutase and glutathione peroxidase, etc., be due to changes in the ratio of type I to type II cells?

Goldstein: Yes. Type II cells appear more granular; however, to correlate enzyme activity with the number of granules is an old-fashioned approach.

Smith: Some enzyme histochemical studies indicate that the type II cells are metabolically much more active than the type I cells. They possibly have more peroxidase than the type I cells.

Winterhalter: You described a curious decrease in fluorescence but claimed that there was no change in 'chemically measurable tryptophan' and that it is not due to denaturation of proteins. Does the emission spectrum give any clue about what causes it?

Goldstein: The emission spectrum does not change—there is no shift in the emission maximum. A shift would be predicted (Burstein *et al.* 1973) if there were a substantial alteration in the ratio of hydrophobic to hydrophilic tryptophans.

Winterhalter: But the excited species must be the same whether ozone is present or not.

Goldstein: I suspect that it is a tryptophan photoproduct that quenches tryptophan. *N*-Formylkynurenine is one possibility. The fluorescence of 8-anilino-1-naphthalenesulphonic acid (ANS) (λ_{max} 370 nm) slightly overlaps that of tryptophan (λ_{max} 330 nm). ANS is one of the 'fluorescent membrane probes' which are supposed to be localized into hydrophobic areas. (Unfortunately, the probes are not as 'pure' as was initially thought.) With ANS present one can excite at 290 nm and observe fluorescence at 470 nm.

Fridovich: That is the Förster transition.

Goldstein: The fluorescence at 330 nm decreases at a different rate than that at 470 nm during exposure to ozone or other oxidants. Making a few assumptions, one can say that the effects are those expected for a somewhat selected decrease in the more hydrophilic tryptophans (Goldstein & McDonagh 1975).

Winterhalter: The fluorescence of tryptophans in hydrophilic surroundings should not be the same as that of the ones in hydrophobic surroundings. Therefore, selective destruction of one should be reflected in the spectrum.

Goldstein: I agree; however, there was no change in the maximum wavelength (within 1–2 nm on an MPF-3 spectrofluorometer). Tryptophan groups have two different lifetimes in the membrane and we noted a tendency toward a specific loss of those with the longer lifetime.

Fridovich: Did you measure depolarization of fluorescence to obtain the lifetimes?

Goldstein: No; this was a straightforward fluorescence lifetime, a form of fluorescence decay. The slight difference we observed was not enough to account for the observed greater decrease in fluorescence than in chemically-measurable tryptophan.

Hill: What chemical method did you use to detect tryptophan?

Goldstein: First, we used *p*-dimethylaminobenzaldehyde (Spies & Chambers 1948). We checked its specificity by treating ribonuclease (which contains no tryptophan) with ozone and u.v. irradiation but found no tryptophan-like product in terms of reaction with this aldehyde. Recently we have been using a fluorescamine technique (Nakamura & Pisano 1976) which is specific for tryptophan — and we get the same result.

Fridovich: Have you tried Koshland's reagent — 2-hydroxy-5-nitrobenzyl bromide (Koshland *et al.* 1964)?

Goldstein: That interferes with the fluorescence.

Fridovich: After you have made your fluorescence measurement there would be no problem.

Presumably, you looked at the fluorescence of the native protein. Did you study it after total denaturation in, say, 6M-guanidinium chloride? If the cause were conformational, a change in the degree of internal quenching or energy transfer by resonance interaction, this treatment would reduce everything into the same state of disorder and eliminate the disparity between the change in fluorescence and the change in chemically detectable tryptophan.

Goldstein: We have done that with guanidinium chloride and with urea, and the disparity is still there. A problem is that we can never be sure that we have 100% denatured ghosts.

Crichton: Several cases are known where urea will not denature proteins although guanidinium chloride will (for example, ferritin) (Hofmann & Harrison 1963; Listowsky *et al.* 1972).

Goldstein: Both urea and guanidine-HCl alter the absolute fluorescence levels but the relative values between control and ozone-exposed membranes do not change (Goldstein & McDonagh 1975).

Fridovich: Dr Cohen's technique of hydrocarbon exhalation (see pp. 177–183) ought to be very useful in studies of ozone toxicity.

Goldstein: The recent results of Dumelin *et al.* (1978) were disappointing. Most of their previous data with Dr Cohen's technique fell between narrow error bands but these results, usually from four animals, varied widely. Also, ozone inhalation will interfere with the method by reacting with the hydrocarbon.

Fridovich: Surely not with pentane and other alkanes?

To what degree did ozone induce the various enzymes — glutathione peroxidase, superoxide dismutase and so on?

Goldstein: The degree of induction varies from enzyme to enzyme: glutathione peroxidase activity increases by about 70% in rats exposed to 0.8 ppm O_3; other enzymes are induced less. The change is not dramatic — it is not like placing anaerobic bacteria in air, or as rapid as Professor Hayaishi's observations (see pp. 199–203).

Smith: Another point is that in measurements of the lung enzyme activity the results are usually expressed as activity per unit weight or unit protein content. The change in a particular cell type in the lung is therefore unknown. Secondly, even if no effects are found in enzyme or co-factor activity in terms of the whole lung the error of measurement may be such that it disguises a very significant effect in a small percentage of lung cells. Consequently, in the interpretation of changes in enzyme or co-factor activity in terms of the whole lung the error of measurement may be such that it disguises a very significant effect in a small percentage of lung cells. The interpretation of whole lung changes in enzyme activities must be interpreted cautiously.

Goldstein: Another complication results from the oedema. Also, red cells brought into lung spaces are rich in glutathione peroxidase and catalase, and the influx of white cells from the bone marrow brings a lot of lysozyme.

Fridovich: Lung is probably the most difficult organ to work with.

Allison: Does ozone poison the prostaglandin synthetase system?

Goldstein: Yes (Menzel *et al.* 1976). This presumably accounts for the increase in lung arachidonic acid after exposure to ozone (Menzel *et al.* 1972).

Chvapil: After chronic exposure of animals to 80% oxygen, the total concentration of arachidonic acid with lung tissue was lower than in controls at ambient air (Chvapil & Peng 1975).

Goldstein: Cortesi & Privett (1972) injected rats intravenously with hydroperoxylinoleic acid. The animals died within a few hours with massive pulmonary oedema. Analysis of the lungs revealed a decrease in the amounts of oleic, linoleic and linolenic acids but an increase in the amount of arachidonic acid.

Slater: Was there a decrease in lung cytochrome P450 after exposure to ozone?

Goldstein: Yes, after a single exposure.

Slater: How did you isolate the microsomes from lung and analyse for cytochrome P450 before and after exposure to ozone? We find that lung microsomes are so easily contaminated with absorbed haemoglobin that cytochrome P450 measurements are very difficult (Benedetto *et al.* 1976).

Goldstein: We were extremely careful to remove the haemoglobin as well as we could. However, 1 ppm ozone for 90 min is not a strongly oedemogenic dose. Not much difference was observed between our control and exposed animals in lung haemoglobin levels (Goldstein *et al.* 1975).

Smith: What happens to benzphetamine metabolism in these animals?

Goldstein: I don't know. Metabolism of benz[*a*]pyrene decreases in hamster lungs after exposure to ozone (Palmer *et al.* 1971).

Smith: Benzphetamine is a better substrate for the pulmonary cytochrome P450 than benz[*a*]pyrene. Secondly, did you start your studies at the completion of exposure?

Goldstein: Some were done on completion of the 90 min exposure and some one or more days after exposure. We kept the animals alive until we needed to analyse the lungs for cytochrome P450.

References

BENEDETTO, C., SLATER, T. F. & DIANZANI, M. U. (1976) *Biochem. Trans. 4*, 1094

BURSTEIN, E. A., VEDENKINA, N. S. & IVKOVA, M. N. (1973) Fluorescence and the location of tryptophan residues in protein molecules. *Photochem. Photobiol. 18*, 263–279

CHVAPIL, M. & PENG, Y. M. (1975) Oxygen and lung fibrosis. *Arch. Environ. Health 30*, 528–532

CORTESI, R. & PRIVETT, O. S. (1972) Toxicity of fatty ozonides and peroxides. *Lipids 2*, 715–721

DUMELIN, E. E., DILLARD, C. J. & TAPPEL, A. L. (1978) Breath ethane and pentane as measures of vitamin E protection of *Macaca radiata* against 90 days of exposure to ozone. *Environ. Res. 15*, 38–43

GOLDSTEIN, B. D. & McDONAGH, E. M. (1975) Effect of ozone on cell membrane protein fluorescence. *Environ. Res. 9*, 179–186

GOLDSTEIN, B. D., SOLOMON, S., PASTERNACK, B. S. & BICKERS, D. R. (1975) Decrease in rabbit lung microsomal cytochrome p-450 levels following ozone exposure. *Res. Commun. Chem. Pathol. Pharmacol. 10*, 759–762

HOFMANN, T. & HARRISON, P. M. (1963) The structure of apoferritin: degradation into and molecular weight of subunits. *J. Mol. Biol. 4*, 256–267

KOSHLAND, D. E., KARBANIS, Y. D. & LATHAM, H. (1964). *J. Am. Chem. Soc. 86*, 1448

LISTOWSKY, I., BLAUER, G., ENGLARD, S. & BETHEIL, J. J. (1972) Denaturation of horse spleen ferritin in aqueous guanidinium chloride solution. *Biochemistry 11*, 2176–2181

MENZEL, D. B., ANDERSON, W. G. & ABOU-DONIA, M. B. (1976) Ozone exposure modifies prostaglandin biosynthesis in perfused rat lungs. *Res. Commun. Chem. Pathol. Pharmacol. 15*, 135–147

MENZEL, D. B., ROEHM, J. N. & LEE, S. D. (1972) Vitamin E: the biological and environmental antioxidant. *J. Agr. Food Chem. 20*, 481–486

NAKAMURA, H. & PISANO, J. J. (1976) Reaction of tryptophan, tryptamines, and some related indoles with fluorescamine: unique fluorescence in strong acid. *Arch. Biochem. Biophys. 172*, 98–101

SPIES, J. R. & CHAMBERS, D. C. (1948) Chemical determination of tryptophan. *Anal. Chem. 20*, 30–39

PALMER, M. S., SWANSON, D. H. & COFFIN, D. L. (1971) Effect of ozone on benzpyrene hydroxylase activity in the Syrian golden hamster. *Cancer Res. 31*, 730–733

The pathology and biochemistry of paraquat

L. L. SMITH, M. S. ROSE and I. WYATT

Central Toxicological Laboratory, ICI, Alderley Park, Cheshire

Abstract After the administration of paraquat to rats the lung is the organ most severely damaged. The pathology in the lung can be divided into two distinct phases: (1) a destruction phase lasting a few days with damage to the type I and type II alveolar epithelial cells, oedema and haemorrhage (most of the rats which die after dosing with paraquat do so during this phase); (2) a reparative phase with regeneration of the epithelium and, in areas of severe damage, a characteristic proliferation of fibroblasts. In both phases of the lesion the death of the rats results from anoxia. Paraquat is selectively accumulated by the rat lung in comparison with other tissues and this accounts, at least in part, for the specific toxic effect in this organ. The accumulation into the lung was shown by *in vitro* studies to depend on energy and is inhibited by various endogenous and exogenous compounds. This uptake process is not that which has been described for 5-hydroxytryptamine and evidence is presented to suggest that the type I and type II alveolar epithelial cells are sites of accumulation. When paraquat is present in lung cells, it undergoes a cyclical reduction and oxidation with the production of superoxide anion. This radical may lead directly or indirectly to the formation of lipid peroxides and hence to cell death. However, paraquat stimulates the pentose-phosphate pathway and both reduces the level of NADPH and inhibits fatty acid synthesis in the lung. These effects occur when there is only minimal ultrastructural damage to the lung cells. It is suggested, therefore, that the primary mechanism of toxicity of paraquat is the extreme oxidation of NADPH which inhibits vital physiological processes and renders the cell more susceptible to attack from lipid hydroperoxides.

Paraquat (1,1′-dimethyl-4,4′-bipyridylium; methyl viologen), a non-selective herbicide marketed by Imperial Chemical Industries Limited, was first synthesized in 1882 (Weidel & Russo 1882) although its herbicidal properties were not discovered until 1955 (Calderbank 1968). The phytotoxic action of paraquat is considered to be related to its ability to be reduced by photosystem 1 of plants. The paraquat radical which is formed then reacts avidly with molecular oxygen to

(1) reform the paraquat cation and (2) generate superoxide anion (O_2^-) which leads to the production of H_2O_2 which damages the chloroplast (Dodge 1971).

Since its introduction as a herbicide in 1963 paraquat has proved safe in use. Although there have been occasional fatalities from the accidental ingestion of the concentrated agricultural product (usually after decanting of the product into improperly labelled containers), nearly all the deaths caused by it result from the deliberate ingestion of large amounts of the concentrate. The severity and rapidity of onset of the toxic effects which result from the ingestion of paraquat depend on the amount taken. In humans this varies with the circumstances of poisoning. From data collected at the Central Toxicological Laboratory (ICI) it appears that doses of paraquat in excess of 3 g per adult are usually fatal if untreated.

In humans the most characteristic effect of paraquat toxicity is damage to the lung. After large doses oedema and haemorrhage develop in the lung (Yoneyama *et al.* 1969) with destruction of the alveolar epithelium (Von der Hardt & Cardesa 1971). Other toxic effects include ulceration of the mouth and oesophagus (McDonagh & Martin 1970; Malone *et al.* 1971), proximal tubular necrosis in the kidney (Oreopoulos *et al.* 1968) and cellular necrosis in the liver (Fennelly *et al.* 1971; Grabensee *et al.* 1971). With acute deaths within two days of ingestion there can be adrenal corticonecrosis (Nagi 1970; Yoneyama *et al.* 1969) and haemorrhage and oedema in the brain (Nienhaus & Ehrenfeld 1971). Clearly, such deaths are associated with damage to several vital organs amongst which the lung is probably most prominent.

Deaths which occur several weeks after the ingestion of paraquat have given rise to a pathology known as 'paraquat lung'. In these cases the mouth, oesophagus, kidney and liver may be damaged initially but this usually resolves or responds to treatment. However, over a period of many days or weeks lung function becomes progressively impaired and the patient dies from anoxia owing to a characteristic fibrosis of the lung parenchyma (Bronkhorst *et al.* 1968; Bony *et al.* 1971; Grabensee *et al.* 1971).

In this paper we shall (1) describe the toxicity of paraquat in experimental animals, (2) explain the selective toxicity of paraquat for the lung, and (3) discuss the possible mechanisms of toxicity which result in damage and death of lung cells.

THE PATHOLOGY OF PARAQUAT IN EXPERIMENTAL ANIMALS

The toxic effects of paraquat in experimental animals were first described by Clark *et al.* (1966) who reported that histologically the effects of paraquat in rats, mice and dogs are similar. The liver, kidney, thymus and lung are the organs affected: the lung most severely. The effects of paraquat in monkeys are similar to those in rats (Murray & Gibson 1972). In contrast, lung lesions do not develop in rabbits after a single lethal dose of paraquat (Butler & Kleinerman 1971; Mehani 1972). However, Restuccia *et al.* (1974) have shown that after the administration of paraquat in small amounts over several consecutive days the rabbit lung develops lesions similar to those seen in rats. Similarly, lung damage cannot be produced in hamsters after a single administration but lung fibrosis does result from a prolonged series of exposures (Butler 1975).

The most extensive investigations of the pathogenesis of paraquat-induced lung damage have been carried out in rats. After giving the animals an approximate LD_{50} dose of paraquat intraperitoneally, Vijeyaratnam & Corrin (1971) examined the lungs of the rats at various times after dosing. Within a day of dosing, the type I and type II alveolar epithelial cells are damaged. The destruction of these epithelial cells continues, such that by 2–4 days after dosing areas of alveolar epithelium are completely destroyed. As the epithelial cells are destroyed considerable alveolar oedema develops and in some areas haemorrhage into the air spaces may occur. During this phase of the lesion there is an extensive infiltration of the alveolar interstitium, air spaces and perivascular areas by inflammatory cells, although the alveolar endothelial capillaries show only minor alterations. Most of the rats which die after an LD_{50} of paraquat do so as a consequence of severe anoxia within a few days of dosing (Smith & Rose 1977). Similar early pathological changes have been described by others (see Kimbrough & Gaines 1970; Modee *et al.* 1972; Wasan & McElligott 1972; Smith *et al.* 1974; Sykes *et al.* 1977).

A few of the rats severely affected by paraquat survive for many days after dosing. These animals develop an extensive hypercellular lesion in the lung dominated by the proliferation of fibroblasts. This hypercellular phase of the lesion is characterized by attempts of the epithelium to regenerate and restore the normal architecture of the alveolar epithelium (Kimbrough & Gaines 1970; Vijeyaratnam & Corrin 1971). The most prominent feature, however, of the later lesion is the proliferation of fibroblasts within the interstitium of the lung which, along with residual oedema and alveolar collapse, results in death due to severe anoxia. It appears that it is the damage to the alveolar epithelium which is the primary event in the development of the 'paraquat lung' and that the proliferative fibrosis should be considered a consequence of extensive and severe acute damage.

THE *IN VIVO* DISTRIBUTION OF PARAQUAT

Most authors agree that paraquat is not metabolized in experimental animals. Sharp *et al.* (1972) first demonstrated that the lungs of rats intravenously dosed

TABLE 1

Paraquat concentrations in rat tissues after oral administration of 680 μmol/kg body weight

Tissue	*Paraquat concentration (nmol/g wet weight tissue)*			
	Time after dosing (hours)			
	2	*4*	*18*	*30*
Brain	6.8 ± 3.2 (4)	0.81 ± 0.08 (4)	1.5 ± 0.1 (4)	3.1 ± 0.3 (4)
Lung	16.5 ± 2.2 (8)	17.0 ± 1.6 (8)	29.6 ± 2.7 (8)	86.6 ± 17.2 (7)
Liver	20.8 ± 6.8 (7)	8.9 ± 1.7 (8)	11.6 ± 1.7 (8)	20.4 ± 3.1 (7)
Kidney	75.0 ± 15.1 (8)	54.9 ± 18.1 (8)	57.7 ± 3.1 (8)	108 ± 22 (7)
Adrenal	12.8 ± 2.5 (8)	30.1 ± 15.4 (8)	16.1 ± 1.5 (8)	26.2 ± 6.5 (7)
Muscle	4.8 ± 0.4 (8)	5.2 ± 2.1 (8)	5.2 ± 1.1 (8)	11.0 ± 2.7 (7)
Plasma	14.0 ± 3.7 (8)	6.6 ± 1.0 (8)	8.0 ± 0.6 (8)	13.8 ± 2.9 (7)

Data are expressed as mean ± S.E.M. with the number of animals in parentheses and are from Rose *et al.* (1976*a*), with permission.

with paraquat had the highest concentration and selectively retained the compound in comparison to other organs. This selective retention was also shown in the lungs of mice after intravenous dosing (Litchfield *et al.* 1973) and in rabbits (Ilett *et al.* 1974).

We have investigated the tissue distribution of paraquat after oral administration of an approximate LD_{100} dose (Table 1). The concentration of paraquat in the plasma remained relatively constant over the period of study (30 h), whereas that in the lung increased with time such that by 30 h after dosing there was 6–7 times more in the lung than in the plasma (Table 1). Of the organs studied no other tissue showed a similar time-dependent increase in paraquat concentration. In the kidney, however, the concentration was high compared with other organs two hours after dosing and remained high but did not increase with time during the study. This observation is consistent with the role of the kidney in the clearance of paraquat from the blood. Thus, the lung, which is the organ most severely damaged by paraquat, selectively retains it after intravenous administration and accumulates it to concentrations several times that present in the plasma after oral dosing. We conclude that, at least in part, its selective uptake accounts for the specific toxicity exhibited by paraquat for the lung.

UPTAKE OF PARAQUAT INTO RAT LUNG SLICES

To study further the mechanism of accumulation, we incubated rat lung slices in a medium containing paraquat (10 μmol/l) and determined the slice/medium ratio as a function of time (Fig. 1). Paraquat accumulated in the lung with time but when KCN was added to the incubation medium the accumulation was inhibited. The accumulation was also inhibited by iodoacetate, rotenone, and when the slices were incubated under nitrogen as opposed to air (Rose *et al.* 1974). Thus, we conclude that the accumulation of paraquat into lung tissue depends on energy.

Increases in the slice/medium ratio as seen in Fig. 1 may result from (1) the binding of paraquat onto or in the lung or (2) the uptake of paraquat into a compartment within the lung slice from which the rate of efflux is slower than the rate of uptake. Ilett *et al.* (1974) were unable to demonstrate any covalent binding of paraquat to the lung. Furthermore, when lung slices which had accumulated paraquat for several hours were gently homogenized, the paraquat was found in the supernatant fraction. When this was passed through a column containing Sephadex G25 gel, the elution volume for 'accumulated' paraquat was the same as that for paraquat dissolved in buffer (Jaques & Smith, unpublished data), a fact which indicates that the 'accumulated' paraquat was not

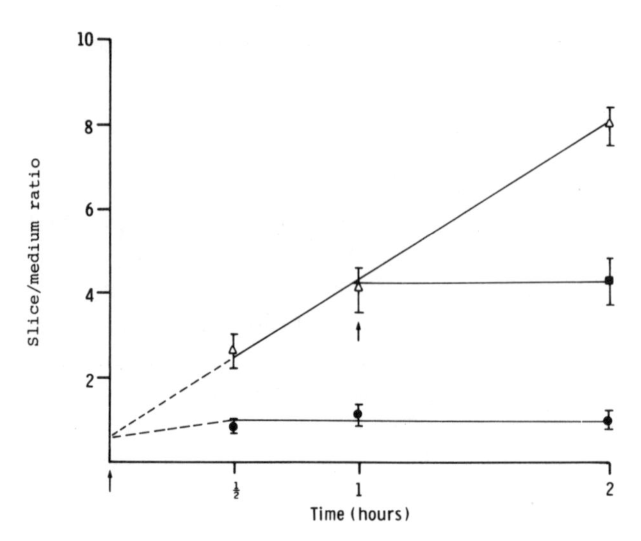

FIG. 1. The effect of metabolic inhibitors on the uptake of paraquat into lung slices. Slices of rat lung were incubated in Krebs–Ringer phosphate medium containing 10μM-paraquat (△), with 1mM-KCN (●) at zero time and 1mM-KCN added (arrow) after 1 h incubation (■). The lung slices were removed at various times and the slice/medium ratio for paraquat was determined. The results are expressed as mean ± S.E.M. with four slices per time point.

bound to any low-molecular-weight material. Therefore, it appears that the accumulation by the lung is unrelated to a binding phenomenon.

The lung plays an important role in the regulation of vasoactive amines (Vane 1968). Studies using isolated perfused lungs have indicated that the endogenous amines 5-hydroxytryptamine (5HT) (Alabaster & Bahkle 1970; Junod 1972) and noradrenalin (Gillis & Iwasawa 1972) are accumulated by the lung. We

TABLE 2

Inhibition of paraquat accumulation in rat lung slices by endogenous amines and amino acids at 0.1 mmol/l concentration

Compound	*% Inhibition after two hours (range)*
Noradrenalin	63 (58–69)
5-Hydroxytryptamine	32 (24–46)
Tryptamine	51 (27–64)
Tyramine	81 (80–82)
Histamine	51 (37–68)
Lysine	35 (33–41)
Spermine	88 (87–89)

Slices of rat lung were incubated in Krebs–Ringer phosphate glucose medium with 10μM-paraquat and 0.1mM-additive. Compounds were added at zero time and the incubation lasted two hours at 37 °C. The results are expressed as mean (with the range of inhibition in parentheses). At least three slices were used per concentration.

TABLE 3

Inhibition of paraquat accumulation in rat lung slices by exogenous compounds at 0.1 mmol/l concentration

Compound	*% Inhibition after two hours (range)*
D-Propranolol	39 (38–41)
Imipramine	58 (45–76)
Betazole	50 (47–53)
Burimimide	39 (24–56)
Promethazine	58 (57–60)

Slices of rat lung were incubated in Krebs–Ringer phosphate glucose medium with 10μM-paraquat and 0.1mM-additive. Compounds were added at zero time and the incubations lasted two hours at 37 °C. The results are expressed as mean (with the range of inhibition in parentheses). At least three slices were used per concentration.

studied the effect of a range of endogenous compounds on the uptake of paraquat into lung slices (Table 2). Among others 5HT and various exogenous compounds, including D-propranolol, imipramine and betazole, inhibited accumulation *in vitro* (Table 3). At present the most effective inhibitor of paraquat uptake that we have found is putrescine.

THE COMPARTMENT INTO WHICH PARAQUAT ACCUMULATES IN THE LUNG

The inhibition of the accumulation of paraquat into lung slices by 5HT raised the possibility that paraquat was accumulated by the same mechanism as that reported for 5HT. We demonstrated that 5HT was accumulated by rat lung slices and confirmed, as had been shown with perfused lung systems (Junod 1972; Steinberg *et al.* 1975), that the process in lung slices also obeyed saturation kinetics, was inhibited by imipramine, the metabolic inhibitors cyanide and iodoacetate and depended on sodium ion concentration (Smith *et al.* 1976). Although paraquat accumulation also obeyed saturation kinetics and was inhibited by imipramine, cyanide and iodoacetate, it was not dependent on sodium; instead, it was markedly stimulated in the absence of sodium and this enhanced accumulation could be inhibited by cyanide (Fig. 2). We conclude, therefore, that the accumulation process for paraquat is not the same as that which has been described for 5HT.

With regard to the cell type(s) in the lung responsible for the accumulation of paraquat, there is at present no reported autoradiographic study of the cellular distribution. However, after the administration of paraquat to rats, it is the type I and type II alveolar epithelial cells of the lung which are first damaged (Kimbrough & Gaines 1970; Vijeyaratnam & Corrin 1971). Sykes *et al.* (1977) showed that, four hours after intravenous dosing, there were minimal

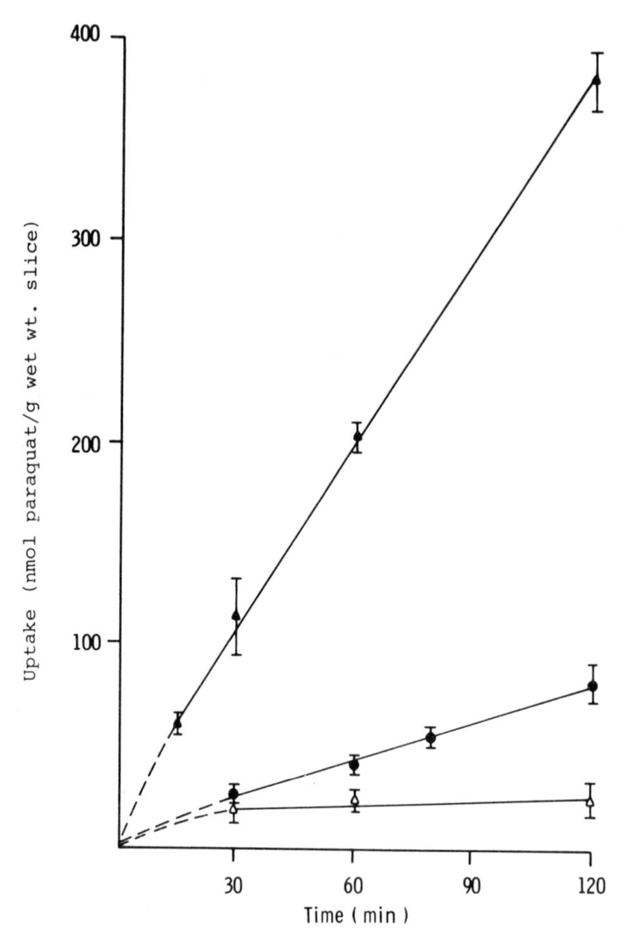

FIG. 2. Uptake of paraquat by rat lung slices incubated in Krebs–Ringer phosphate and sodium-deficient medium. Slices of rat lung were incubated with 10μM-paraquat in Krebs–Ringer phosphate medium (●) or in sodium-deficient medium (▲) or in sodium-deficient medium containing 1mM-KCN (△). The amount of paraquat present in the lung is expressed as mean ± S.E.M. with four slices per time point.

ultrastructural changes in the type I and type II alveolar cells which progressed with time such that by 16 h after dosing there was significant destruction of these cells. We then determined the accumulation of paraquat by lung slices taken from rats at various times after intravenous dosing with paraquat. By 16 h after dosing, the lung slices with damaged type I and type II cells no longer accumulated paraquat as effectively as control lung slices or indeed lung slices taken from rats at earlier times when there was only minimal epithelial damage (Table 4). That this inhibition of accumulation is not the result of non-specific lung damage was shown by the ability of lung slices taken from rats 16 h after

TABLE 4

Accumulation of paraquat into normal lung slices and lung slices taken from rats treated with paraquat

Time after administration of paraquat (h)	*Accumulation of paraquat (nmol[g wet wt]$^{-1}$[2 h]$^{-1}$)*
0 (Control; untreated)	73.8 ± 2.9 (7)
2	71.0 ± 2.5 (10)
4	53.6 ± 1.8 (10)
8	63.1 ± 2.1 (10)
16	36.2 ± 2.2 (10)

Treated rats were given 65 μmol paraquat/kg body weight intravenously and killed at the times indicated. The lungs were removed and slices cut from the left lobe. The slices were incubated in 10μM-paraquat for two hours and the concentration accumulated into the lung was determined. Results are expressed as mean ± S.E.M. with the number of rats used in parentheses.

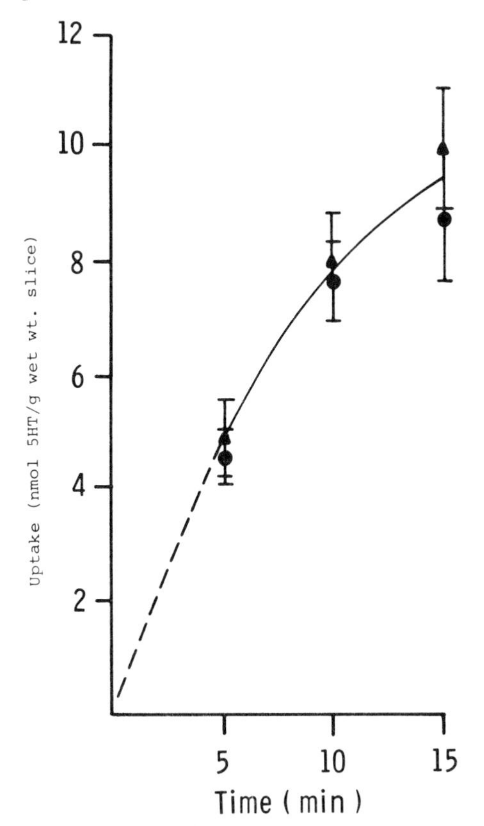

FIG. 3. Uptake of 5-hydroxytryptamine (5HT) by lung slices from rats given paraquat and by control slices: accumulation of 5HT into control slices (●) or lung slices from rats given 65 μmol paraquat/kg body weight intravenously 16 h previously (▲). The slices were incubated in Krebs–Ringer phosphate medium containing 1μM-5HT. Results are expressed as mean ± S.E.M. with four slices per time point.

dosing with paraquat to accumulate 5HT as effectively as control lung slices (Fig. 3). This is consistent with the observation that the capillary endothelial cells are not damaged by paraquat and this cell type is responsible for the uptake of 5HT (Strum & Junod 1972).

In conclusion, therefore, paraquat is accumulated by an energy-dependent process into lung by a mechanism different from that described for 5HT. The sites of paraquat accumulation in the lung are at least in part the type I and type II alveolar epithelial cells, and the transport system responsible for this uptake has not previously been identified.

MECHANISMS OF TOXICITY

Paraquat can be reduced to form a free radical which is stable in aqueous solution in the absence of oxygen (Michaelis & Hill 1933). Oxygen extremely rapidly reoxidizes this radical to the cation with the concomitant production of superoxide anion, O_2^- (Farrington *et al.* 1973). Paraquat can be reduced to its free radical by homogenates of liver, kidney or lung (Gage 1968; Baldwin *et al.* 1975). This reduction depends on NADPH and is enzymically catalysed (Gage 1968), for instance by cytochrome *c* reductase (Bus *et al.* 1974). Therefore, one probable consequence of the accumulation of paraquat in lung cells is the cyclic reduction and oxidation of paraquat with the production of O_2^- and the oxidation of NADPH. In mammalian cells the pentose-phosphate pathway is considered the major process for the reduction of $NADP^+$ to NADPH. We have shown that in lung slices incubated with paraquat *in vitro* and in lung slices taken from paraquat-poisoned rats the activity of the pentose-phosphate pathway was markedly stimulated (Rose *et al.* 1976*b*). This suggests that the cycle of oxidation and reduction of paraquat in broken-cell preparations also occurs in intact lung cells. Fig. 4 summarizes the initial reactions in the lung. The involvement of O_2^- in paraquat toxicity leads naturally to a consideration of the relationship of paraquat in oxygen toxicity.

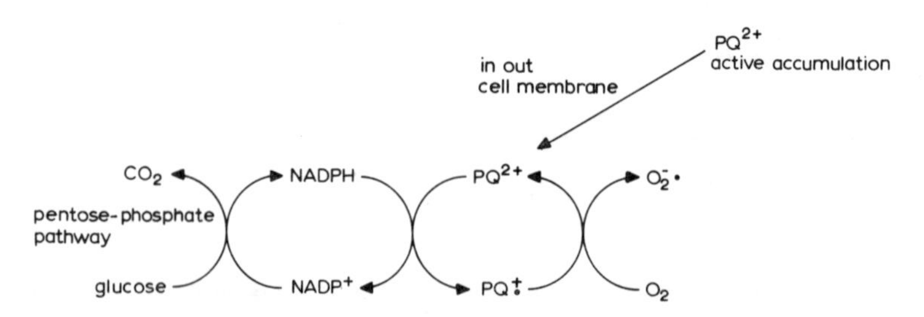

FIG. 4. Initial reactions in the lung on active accumulation of paraquat dication (PQ^{2+}) across the cell membrane.

Fisher *et al.* (1973) demonstrated that rats given a lethal dose of paraquat and placed in a 100% oxygen atmosphere die more rapidly than those given the same dose of paraquat and left in ambient air. Bus *et al.* (1976) demonstrated that rats pre-exposed to 85% oxygen for seven days were more resistant to the effects of paraquat compared to those pre-exposed to ambient air. This is consistent with the observation that pre-exposure of rats to 85% oxygen for seven days induces superoxide dismutase in the lung (Crapo & Tierney 1974). This enzyme is considered to be of prime importance in the defence against attack by O_2^- (Fridovich 1975).

Bus *et al.* (1974, 1975, 1976) have tested the hypothesis that the continued reduction and oxidation of paraquat with concomitant production of O_2^- leads, directly or indirectly, via the production of hydroxyl radicals or singlet oxygen, to lipid peroxidation and hence cell death. Although this is an attractive hypothesis, we consider that the primary mechanism of toxicity may not be the production of lipid peroxides but involves the oxidation of NADPH (Rose *et al.* 1976*b*). In those cells in which paraquat is accumulated, the concentration may be very high and result in very fast rates of NADPH oxidation. If this rate is greater than the rate of NADPH generation by the pentose-phosphate pathway, the concentration of NADPH will fall below that required to sustain vital physiological processes.

The stimulation of the activity of the pentose-phosphate pathway in the lung by paraquat indicates a requirement for NADPH. To test the 'NADPH depletion' hypothesis further we have examined the ability of paraquat-treated lungs to synthesize fatty acids since this system depends on NADPH (Wakil 1962) and is probably of particular importance in type II alveolar epithelial cells since these are active in surfactant synthesis (Macklin 1954). When lung slices were incubated with increasing concentrations of paraquat, fatty acid synthesis in the lung was correspondingly inhibited. Since putrescine, which inhibits the accumulation of paraquat into lung slices, abolished the inhibition of fatty acid synthesis when added to the incubation medium (Table 5), this inhibition must have been caused by the paraquat that had been accumulated into the lung.

After intravenous dosing of paraquat to rats, fatty acid synthesis was inhibited within four hours of dosing and practically abolished by 16 h (Table 6). This system was inhibited at times when there was only minimal ultrastructural damage to some of the alveolar epithelial cells (Sykes *et al.* 1977).

Recently, Witschi *et al.* (1977) have shown that the ratio of NADPH to NADP is reduced in lungs taken from paraquat-treated rats. We found that lung slices incubated in a medium containing paraquat also showed a reduction in the NADPH concentrations and in the NADPH/NADP ratio (Table 5). As with

TABLE 5

The effect of paraquat on fatty acid synthesis, NADPH concentrations and NADPH/NADP ratios in lung slices

Treatment	*Paraquat concentration in the lung slices (nmol/g wet wt.)*	*[^{14}C]Acetate incorporation into fatty acids (% of control)*	*NADPH concentration in the lung slices (nmol/100 mg)*	*NADPH/NADP ratio in the lung slices*
Control		100	1.48 ± 0.04 (17)	1.48 ± 0.06 (17)
100μM-Paraquat	418 ± 61 (4)	37 ± 4 (5)	0.85 ± 0.05 (6)	0.685 ± 0.03 (6)
10μM-Paraquat	109.6 ± 7.9 (4)	53 ± 6 (19)	1.00 ± 0.1 (6)	1.01 ± 0.09 (5)
1μM-Paraquat	17.0 ± 1.2 (4)	96 ± 9 (8)	1.22 ± 0.05 (6)	1.15 ± 0.09 (5)
100μM-Putrescine		103 ± 6 (5)	1.57 ± 0.14 (9)	1.49 ± 0.15 (9)
5μM-Paraquat	65.8 ± 4.6 (4)	77 ± 2 (5)	1.07 ± 0.13 (9)	1.08 ± 0.12 (9)
100μM-Putrescine + 5μM-Paraquat	8.6 ± 0.9 (4)	101 ± 8 (5)	1.49 ± 0.17 (9)	1.53 ± 0.15 (9)

Lung slices (about 300 mg) from the left lobe of male rats were incubated for five hours with various concentrations of paraquat in Krebs–Ringer phosphate (7.0 ml) with 11mM-glucose at 37 °C and 140 spm. For fatty acid synthesis [^{14}C]acetate (4 μCi, 40 μmol) was added for the last 90 min of incubation; for determinations of paraquat, NADPH and NADP, cold acetate (40 μmol) was added for the last 90 min. The results are expressed as mean ± S.E.M. (with the number of determinations in parentheses). From control lung tissue the radioactivity obtained was about 5×10^3 d.p.m./mg fatty acid. [^{14}C]Paraquat was used to determine paraquat concentrations in lung slices. The slices from the incubation were divided in half so that NADPH and NADP concentrations could be determined on the same sample so that the ratios could be obtained.

TABLE 6

The effect of paraquat on fatty acid synthesis in lung tissue taken from rats given paraquat intravenously

Time after intravenous dosing (h)	*[^{14}C]Acetate incorporation into fatty acids (% of control)*
2	75 ± 9 (5)
4	31 ± 3 (5)
16	15 ± 4 (8)

Male rats were injected intravenously (via the tail vein) with 65 μmol/kg of paraquat and killed at various time intervals. The lungs were removed, slices were prepared (300 mg), placed in Krebs–Ringer phosphate (7.0 ml) with 11mM-glucose, and the incorporation of [^{14}C]acetate into fatty acids was measured over 90 min incubation at 37 °C and 140 spm. The results represent the mean ± S.E.M. (with the number of determinations in parentheses). From control lung tissue the radioactivity obtained was about 5 × 10^3 d.p.m./mg fatty acid.

the inhibition of fatty acid synthesis, this inhibition could be prevented by the addition of putrescine to the medium. It, therefore, appears that in both *in vivo* and *in vitro* studies paraquat reduced NADPH concentrations in the lung and inhibited fatty acid synthesis which depends on NADPH. Also, the effects of paraquat on NADPH may be much more severe than suggested by our studies. If the type I and type II alveolar cells are mostly responsible for the accumulation of paraquat, the observed reduction in NADPH concentrations on a lung (wet weight) basis may result from considerable reductions in only a small percentage of the total cell population.

A fall in the concentration of NADPH in lung cells may not only debilitate

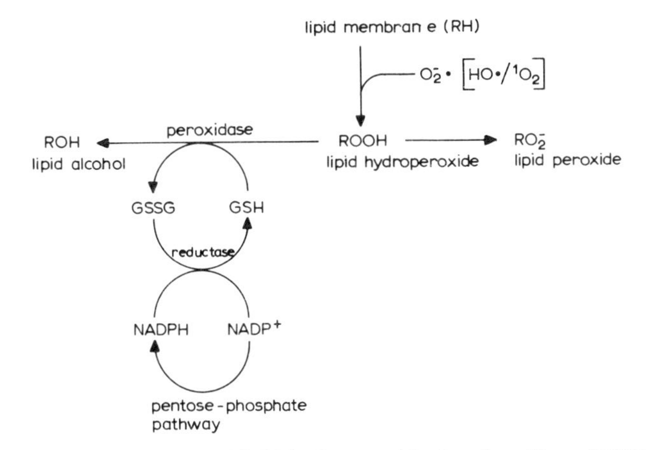

FIG. 5. Reduction of lipid hydroperoxides by glutathione (GSH) peroxidase.

vital physiological processes but render them more susceptible to lipid peroxidation. When lipid hydroperoxides are formed as a result of the interaction of lipids with reactive radicals, glutathione peroxidase mediates the formation of the less toxic lipid alcohols (Fig. 5). This enzyme has, as an essential cofactor, reduced glutathione (GSH) which is oxidized in the reaction. The reduction of oxidized glutathione (GSSG) by glutathione reductase depends on NADPH.

Nishiki *et al.* (1976) suggested that oxygen toxicity in the lung results from the extreme oxidation of NADPH which leads to a decrease in respiration and to formation of lipid peroxides. If this were the case in oxygen toxicity, then in the toxicity of paraquat the cell would be even further compromised. The concentration of NADPH would be lowered first by the formation of the paraquat radical and then further in the defence against lipid hydroperoxides. Furthermore as the pentose-phosphate pathway is stimulated to regenerate NADPH, so this cofactor may become available not only for the conversion of GSSG into GSH but also for the formation of paraquat radical. In the event of the latter, attempts by the cell to replace NADPH may contribute to the further formation of lipid hydroperoxides and hence oxidation of NADPH.

References

ALABASTER, V. A. & BAHKLE, Y. S. (1970) Removal of 5-hydroxytryptamine in the pulmonary circulation of isolated lungs. *Br. J. Pharmacol. 40*, 468–482

BALDWIN, R. C., PASI, A., MACGREGOR, J. T. & HINE, C. H. (1975) The rates of radical formation from the dipyridilium herbicides, paraquat, diquat and morfamquat in homogenates of lung, kidney and liver: an inhibitory effect of carbon monoxide. *Toxicol. Appl. Pharmacol. 32*, 298–304

BONY, D., FAVAREL-GARRIGUES, J. C., CLEDES, J. CAMBEILH, J. & CASTAING, R. (1971) Paraquat poisoning. *Eur. J. Toxicol. 4*, 406–411

BRONKHORST, F. B., VAN DAAL, J. M. & TAN, H. D. (1968) Fatal poisoning with paraquat. *Ned. Tijdschr. Geneeskd. 112*, 310–313

BUS, J. S., AUST, S. D. & GIBSON, J. E. (1974) Superoxide and singlet oxygen catalysed lipid peroxidation as a possible mechanism for paraquat toxicity. *Biochem. Biophys. Res. Commun. 58*, 749–755

BUS, J. S., AUST, S. D. & GIBSON, J. E. (1975) Lipid peroxidation: a possible mechanism for paraquat toxicity. *Res. Commun. Chem. Pathol. Pharmacol. 11*, 31–38

BUS, J. S., CAGEN, S. Z., OLGAARD, M. & GIBSON, J. E. (1976) A mechanism of paraquat toxicity in mice and rats. *Toxicol. Appl. Pharmacol. 35*, 501–513

BUTLER, C. (1975) Pulmonary interstitial fibrosis from paraquat in the hamster. *Arch. Pathol. 99*, 503–507

BUTLER, C. & KLEINERMAN, J. (1971) Paraquat in the rabbit. *Br. J. Ind. Med. 28*, 67–71

CALDERBANK, A. (1968) The bipyridilium herbicides. *Adv. Pest Control Res. 8*, 127–235

CLARK, D. G., MCELLIGOTT, T. F. & WESTONHURST, E. (1966) The toxicity of paraquat. *Br. J. Ind. Med. 23*, 126–132

CRAPO, J. D. & TIERNEY, D. F. (1974) Superoxide dismutase and pulmonary oxygen toxicity. *Am. J. Physiol. 226*, 1401–1407

DODGE, A. D. (1971) The mode of action of the bipyridilium herbicides, paraquat and diquat. *Endeavour 30*, 130–135

FARRINGTON, J. A., EBERT, M., LAND, E. J. & FLETCHER, K. (1973) Bipyridilium quaternary salts and related compounds. V. Pulse radiolysis studies on the reaction of paraquat radical with oxygen. Implications for the mode of action of bipyridilium herbicides. *Biochim. Biophys. Acta 314*, 372–381

FENNELLY, J. J., FITZGERALD, M. X. & FITZGERALD, O. (1971) Recovery from severe paraquat poisoning following forced diuresis and immunosuppressive therapy. *J. Ir. Med. Assoc. 64*, 69–71

FISHER, H. K., CLEMENTS, J. A. & WRIGHT, R. R. (1973) Enhancement of oxygen toxicity by the herbicide paraquat. *Am. Rev. Resp. Dis. 107*, 246–252

FRIDOVICH, I. (1975) Superoxide dismutases. *Annu. Rev. Biochem. 44*, 147–159

GAGE, J. C. (1968) The action of paraquat and diquat on the respiration of liver cell fractions. *Biochem. J. 109*, 757–761

GILLIS, C. N. & IWASAWA, Y. (1972) Technique for measurement of norepinephrine and 5-hydroxytryptamine uptake by rabbit lung. *J. Appl. Physiol. 33*, 404–408

GRABENSEE, B., VELTMANN, G., MÜRTZ, R. & BORCHARD, F. (1971) Poisoning by paraquat. *Dtsch. Med. Wochenschr. 96*, 498–506

ILETT, K. F., STRIPP, B., MENARD, R. H., REID, N. D. & GILLETTE, J. R. (1974) Studies on the mechanism of the lung toxicity of paraquat: comparison of tissue distribution and some biochemical parameters in rats and rabbits. *Toxicol. Appl. Pharmacol. 28*, 216–226

JUNOD, A. F. (1972) Uptake, metabolism and efflux of ^{14}C-5-hydroxytryptamine in isolated perfused rat lungs. *J. Pharmacol. Exp. Ther. 183*, 341–355

KIMBROUGH, R. D. & GAINES, T. B. (1970) Toxicity of paraquat to rats and its effect on rat lung. *Toxicol. Appl. Pharmacol. 17*, 679–690

LITCHFIELD, M. H., DANIEL, J. W. & LONGSHAW, S. (1973) The tissue distribution of the bipyridilium herbicides paraquat and diquat in rats and mice. *Toxicology 1*, 155–163

MACKLIN, C. C. (1954) The pulmonary alveolar mucoid film and the pneumonocytes. *Lancet 1*, 1099–1104

MALONE, J. D. G., CARMODY, M., KEOGH, B. & O'DWYER, W. F. (1971) Paraquat poisoning — a review of nineteen cases. *J. Ir. Med. Assoc. 64*, 59–68

MCDONAGH, B. J. & MARTIN, J. (1970) Paraquat poisoning in children. *Arch. Dis. Childh. 45*, 425–427

MEHANI, S. (1972) The toxic effect of paraquat in rabbits and rats. *Ain Shams Med. J. 23*, 599–601

MICHAELIS, L. & HILL, E. S. (1933) Potentiometric studies on semiquinones. *J. Am. Chem. Soc. 55*, 1481–1494

MODÉE, J., IVEMARK, B. I. & ROBERTSON, B. (1972) Ultrastructure of the alveolar wall in experimental paraquat poisoning. *Acta Pathol. Microbiol. Scand. 80*, 54–60

MURRAY, R. E. & GIBSON, J. E. (1972) A comparative study of paraquat intoxication in rats, guinea pigs and monkeys. *Exp. Mol. Pathol. 17*, 317–325

NAGI, A. H. (1970) Paraquat and adrenal cortical necrosis. *Br. Med. J. 2*, 669

NIENHAUS, H. & EHRENFELD, M. (1971) Pathogenesis of lung disease in paraquat poisoning. *Beitr. Pathol. 142*, 244–267

NISHIKI, K., JAMIESON, D., OSHINO, N. & CHANCE, B. (1976) Oxygen toxicity in the perfused rat liver and lung under hyperbolic conditions. *Biochem. J. 160*, 343–355

OREOPOULOS, D. G., SOYANNWO, M. A. O., SINNIAK, R., FENTON, S. S. A., MCGEOWN, M. G. & BRUCE, J. H. (1968) Acute renal failure in cases of paraquat poisoning. *Br. Med. J. 1*, 749–750

RESTUCCIA, A., FOGLINI, A. & DE ALENTIS NANNINI, D. (1974) Paraquat toxicity for rabbits. *Vet. Ital. 25*, 555–565

ROSE, M. S., SMITH, L. L. & WYATT, I. (1974) Evidence for the energy dependent accumulation of paraquat into rat lung. *Nature (Lond.) 252*, 314–315

ROSE, M. S., LOCK, E. A., SMITH, L. L. & WYATT, I. (1976*a*) Paraquat accumulation: tissue and species specificity. *Biochem. Pharmacol. 25*, 419–423

ROSE, M. S., SMITH, L. L. & WYATT, I. (1976*b*) The relevance of pentose phosphate pathway stimulation in rat lung to the mechanism of paraquat toxicity. *Biochem. Pharmacol. 25*, 1763–1767

SHARP, C. W. M., OTTOLENGHI, A. & POSNER, H. S. (1972) Correlation of paraquat toxicity with tissue concentration and weight loss of the rat. *Toxicol. Appl. Pharmacol. 22*, 241–251

SMITH, L. L. & ROSE, M. S. (1977) A comparison of the effects of paraquat and diquat on the water content of rat lung and the incorporation of thymidine into lung DNA. *Toxicology 8*, 223–230

SMITH, L. L., LOCK, E. A. & ROSE, M. S. (1976) The relationship between 5-hydroxytryptamine and paraquat accumulation into rat lung. *Biochem. Pharmacol. 25*, 2485–2487

SMITH, P., HEATH, D. & KAY, J. M. (1974) The pathogenesis and structure of paraquat induced pulmonary fibrosis in rats. *J. Pathol. 114*, 57–67

STEINBERG, H., BASSETT, D. J. & FISHER, A. B. (1975) Depression of pulmonary 5-hydroxytryptamine uptake by metabolic inhibitors. *Am. J. Physiol. 228*, 1298–1303

STRUM, J. & JUNOD, A. F. (1972) Radioautographic demonstration of 5-hydroxytryptamine-^3H uptake by pulmonary endothelial cells. *J. Cell Biol. 54*, 456–467

SYKES, B. I., PURCHASE, I. F. H. & SMITH, L. L. (1977) Pulmonary ultrastructure after oral and intravenous dosage of paraquat to rats. *J. Pathol. 121*, 233–241

VANE, J. R. (1968) The alteration or removal of vaso-active substances by the pulmonary circulation, in *The Importance of Fundamental Principles in Drug Evaluation*, pp. 217–235, Raven Press, New York

VIJEYARATNAM, G. S. & CORRIN, B. (1971) Experimental paraquat poisoning: a histological and electron-optical study of the changes in the lung. *J. Pathol. 103*, 123–129

VON DER HARDT, H. & CARDESA, A. (1971) Early histopathological alterations following paraquat intoxication. *Klin. Wochenschr. 49*, 544–550

WAKIL, S. J. (1962) Lipid metabolism. *Annu. Rev. Biochem. 31*, 369–406

WASAN, S. M. & MCELLIGOTT, T. F. (1972) An electron microscopic study of experimentally induced interstitial pulmonary fibrosis. *Am. Rev. Resp. Dis. 105*, 276–282

WEIDEL, H. & RUSSO, M. (1882) *Mh. Chem. 3*, 850–885

WITSCHI, H., KACEW, S., HIRAI, K. & CÔTE, M. G. (1977) In vivo oxidation of reduced nicotinamide-adenine dinucleotide phosphate by paraquat and diquat in rat lung. *Chem.-Biol. Interact. 19*, 143–160

YONEYAMA, M., KAINUMA, T. & TAKEUCHI, I. (1969) Swift death by poisoning caused by the new herbicide gramoxone. *Nihon Iji Shimpo (Jpn. Med. Affairs News) No. 2374*, 32–34

Discussion

Bielski: Might O_2^- run down the reduced nucleotide concentration directly? I ask because we found that NADH attached to lactate dehydrogenase or glycerophosphate dehydrogenase (Bielski & Chan 1977) is oxidized by O_2^- four to five orders of magnitude faster than free NADH.

Smith: We are currently investigating whether *in vitro* O_2^- reacts directly with nucleotide(s).

Winterhalter: According to Crapo & Tierney (1974), one can increase the SOD content of lung cells (presumably the lining cells). By pre-conditioning the rats with 85% oxygen could you increase their resistance to paraquat?

Smith: Yes. However, when rats are pre-exposed to 85% oxygen for eight days the SOD levels in the whole lung are increased by only 50% (Crapo & Tierney 1974). Also the work of Kimball *et al.* (1976) besides our own has shown that various other enzymes, notably glucose-6-phosphate dehydrogenase is increased. Therefore, the importance of the induction of SOD in the protection against both oxygen and paraquat needs careful investigation. I should especially draw your attention to the replacement of type I alveolar epithelial cells by type II cells as a result of exposure to 85% oxygen.

Fridovich: That is why we work with *E. coli*! Depletion of NADPH in *E. coli* cannot explain paraquat toxicity. Cells suspended aerobically in a medium devoid of substrate were not affected by paraquat.

Smith: I am merely suggesting that, in the mammalian system, the defence mechanisms have a different emphasis.

Klebanoff: Presumably, the inflammatory response to paraquat consists of neutrophils. On degeneration, these cells release peroxidase which is powerfully toxic to mammalian cells in the presence of hydrogen peroxide (which could be generated by the oxidation and reduction of paraquat) and a halide. This could be an additional source of toxicity to the lining cells.

Smith: Various agents have been given to rats treated with paraquat in an attempt to reduce the inflammatory lesion. I should point out that the acute inflammatory lesion in rats appears to be more severe than that which occurs in human poisoning. Dr Cross's group at University of California (Davis) has claimed some protection with anti-inflammatory agents in rats. Autor (1974) administered superoxide dismutase to paraquat-treated rats and claimed that the treated rats survived longer although the overall mortality was not affected. We have certainly confirmed the lack of efficacy of superoxide dismutase in protecting against lethality but have been unable to demonstrate a delay in the time to death.

Klebanoff: What about catalase?

Smith: That has not been investigated, as far as I know.

Allison: The use of anti-inflammatory agents is ambiguous because they could act at several different levels but it would be useful to reduce the number of circulating leucocytes. Leucocyte production could be prevented by various means, for example, irradiation or anti-neutrophil serum.

Klebanoff: Nitrogen mustards also could be used.

Dormandy: The toxicity of oxygen to humans is puzzling by its unpredictability. After exposure of many people to apparently similar oxygen pressures for similar lengths of time, only a few develop untoward complications, such as pulmonary fibrosis or blindness. Could there be other substances like paraquat that potentiate oxygen as a toxic agent?

Smith: It is possible. The major exposure of the lung to toxic agents is by inhalation and in most cases particles are inhaled and deposited in the upper respiratory airways. Paraquat is unusual; it belongs to a relatively small group of drugs and compounds which, when given systemically, damage the parenchyma of the lung.

Willson: Bleomycin has certain side-effects in the lung. How does this damage compare with that caused by paraquat?

Smith: We had thought that bleomycin would have to be administered for several weeks but Dr Chvapil informs me that one dose will cause fibrosis.

Chvapil: We compared the effect of silica and paraquat on the development of lung fibrosis in rats. We were amazed that, even after exposing animals for 2–3 months to high doses of paraquat, they survived but did not grow and there was no sign of lung fibrosis. The activity of proline hydroxylase and rate of collagen synthesis increased enormously, but no collagen was deposited (E. Hollinger & M. Chvapil, unpublished results). We concluded that the collagen synthesized was digested by the collagenase produced by inflammatory cells and that this prevented its deposition. A single injection of bleomycin does not induce such an acute inflammatory response and the amount of collagen deposited is significantly higher. The question then is whether stimulated macrophages pass on the message by secreting some fibroblast-stimulating factors to the fibroblasts to form more collagen or whether, by activation of these cells, the collagenase is activated and degrades the deposited collagen. It seems that the more acute the response, the less the fibrosis owing to compensation and degradation of the deposited collagen.

Willson: I am interested in the reduction of NADP to NADPH, which is a two-electron process. The paraquat reduction is a one-electron process. Could the enzyme involved be a flavoprotein or an iron-sulphur system? I am reminded of the work of Daniel & Gage (1966) on gut flora; when paraquat is incubated with anaerobic bacteria, the one-electron product — the cation — can be detected.

Smith: The prosthetic group is probably FAD (FADH) (Gage 1968). It seems that the two-electron donor, NADPH, provides reducing equivalents for the reduction of FAD to FADH and hence a one-electron addition to paraquat. In suitable conditions addition of paraquat to lung microsomes with NADPH will generate the paraquat radical, which is relatively stable, provided that it cannot react with oxygen. The system turns blue as the radical is formed.

Fridovich: We have observed the same colour change with anaerobic *E. coli.*

Cohen: Apparently, NADPH serves a double role: (1) a toxicological role, in that it generates superoxide, and (ii) a protective role, in that it reduces oxidized glutathione (via glutathione reductase). If the NADPH were shunted towards superoxide formation, with, consequently, less reduction of oxidized glutathione, the toxicity might be accelerated. Let us consider the converse: if the electrons were shunted towards reduction of oxidized glutathione with, consequently, less formation of superoxide, this might protect against the toxicity. Leaving aside for the moment questions about compartmentalization of the two enzyme systems, I wonder whether prior conversion of reduced glutathione into GSSG by, for example, incubating cells with diamide (e.g. Ammon *et al.* 1977) might not protect cells for a period of time. Of course, oxidizing the glutathione might be a toxic phenomenon in itself, but, by lowering the NADPH availability, might it not protect in the short-term?

Smith: I suspect that would lead to supertoxicity. The nearest experiment is the administration of paraquat to selenium-deficient rats which did not have such a functional glutathione peroxidase system (Gibson & Cagen 1977). Instead of lung damage they found liver damage and argued that removal of that defence mechanism renders the animals more susceptible.

Fridovich: That is not the same thing. Dr Cohen was suggesting the same experiment that we performed: taking away the nutrients from *E. coli* — that is to say, if one cannot have NADPH, one cannot reduce paraquat to the monocation radical and one cannot then make O_2^-. We depleted the stores of NADPH by removing the nutrient — glucose. Although one cannot do that in liver or lung slices, one can do it in cells by accelerating the electron outflow from NADPH to glutathione reduction, by causing there to be a great deal of oxidized glutathione through use of diimide.

Roos: In a similar experiment with neutrophils we observed that the production of superoxide was greatly reduced after incubation of the cells with diamide, but, also, all the other functions of the neutrophil were reduced (Voetman & D. Roos, unpublished work, 1977).

Goldstein: We are all guilty of drawing these loaded diagrams with cyclic reaction paths, but obviously the cells are not perpetual-motion machines. Eventually the paraquat is disposed of. In doing so, does it form another radical which might be responsible for the toxicity such as lipid peroxidation and membrane damage? Also, could the toxic agent that is damaging the membranes be produced elsewhere in the metabolic pathway?

Smith: The paraquat merely cycles between the cation and the reduced form. As far as I know nobody has detected other paraquat radicals in biological systems. The major loss of paraquat from these systems is its efflux from the lung; it has a half-life in the lung of about 24 h. Most of the paraquat in the lung will be in the cationic form. Provided oxygen is present, the small amount of paraquat radical formed will be converted immediately back into the cation.

Dormandy: Paraquat is eventually excreted by the kidneys.

Winterhalter: Against all predictions paraquat accumulates in the lung after oral administration even though, for geometric reasons, it reaches the liver first but is not accumulated there. Is this accumulation in the lung due to the fact that the contact between the blood stream and the cells in the lung is particularly intimate, as far as surfaces are concerned, or to a specific property of the lung cells?

Smith: Tissue slices taken from various organs of the rat have been examined for their ability to accumulate paraquat *in vitro*. No other organ accumulates paraquat to the same extent as the lung. Of the organs studied, brain slices are second to lung although we can see no uptake into the brain with *in vivo* studies — presumably because of the blood–brain barrier. Different species show differences in the ability of the lung to accumulate paraquat. *In vitro* the apparent K_m and V_{max} for accumulation are similar in rat and human lungs.

Kerberle: The accumulation of paraquat in the lung is supposed to be an active process. What physiological purpose does this active transport serve in the absence of paraquat?

Many drugs belonging to the category of lipophilic bases (e.g. the tricyclic antidepressants or the β-blockers) also accumulate in the lung in considerable concentrations, yet it has never been assumed that this accumulation is an active process. Is the accumulation of paraquat in the lung blocked by metabolic inhibitors or by putrescine?

Smith: We do not know about the natural endogenous compound or the process. However, I should remind you that it is a slow process. With *in vivo* studies on rats we have to wait for more than 30 h to detect a concentration in the lung some seven times that in the plasma. *In vitro* events are much faster. Also *in vitro* several compounds, including putrescine, block the uptake. I take issue with you if you are suggesting that we should have expected a compound like paraquat to be actively accumulated into the lung, especially as we have discovered that various 4,4′-bipyridyl substitutes, with similar chemical redox properties to paraquat, are not taken up.

Cohen: The lung serves a function like that of liver: it is a site of metabolism of xenobiotic substances. Perhaps the question should not be what is the endogenous substance that is normally transported but rather is this the mechanism by which foreign compounds are transported for metabolism? The lung, being exposed to materials in the air, may develop these mechanisms for protection against aerobically derived benzenoid substances. That may be why paraquat accumulates there: it triggers the transport mechanism that is normally concerned with the catabolism of foreign substances.

Smith: When paraquat is directly instilled into perfused lung preparations (Charles *et al.* 1978), it is retained in the lung tissue for a considerable time. These authors suggested that this prolonged retention is the result of the accumulation of paraquat into alveolar epithelial cells which we have described both from *in vitro* studies using lung slices and after oral dosing to rats.

Hill: Does paraquat affect phagocytosis?

McCord: This has not been investigated.

References

AMMON, H. P. T., AKHTAR, M. S., NILAS, H. & HEGNER, D. (1977) Inhibition of *p*-chloromercuribenzoate- and glucose-induced insulin release *in vitro* by methylene blue, diamide and tert-butyl hydroperoxide. *Mol. Pharmacol. 13*, 598–605

AUTOR, A. P. (1974) Reduction of paraquat toxicity by superoxide dismutase. *Life Sci. 14*, 1309–1319

BIELSKI, B. H. J. & CHAN, P. C. (1977) Enzyme-catalysed chain oxidation of nicotinamide adenine dinucleotide by superoxide radicals, in *Superoxide and Superoxide Dismutases* (Michelson, A. M., McCord, J. M. & Fridovich, I., eds.), pp. 409–416, Academic Press, London & New York

CHARLES, J. M., ABOV-DONIA, M. B. & MENZEL, D. B. (1978) Absorption of paraquat and diquat from the airways of the perfused rat lung. *Toxicology 9*, 59–67

CRAPO, J. D. & TIERNEY, D. F. (1974) Superoxide dismutase and pulmonary oxygen toxicity. *Am. J. Physiol. 226*, 1401–1407

DANIEL, J. W. & GAGE, J. C. (1966) Absorption and excretion of diquat and paraquat in rats. *Br. J. Ind. Med. 23*, 133–136

GAGE, J. C. (1968) The action of paraquat and diquat on the respiration of liver cell fractions. *Biochem. J. 109*, 757–761

GIBSON, J. E. & CAGEN, S. Z. (1977) Paraquat induced functional changes in kidney and liver, in *Biochemical Mechanisms of Paraquat Toxicity* (Autor, A., ed.), pp. 117–136, Academic Press, London & New York

KIMBALL, R. E., REDDY, K., PEIRCE, T. H., SCHWARTZ, L. W., MUSTAFA, M. G. & CROSS, C. E. (1976) Oxygen toxicity: augmentation of antioxidant defense mechanisms in rat lung. *Am. J. Physiol. 230*, 1425–1431

Phagocyte-produced free radicals: roles in cytotoxicity and inflammation

JOE M. McCORD and KENNETH WONG

Department of Biochemistry, University of South Alabama, Mobile

Abstract The production of superoxide free radical, O_2^-, by metabolically activated phagocytes results in damage to the phagocyte which is manifested by the premature death of the cell *in vitro*. The cytotoxic agent appears to be formed by the reaction of superoxide with hydrogen peroxide, and is thought to be hydroxyl radical or a secondary radical thereof. *In vivo* two animal models of induced inflammation also appear to be largely dependent on superoxide production by phagocytes for the development of tissue damage manifested as oedema. Intravenously administered superoxide dismutase shows anti-inflammatory activity in these models, but only when so derivatized that it can remain in the circulation for longer periods of time. Catalase, or a catalase derivative, on the other hand, shows no anti-inflammatory activity *in vivo*.

The recognition that bovine cuprozinc superoxide dismutase (EC 1.15.1.1) possesses an anti-inflammatory activity (see Huber & Saifer 1977) predates the recognition that the metalloprotein possesses the enzymic ability to catalyse the dismutation of superoxide (McCord & Fridovich 1969). These two seemingly unrelated properties of the protein stood side by side, incongruously, for several years. Then a third key observation provided an important link. Babior *et al.* (1973) noted that phagocytosing leukocytes liberate reasonably large quantities of superoxide radical into the medium in which the cells are suspended. (Babior postulated a bactericidal role for the radical, and this notion has been strongly supported [Johnston *et al.* 1975].) The three pieces of information (1) that phagocytosing leukocytes liberate superoxide radical, (2) that superoxide dismutase efficiently scavenges and eliminates the radical, and (3) that superoxide dismutase has the pharmacological activity of an anti-inflammatory agent, provided the basis for our hypothesis that superoxide radical is a significant chemical mediator of the inflammatory process *in vivo*. Our goals are to elucidate the mechanism of this role of superoxide for a two-fold purpose: first, to understand more fully the biochemistry of the inflammatory process and the sequence and nature of events leading to tissue damage and, secondly, to apply this understanding toward the development of new pharmacological agents aimed at controlling the inflammatory process in clinical situations. Regarding the latter, there is a real need for improved anti-inflammatory agents. Most of the ones in current use are ill-suited for the treatment of chronic inflammatory conditions owing to the accompaniment of undesirable and, at times, dangerous side-effects. Their mechanisms of action are, at best, only vaguely understood. Furthermore, many diseases not normally thought of as inflammatory diseases (e.g. multiple sclerosis and coronary infarction) do have inflammatory components of great clinical significance.

THE BIOLOGICAL MANIFESTATIONS OF SUPEROXIDE PRODUCTION

We began our studies by examining whether superoxide was responsible for, or at least capable of, damage to extracellular components such as accompanies chronic inflammatory conditions. A model was chosen to mimic the conditions which might exist in an inflamed rheumatoid joint. Bovine synovial fluid was exposed *in vitro* to a source of superoxide (generated by xanthine oxidase) equivalent to that which might be produced by the numbers of metabolically activated polymorphonuclear leukocytes commonly found in such a joint (McCord 1974). Extensive degradation of the polysaccharide hyaluronic acid occurred within 30 min, with concomitant loss of viscosity and lubricating quality of the joint fluid. The changes observed in synovial fluid exposed to superoxide *in vitro* appear identical to the changes synovial fluid undergoes *in vivo* as a result of rheumatoid arthritis. Although normal synovial fluid (as well as other extracellular fluids) was found to contain a trace of superoxide dismutase activity, it was insufficient to protect fully the fluid against a flux of the radical. The addition of exogenous superoxide dismutase, however, provided complete protection against superoxide-induced degradation of the fluid.

We were already aware that certain superoxide-dependent phenomena were the result of a species generated secondarily through the reaction of superoxide radical with hydrogen peroxide. Such phenomena are inhibited by either superoxide dismutase or by catalase, as well as by radical-scavenging agents such as mannitol, ethanol, and benzoate. The first example was the production of ethylene from methional by xanthine oxidase-generated superoxide (cited by Beauchamp & Fridovich 1970). They invoked a reaction (1) proposed by Haber & Weiss (1934) to account for their observations, proposing the very reactive

$$O_2^- + H_2O_2 \rightarrow O_2 + OH^- + OH^\bullet \qquad (1)$$

hydroxyl radical, $OH^\bullet$, as the responsible agent.

Our degradation of synovial fluid with a superoxide flux could also be prevented by catalase and by mannitol. Hence, the conclusion drawn was that the Haber–Weiss reaction could occur in a physiological fluid as a result of superoxide generation (McCord 1974).

Soon after the work of Beauchamp & Fridovich, many independent investigators began to obtain results interpretable by the Haber–Weiss reaction. At the same time the reaction began to be sharply criticized, first on thermodynamic grounds, later on kinetic grounds, as a 'non-reaction' (Fee & Valentine 1977). For several years this remained an unhappy dilemma. On the one hand, the direct reaction between superoxide and hydrogen peroxide could not be demonstrated in chemically well-defined conditions; on the other hand, there existed an abundance of data which could be satisfactorily explained only by the Haber–Weiss reaction, or by something very close to it.

A solution to the dilemma appears to be emerging. Koppenol (1976) showed the reaction to be thermodynamically feasible. He later showed that molecular-orbital considerations predict that iron salts should be effective catalysts of the 'non-reaction' (Koppenol & Butler 1977), as we were demonstrating that iron-EDTA is, in fact, an efficient catalyst of the reaction in a model system (McCord & Day 1978). The identity of the biological catalyst(s) of this reaction will be a matter of great interest, and one that we are currently pursuing.

SUPEROXIDE-PRODUCED DAMAGE TO LEUKOCYTES

The next step in exploring the potential toxicity of superoxide was to examine its effects on the cells which produce it, the phagocytosing polymorphonuclear leukocytes. Since these cells produce the radical in order to kill ingested micro-organisms, it seemed possible that in the process they might damage themselves to a greater or lesser extent. We found this to be the case when isolated human peripheral polymorphonuclear leukocytes were induced to produce superoxide *in vitro* (Salin & McCord 1975; McCord & Salin 1977). Although ingested bacteria are killed quickly, largely within the first few minutes, the self-inflicted damage to the leukocytes was manifested by cell death which began only after a delay of about 10 h. The burst of superoxide production lasts for only about the 20 min following the stimulation of the leukocytes, but it was during this period of the incubation that protection by exogenous superoxide dismutase was crucial. The reason for the delayed expression of the radical-induced damage is not understood.

The viability of phagocytosing leukocytes could be protected not only by superoxide dismutase, but equally well by catalase or mannitol. This once again implicates the Haber–Weiss reaction, and the hydroxyl radical is presumed to be the actual cytotoxic species.

This observance of the cytotoxicity of superoxide-radical generation to human leukocytes *in vitro* provides a rational link to the potential anti-inflammatory activity of superoxide dismutase. The key features are (1) that the radical is produced by the leukocytes, (2) that it is released into the extracellular milieu, (3) that superoxide or secondary products can damage the leukocyte, and, therefore, presumably can inflict similar damage *in vivo* on adjacent tissue cells, capillary walls, etc., and (4) that exogenous superoxide dismutase added to the extracellular milieu can protect cells against superoxide-induced damage.

SUPEROXIDE DISMUTASE AS A PROBE INTO THE MECHANISM OF INFLAMMATION

Although there were reports in the literature of the empirical observance of the anti-inflammatory activity of superoxide dismutase (as orgotein) (for a review, see Huber & Saifer 1977 and Menander-Huber & Huber 1977), there had been no systematic probing of the biochemical mechanisms involved at the time our studies began. We had evolved a hypothetical mechanism for the involvement of superoxide in phagocyte-mediated inflammation (McCord 1974; Salin & McCord 1974; McCord & Salin 1975; Salin & McCord 1975; McCord & Salin 1977), but the hypothesis remained to be tested.

The proposition that leukocyte-generated superoxide plays a major role in the development of tissue damage which accompanies inflammation rests on the fact that superoxide dismutase is an intracellular enzyme, and that extracellular fluids are nearly devoid of the activity (McCord 1974). To obtain therapeutic benefit, then, one must introduce the enzyme to these extracellular fluids and, ideally, maintain a functionally-efficient concentration of the enzyme during the course of the inflammatory process. By far the easiest and most economical source of superoxide dismutase is bovine liver. This copper-zinc enzyme, however, is small (molecular weight 31 000) and is quickly cleared from the circulation of laboratory animals by the kidneys, with typical half-lives of 4–6 min. Maintenance of reasonably high concentrations of native bovine enzyme in the circulation, therefore, requires large amounts of the enzyme and either continuous infusion or frequent multiple injections. These factors render the use of native bovine superoxide dismutase experimentally undesirable except in highly specialized applications (e.g. when narrow 'windows' of protection might be desired to study the temporal development of a model, or when the kidney is the target organ, as in immune complex-induced glomerulonephritis). Thus, we set out to modify chemically bovine superoxide dismutase with the objective of producing enzymically-active preparations with substantially increased survival times in the circulation. Our approach was to attach covalently polymers such as polyethylene glycol monomethyl ether (average molecular weight 1900), dextran T-70 (a linear polymer of average molecular weight 70 000) or Ficoll 70 (a branched polymer of average molecular weight 70 000) to native superoxide dismutase. The details regarding synthesis and characterization of these derivatives will be published elsewhere (Wong & McCord, unpublished work). Table 1 summarizes the properties of several such preparations. The increase in survival times is dramatic, ranging from 70- to 350-fold. The loss of enzymic activity on derivatization varied from 10 to 40%.

Because catalase was as effective as superoxide dismutase in protecting phagocytosing leukocytes *in vitro* (implicating the Haber–Weiss mechanism), we also synthesized a derivative of catalase with a long circulating half-life for assessment as an anti-inflammatory agent *in vivo*. As shown in Table 1, native beef-liver catalase was rapidly cleared from the circulation in the rat, whereas the Ficoll derivative showed a clearance half-life of 2.3 h, an increase of 35-fold. The loss of catalase activity on derivatization was 40%.

Our initial choice of an inflammatory model in which to assess the pharmacological properties of these derivatives was the reverse passive Arthus reaction, brought about by intravenous injection of antigen (e.g. human serum albumin) and the concurrent intradermal injection of antibody (e.g. rabbit anti-human serum albumin). This model is thought to be nearly totally mediated by polymorphonuclear leukocytes, is rapid in its development, and is characterized by reasonably-quantifiable parameters such as erythema and oedema. Although others have attempted to quantify lesion diameter or thickness, we found that the most reliable numbers resulted from excision of the shaved area of skin containing the lesion, punching out a half-inch circle of skin centred on the lesion using a steel punch and mallet, and then weighing the lesion.

The efficacy of Ficoll-superoxide dismutase in preventing the development of immune complex-induced oedema in the reverse passive Arthus reaction is shown in Table 2. In the conditions specified, native superoxide dismutase had no effect on the development of oedema, whereas an equivalent amount of the

TABLE 1

Circulating half-lives of native and derivatized superoxide dismutase and catalase

Enzyme type	*Circulating half-life*
Native bovine superoxide dismutase	6 min
Polyethylene glycol-superoxide dismutase	35 h
Dextran-superoxide dismutase	7 h
Ficoll-superoxide dismutase (high molecular weight)	24 h
Ficoll-superoxide dismutase (low molecular weight)	14 h
Native bovine catalase	4 min
Ficoll-catalase	2.3 h

TABLE 2

Inhibition of the reverse passive Arthus reaction in the rat by Ficoll-superoxide dismutase

Treatment	*Oedema (as indicated by weight in g ± S.E.M.)*[a]	*Inhibition (%)*	*Significance*
Saline control	0.20 ± 0.01	0	
Ficoll, 30 mg/kg	0.20 ± 0.02	0	N.S.
Native superoxide dismutase 24 000 units/kg	0.21 ± 0.01	0	N.S.
Ficoll-superoxide dismutase 24 000 units/kg	0.01 ± 0.02	95	$P < 0.001$
Ficoll-superoxide dismutase 10 000 units/kg	0.10 ± 0.02	50	$P < 0.005$
Ficoll-superoxide dismutase 24 000 units/kg (24 h predose)	0.06 ± 0.02	70	$P < 0.001$

The reaction was elicited by intravenous injection of human serum albumin (5 mg/kg), followed by the intradermal injection of rabbit anti-human serum albumin (50 μl). Lesion weight was determined after 3.5 h.

[a] Oedema weight was determined by subtraction of the average weight of uninflamed skin punches (from animals which received antibody, but no antigen). In each experiment, groups of at least seven animals were used, and treatment was administered one hour before elicitation of the inflammatory reaction except as noted otherwise.

Ficoll derivative produced an impressive 95% inhibition. In fact, such a dose was still effective after 24 h. An equivalent amount of free Ficoll, unattached to superoxide dismutase, was also without effect on the development of oedema.

A second inflammatory model we have examined is carrageenan-induced foot oedema in the rat. In this model, carrageenan was administered by subplantar injection, as described by Vinegar *et al.* (1969). Oedema formation was assessed by either of two methods. The first consisted of killing the animals after four hours, amputating both rear feet at the ankles, and weighing the feet. Comparison of the carrageenan-injected foot to its saline-injected mate revealed the magnitude of the oedema. The second method consisted of estimating foot volume by a water-displacement method (Vinegar 1968) which allowed multiple measurements with time for kinetic studies. With the latter method, each foot served as its own control.

The development of carrageenan-induced oedema is biphasic, the early phase being referred to as the histamine and serotonin (5-hydroxytryptamine) phase, the later phase as the prostaglandin phase. Vinegar *et al.* (1969) showed the late phase to be inhibited by hydrocortisone, phenylbutazone, and indomethacin. Oyanagui (1976) recently showed that native bovine superoxide dismutase, administered by frequent multiple intravenous injections, significantly inhibited the late phase of this model.

Table 3 shows the effects of native and derivatized superoxide dismutase on the extent of swelling four hours after carrageenan injection. The degree of oedema suppression obtained with the low dose of Ficoll-superoxide dismutase reflects nearly complete inhibition of the prostaglandin phase of the reaction. Note that native enzyme was without effect at five times this dose rate. All treatments were administered one hour before carrageenan injection.

TABLE 3

Gravimetric analysis of the effects of native and derivatized superoxide dismutase on carrageenan-induced foot oedema in the rat

Treatment	*Increase in weight (%)* ± S.E.M.	*Inhibition (%)*
Saline control	24 ± 3	
Native superoxide dismutase (24 00 units/kg)	24 ± 3	0
Ficoll-superoxide dismutase (9600 units/kg)	10 ± 2	58
Ficoll-superoxide dismutase (4800 units/kg)	11 ± 1	54

Treatment was administered by intravenous injection one hour before the administration of carrageenan. Each experiment represents the averaged results of 9 or 10 rats. Oedema was determined 4 h after carrageenan injection.

More extensive data were collected using the foot volume method for following the development of oedema at Ficoll-superoxide dismutase doses ranging from 1000 to greater than 15 000 units/kg. A maximal inhibition of 55–60% at three hours after carrageenan injection was attained at a dose of about 10 000 units/kg. Half-maximal inhibition was seen at a dose of 1800 units/kg. This is equivalent in activity to about 0.5 mg of native superoxide dismutase per kg.

In vitro results have suggested that it is not superoxide radical *per se* that is cytotoxic to phagocytosing neutrophils, but rather the product of a reaction between superoxide and hydrogen peroxide. This product is thought to be the hydroxyl radical. Thus, the cells could be protected not only by superoxide dismutase, but also by catalase or by scavengers of hydroxyl radical such as mannitol (Salin & McCord 1975). It seemed likely that the cytotoxicity resulting from the inflammatory process *in vivo* is explicable by the same cytotoxic mechanism observed *in vitro*. That is, we expected catalase and perhaps even hydroxyl-radical scavengers such as mannitol to display anti-inflammatory activity. This was not the case. Oyanagui (1976) had already reported the failure of native beef-liver catalase to inhibit carrageenan-induced foot oedema, but native catalase is cleared from the rat's circulation even faster than native superoxide dismutase is (see Table 1), so we were not convinced by this negative report. A Ficoll-derivative of catalase, however, which had a plasma half-life of 2.3 h (compared to 4 min for native enzyme) was totally without effect on carrageenan-induced foot oedema at a dose of 200 000 units/kg. Likewise, mannitol at a dose of 470 mg/kg was totally without effect on oedema development. The Ficoll-catalase derivative was also assessed in the reverse passive Arthus model at a dose of 180 000 units/kg. The lesion weights at 3.5 h were identical to control values.

In summary, data now exist which strongly implicate a major role for superoxide in the oedema formation which accompanies the phagocyte-mediated acute inflammatory response. The biochemical nature of this role remains to be elucidated. It appears not to involve the formation of hydroxyl radical to any great extent. Superoxide dismutase possesses anti-inflammatory activity which can be dramatically enhanced by keeping the enzyme in the circulation during the development of the inflammatory event. We have accomplished this end by the chemical synthesis of various high molecular weight derivatives of the enzyme.

ACKNOWLEDGEMENTS

This work was supported in part by Grant AM 20527 from the US Public Health Service, National Institutes of Health. K.W. was supported by a fellowship from the Canadian Medical Research Council.

References

BABIOR, B. M., KIPNES, R. S. & CURNUTTE, J. T. (1973) Biological defense mechanisms. The production by leukocytes of superoxide, a potential bactericidal agent. *J. Clin. Invest. 52*, 741–744

BEAUCHAMP, C. & FRIDOVICH, I. (1970) A mechanism for the production of ethylene from methional: the generation of hydroxyl radical by xanthine oxidase. *J. Biol. Chem. 245*, 4641–4646

FEE, J. A. & VALENTINE, J. S. (1977) Chemical and physical properties of superoxide, in *Superoxide and Superoxide Dismutases* (Michelson, A. M., McCord, J. M. & Fridovich, I., eds.), pp. 19–60, Academic Press, London & New York

HABER, F. & WEISS, J. (1934) The catalytic decomposition of hydrogen peroxide by iron salts. *Proc. R. Soc. Lond. A 147*, 332–351

HUBER, W. & SAIFER, M. G. P. (1977) Orgotein, the drug version of bovine Cu–Zn superoxide dismutase. I. A summary account of safety and pharmacology in laboratory animals, in *Superoxide and Superoxide Dismutases* (Michelson, A. M., McCord, J. M. & Fridovich, I., eds.), pp. 517–536, Academic Press, London & New York

JOHNSTON, R. B. JR., KEELE, B. B. JR., MISRA, H. P., WEBB, L. S., LEHMEYER, J. E. & RAJAGOPALAN, K. V. (1975) Superoxide anion generation and phagocytic bactericidal activity, in *The Phagocytic Cell in Host Resistance* (Bellanti, J. A. & Dayton, D. H., eds.), pp. 61–75, Raven Press, New York

KOPPENOL, W. H. (1976) Reactions involving singlet oxygen and the superoxide anion. *Nature (Lond.) 262*, 420–421

KOPPENOL, W. H. & BUTLER, J. (1977) Mechanism of reactions involving singlet oxygen and the superoxide anion. *FEBS (Fed. Eur. Biochem. Soc.) Lett. 83*, 1–6

MCCORD, J. M. (1974) Free radicals and inflammation: protection of synovial fluid by superoxide dismutase. *Science (Wash. D.C.) 185*, 529–531

MCCORD, J. M. & DAY, E. D. JR. (1978) Superoxide-dependent production of hydroxyl radical catalyzed by iron-EDTA complex. *FEBS (Fed. Eur. Biochem. Soc.) Lett. 86*, 139–142

MCCORD, J. M. & FRIDOVICH, I. (1969) Superoxide dismutase, an enzymic function for erythrocuprein. *J. Biol. Chem. 244*, 6049–6055

MCCORD, J. M. & SALIN, M. L. (1975) Free radicals and inflammation: studies on superoxide-mediated NBT reduction by leukocytes. *Prog. Clin. Biol. Res. 1*, 731–752

MCCORD, J. M. & SALIN, M. L. (1977) Self-directed cytotoxicity of phagocyte-generated superoxide free radical, in *Movement, Metabolism, and Bactericidal Mechanisms of Phagocytes* (Rossi, F., Patriarca, P. L. & Romeo, D., eds.), pp. 257–264, Piccin, Padua

MENANDER-HUBER, K. B. & HUBER, W. (1977) Orgotein, the drug version of bovine Cu-Zn superoxide dismutase. II. A summary account of clinical trials in man and animals, in *Superoxide and Superoxide Dismutases* (Michelson, A. M., McCord, J. M. & Fridovich, I., eds.), pp. 537–549, Academic Press, London & New York

OYANAGUI, Y. (1976) Participation of superoxide anions at the prostaglandin phase of carrageenan foot-edema. *Biochem. Pharmacol. 25*, 1465–1472

SALIN, M. L. & MCCORD, J. M. (1974) Superoxide dismutases in polymorphonuclear leukocytes. *J. Clin. Invest. 54*, 1005–1009

SALIN, M. L. & MCCORD, J. M. (1975) Free radicals and inflammation: protection of phagocytosing leukocytes by superoxide dismutase. *J. Clin. Invest. 56*, 1319–1323

VINEGAR, R. (1968) Quantitative studies concerning kaolin-edema formation in rats. *J. Pharmacol. Exp. Ther. 161*, 389–395

VINEGAR, R., SCHREIBER, W. & HUGO, R. (1969) Biphasic development of carrageenan edema in rats. *J. Pharmacol. Exp. Ther. 166*, 96–103

Discussion

Klebanoff: The anti-inflammatory action of superoxide dismutase may be due either to a decrease in the amount of superoxide or an increase in the amount of hydrogen peroxide. Hydrogen peroxide, particularly in the presence of peroxidase, could prevent inflammation by inactivating some inflammatory agent. To determine whether such a mechanism is operative, one could administer catalase with the superoxide dismutase, to destroy any generated peroxide.

McCord: We could easily try simultaneous administration.

Winterhalter: In the plasma clearance experiments with non-modified superoxide dismutase, did all the enzyme that disappeared appear in the urine?

McCord: We did not keep a balance sheet.

Michelson: Certainly, in rats it enters the kidneys and within 24 h is all excreted in urine, but that is not necessarily so in humans.

Fridovich: Were there significant differences in the level of residual activity among these different derivatives?

McCord: That depends on how extensive the modification is. Typically, the enzymic activities fell by about 30% on formation of the derivatives.

Winterhalter: The polyethylene glycol–superoxide dismutase appears to exhibit a redistribution between extravascular space and intravascular space and has the longest circulating half-life in the plasma (Table 1), but you used the Ficoll-enzyme for all the subsequent experiments. Why?

McCord: The Ficoll derivative is the easiest to produce. The polyethylene glycol derivative was the first we made and we used cyanuric acid to attach it. We shall probably return to the polyethylene glycol derivative but with a better linking agent than cyanuric acid, such as cyanogen bromide.

Crichton: In the carrageenan-induced oedema experiments why did you administer the modified, non-immobilized, superoxide dismutase one hour beforehand when its half-life is so short? Why not administer the native enzyme at the same time as the carrageenan?

McCord: These data were assembled to show that the derivatives are much more effective than native enzyme; we were not looking for an effect of native enzyme under these circumstances.

Crichton: The derivative of the enzyme might not be able to reach some of the places where superoxide is being produced whereas the native enzyme might be able to.

McCord: That is possible. In probing the mechanism of the inflammatory process, we plan to use native enzyme for that very reason. It is important to have derivatives that can and cannot get out of the capillary lumen because in some cases one derivative may be effective but another ineffective.

Allison: Are these materials endocytosed by phagocytic cells? Often the formation of derivatives alters that process as well.

McCord: The native enzyme is not taken up significantly by phagocytosing cells. I guess that the derivatives, which are quite soluble, are also not taken up but we do not have any data on that.

Winterhalter: Is that true for both the bovine and the pig enzyme? I should have expected the pig enzyme to be taken up better.

McCord: I don't know.

Flohé: Is the antigenicity of superoxide dismutase altered in the derivatives?

McCord: According to Abuchowski *et al.* (1977) the antigenicity of the polyethylene glycol derivative of albumin is markedly decreased. We have not yet looked at the dextran or Ficoll derivatives.

Reiter: Are the superoxide dismutases from different animals immunologically cross-reactive?

McCord: Not to a great extent: for example, about 10% between human and bovine enzymes.

Reiter: The derivatives might have the same sort of effect as the native enzyme in staphylococcal infections. Staphylococci injected intradermally produce skin lesions of varying severity; a strain of high virulence which is lethal when injected intramammary even at low doses produced necrosis intradermally due to a non-antigenic extracellular slime (Brock *et al.* 1973). Superoxide dismutase may be involved (preventatively?).

Roos: Does the prevention of oedema by superoxide dismutase have anything to do with the protection of the leukocytes by the enzyme? The effect on prostaglandin synthesis, for instance, might play a role.

McCord: It may. In the carrageenan model the phase of the inflammatory process that we are inhibiting is referred to by others as the prostaglandin phase but I am not sure by what evidence it is termed that.

Allison: It is sensitive to indomethacin.

Segal: The evidence that neutrophils have a *kamikaze* action is non-existent; there is no direct evidence that when a neutrophil *in vivo* phagocytoses a bacterium it dies. One problem with the assumption that the superoxide produced kills the neutrophil is the extreme resistance of these cells to radiation: they tolerate at least 50 000 rad without any abnormality of function. Compare the lymphocyte, for instance, which is killed by about 1000 rad. The chances of the cells being functional (by other criteria) 36 h after the addition of bacteria in the presence of superoxide dismutase are slight.

McCord: The evidence that neutrophils have a *kamikaze* action *in vitro* is abundant (Salin & McCord 1975; McCord & Salin 1977). Whatever criterion of cell viability (whether cell lysis — an obvious indication of cell death — or something more subtle) one uses, the cells are going downhill. The question is, when does it happen and does SOD prevent it?

Segal: Neutrophils have an important secondary function after killing, namely digestion of the ingested materials and this takes a considerable length of time. Has this been studied?

McCord: I don't know. Exocytosis of indigestible material from *macrophages* has been observed hours after phagocytosis.

Fridovich: Dr Segal, what is the connection between radiation resistance and the lack of effect of O_2^- on the cell?

Segal: According to the general consensus, radiation damages tissues through the generation of free radicals and the damage is a consequence of the production of oxygen radicals.

Fridovich: But which oxygen radical? That is an important point. The culprit which does most of the damage is the hydroxyl radical.

Allison: But wouldn't subsequently formed peroxides do what superoxide does, only to a greater extent?

Willson: Yes; this is one mechanism suggested for the degradation of hyaluronic acid in synovial fluid (Matsumara *et al.* 1966).

Fridovich: Apparently, the cells, including neutrophils, contain superoxide dismutase to protect themselves against superoxide whether made as a consequence of ionizing radiation or of metabolic processes.

Willson: Surely Dr Segal's point is that the resistance he claims of neutrophils to radiation strongly implies that they would also be resistant to free-radical processes.

McCord: Perhaps they are resistant, relative to other cells. We have not compared them.

Fridovich: We have some relevant data: we compared *E. coli* that had a high level of superoxide dismutase with *E. coli* that had a low level, in terms of resistance to the lethality of ionizing radiation and to the oxygen enhancement of that lethality. There was no difference.

Allison: It is interesting that mutants of *E. coli* lacking the capacity to make glutathione are much more sensitive to ionizing radiation.

Fridovich: That makes good sense: $OH^{\bullet}$ will abstract a proton from RH to give $R^{\bullet}$ — a target radical. Glutathione can reduce $R^{\bullet}$, thereby repairing the damage. In the absence of such a healing reaction, O_2 will react with the radical to give a peroxy radical which can develop into other compounds that make the damage permanent. The glutathione data do not necessarily bear on O_2^- involvement.

Willson: If I may change the subject slightly, discussion of inflammation reminds me that zinc has been used for the treatment of rheumatoid arthritis, with some claims of success. As leukocytes are known to contain the highest concentration of zinc of any cells in the body, I wonder whether zinc could be the first line of defence, in the sense that, although superoxide dismutase, catalase, etc., are indeed defence mechanisms, the best defence is to prevent the superoxide radical being formed in the first place. Alternatively under some circumstances zinc might also act as a trigger; once it is removed from sites catalytic processes may begin. Dr P. Beswick (Brunel University) has been isolating neutrophils and studying the respiratory burst. He has concluded that care must be taken in the interpretation of the results of this type of experiment because the final cellular zinc concentration of preparations may vary: zinc can be readily lost from neutrophils during handling and this may affect the results. What is the zinc content of the neutrophils you study?

McCord: I don't know. Many factors probably are responsible for the variation between individuals. We discovered with leukocytes that the time-course of phagocytosis-induced suicide, the capacity for reduction of NBT, and the magnitude of the respiratory burst varied considerably between individuals.

Flower: Given that superoxide dismutase protects leukocytes from self-destruction, does it not follow that it should also drastically reduce their ability to kill bacteria?

McCord: No, because the soluble enzyme is not taken into the phagocytic vacuoles. The enzyme will protect if it is attached to a particle such as latex (Johnston *et al.* 1975).

Cohen: Why isn't the soluble enzyme taken up?

McCord: Diagrams usually depict the bacterium floating around in the phagocytic vacuole, but the membrane probably adheres tightly to the bacterial cell wall, squeezing out the water as the leukocyte flows around the microorganism.

Dormandy: Most effusions, certainly synovial fluid, contain traces of blood — in humans and presumably cows as well. Some of the blood cells haemolyse, as shown by the iron–pigment deposits. Superoxide dismutase, too, is presumably liberated from these cells. Both the amount of blood and the amount of haemolysis are variable. Do you take this into account?

McCord: That may be the source of the trace amounts of superoxide dismutase found in extracellular fluid. In our work on superoxide dismutase in cerebrospinal fluid we reject any fluids that are contaminated with haemolysed blood which will release significant amounts of superoxide dismutase into the fluid.

Dormandy: It is interesting, however, how fully and rapidly many haemorrhagic effusions resolve. One of the reasons may be that there are substances in haemolysed blood suppressing the inflammatory reaction. One might expect a much more vigorous inflammatory reaction from bleeding than is often observed.

McCord: Release of superoxide dismutase from damaged tissue may break the inflammatory cycle and lead to the resolution of the inflammatory process.

Klebanoff: I want to emphasize again that the myeloperoxidase–halide–hydrogen peroxide system has tremendous cytotoxic properties. Various cell types are killed within minutes. Peroxidase-mediated toxicity also can occur with intact neutrophils as effector cells. When neutrophils ingest a particle, adjacent tumor cells are rapidly killed; this is not observed with myeloperoxidase-deficient cells or chronic granulomatous disease cells (which lack hydrogen peroxide) or when the halides are removed from the medium (Clark & Klebanoff 1975). This suggests that, in these conditions, myeloperoxidase and hydrogen peroxide can leak out of the cell and be toxic to adjacent cells.

McCord: If one were looking at parameters other than oedema, one might expect catalase to have some anti-inflammatory activity. Oedema may be chiefly mediated by superoxide whereas the damage such as you describe and other cytotoxic phenomena may be mediated by myeloperoxidase.

Klebanoff: C5a, a potent chemotactic product of complement activation and an important mediator of the inflammatory response, is inactivated by myeloperoxidase, hydrogen peroxide and a halide (Clark & Klebanoff 1978).

Flohé: If the production of O_2^- by leukocytes is a major factor in the development of inflammation, one would expect patients with chronic granulomatous disease to be less prone to inflammatory diseases. Is there any clinical evidence that favours such an assumption?

McCord: I don't think so. I hasten to point out that this is one aspect of the overall inflammatory process that we have dissected out as our measured parameter; I do not suggest that the whole inflammatory process is due to superoxide production — many other mechanisms contribute. Dilworth & Mandell (1977) described a family of patients with chronic granulomatous disease (probably not the classical disease) who have various inflammatory diseases: arthritis, etc. They are also older than most patients with the disease and their symptoms are not quite so severe.

Flower: It seems rather wasteful for leukocytes to produce superoxide radicals all round the perimeter of this membrane when they are required only in the phagosome itself. Is there a reason why they should do this?

McCord: It may be difficult for the leukocyte to tell which part of the membrane is the forming vacuole (and which is still on the outside). On the other hand, perhaps we are being naive in assuming that nature has not put the production of superoxide by inflammatory cells to some good use. It may be a signal to other parts of the inflammatory response. For example, superoxide may react with inactive plasma factors to produce chemotactic factors that would signal other leukocytes to come to the area where one cell has found something that turns it on.

Hayaishi: Is tyramine or any other molecularly small scavenger of O_2^- effective in your system?

McCord: We have not looked at tyramine. That is a good suggestion.

Dormandy: Your mention of leukotactic agents brings to mind the systemic components of inflammation. You discussed local changes influenced by superoxide dismutase; but are there parallel changes in the systemic inflammatory response, such as the rise in body temperature and the F.S.R.?

McCord: We have not monitored any of these parameters.

Michelson: I want to comment on outer-cell membrane charge and the acidity or basicity of superoxide dismutase. We studied a series of the natural enzymes, with p*I*s ranging from 4.2 to 8.7, with respect to their attachment to or penetration of erythrocytes. We have also modified the outer charge on the enzyme by dimethylating the ϵ-NH_2 groups of lysine (to make it more basic) or by methylating COOH groups (to make the enzyme less acidic). We found a seven-fold difference in uptake by the erythrocytes but the amounts were so small as to be insignificant: minimum uptake of superoxide dismutase was about 240 molecules and the maximum was about 2000 molecules per erythrocyte. This is to be compared with the normal human erythrocyte content of about 260 000 molecules of SOD.

Hill: Dr McCord, how significant are the reports of the clinical use of superoxide dismutase?

McCord: The native bovine enzyme is used clinically as a veterinary product (frequent doses by intramuscular injection) and is claimed to be an effective anti-inflammatory agent. The progress with human clinical use has recently been reviewed (Huber & Saifer 1977; Menander-Huber & Huber 1977). (We administered it intravenously.)

Winterhalter: Weren't intra-articular injections also used in horses?

McCord: Yes, but the suggested way is intramuscularly.

Allison: For what diseases is it used and how can the observations be properly controlled?

McCord: A variety of diseases has been treated. In horses bony exostoses involving the carpal joint have been reported to be susceptible to treatment by superoxide dismutase; also inflammation of the spinal cord of dogs, corneal ulceration, etc.

Fridovich: It has been used in humans against urinary tract inflammations (Marberger *et al.* 1974).

McCord: Specifically, radiation-induced cystitis in humans has been treated by intramural injection of the bladder wall (Marberger *et al.* 1974).

Flohé: Clinical trials are continuing. The results of local intra-articular application are promising (Lund-Olesen & Menander 1974; G. Biehl & K. Menander-Huber, unpublished work, 1978).

Chvapil: Topical application (into the peritendineous sheet) in tendinitis in horses has a dramatic beneficial effect within 12 h — almost as good as dimethyl sulphoxide with butazolin (which is commonly used to treat pulled tendons in horses).

Allison: A point that still disturbs me is whether these effects are due to the damaging actions of superoxide itself or are indirectly mediated through prostaglandins or other mediators.

Flower: I was wondering about that, too. Production of prostaglandins is one of the first detectable events in an injured cell. A spectacular property of prostaglandins is the way that they can potentiate enormously the effects of other mediators (Vane 1976).

Allison: In the G_2 phase isn't the endoperoxide the most effective in that respect?

Flower: Prostaglandins of the E series seem to be most effective.

Allison: Prostaglandins and superoxide could be synergistic.

Fridovich: The tool that has been applied again and again, superoxide dismutase, does not tell one whether the O_2^- is the initiator of a chain of events or the proximal agent. There is no way of knowing that at present.

Willson: Aspirin is the most often used anti-inflammatory drug. I mentioned (p. 55) the work of Weser but, on reflection, I don't believe the reaction rate constants he obtained can be explained as Professor Fridovich suggested by the reaction of free copper — they are surely due to a copper salicylate complex. Is copper salicylate acting as an artificial dismutase as has been suggested (p. 55) and is this one of the mechanisms of aspirin action?

McCord: I don't know. Aspirin does not prevent the depolymerization of synovial fluid by superoxide *in vitro*, but we did not try an aspirin-copper chelate.

Fridovich: In vivo efficient chelating compounds abound: ATP, citrate, proteins of many kinds and so on. Even were one to administer copper salicylate, except in sufficiently massive doses to saturate all natural chelating capacities, copper will be taken from the salicylate by other chelating agents.

Hill: The reactions of most copper complexes are rapid.

Bielski: We experimented with O_2^- in citrate and tartrate buffers and found that they effectively complexed metal impurities thus rendering them harmless as catalysts for the acceleration of O_2^- decay.

Fridovich: EDTA, of course, does that very well.

Keberle: Dr McCord, in the Arthus model if you were to use sub-optimal doses of classical anti-inflammatory drugs and a sub-optimal dose of the Ficoll derivative, would you expect to observe an effect, i.e. to demonstrate an additive effect?

McCord: I would expect an additive, if not synergistic, effect because I believe the mechanisms of action would be different. We have not done this kind of experiment.

Michelson: So far I have gained the impression that everything is black for superoxide and white for superoxide dismutase. May I propose that sometimes O_2^- is good? Chemically, it should be a good radical-chain terminator. Reaction of superoxide with RH$^\bullet$ or lipid radicals to give simply the hydroperoxide would be less dangerous than the generation of OH$^\bullet$ (or even singlet oxygen). On some occasions, addition of superoxide dismutase should increase the damage because radical-chain terminators will have been removed. A few experiments in the literature seem to support this view.

McCord: If it were always beneficial to have superoxide dismutase in the extracellular fluid, surely it would have been a simple matter for nature to have put it there! Perhaps we should be cautious about naively thinking that putting it there will only result in good.

Fridovich: In our experiments with the simpler (*E. coli*) system the correlation is clear and leads one to conclude that superoxide is bad and entirely without redeeming features. That may be totally different in parts of the mammalian system.

Michelson: In a simple human instance, trisomy 21, the 50% excess of superoxide dismutase is certainly not good. Results on partial trisomies indicate that some of the major symptoms are attenuated when the trisomy 21 has a normal complement of superoxide dismutase.

References

ABUCHOWSKI, A., VAN ES, T., PALCZUK, N. C. & DAVIS, F. F. (1977) Alteration of immunological properties of bovine serum albumin by covalent attachment of polyethylene glycol. *J. Biol. Chem. 252*, 3578–3581

BROCK, J. H., TURVEY, A. & REITER, B. (1973) Virulence of two mastitis strains of *Staphylococcus aureus* in bovine skin: enhancement by growth in high carbohydrate-high salt medium or in raw milk. *Infect. Immun. 7*, 865–872

CLARK, R. A. & KLEBANOFF, S. J. (1975) Neutrophil-mediated tumor cell cytotoxicity: role of the peroxidase system. *J. Exp. Med. 141*, 1442–1447

CLARK, R. A. & KLEBANOFF, S. J. (1978) Chemotactic factor inactivation by the myeloperoxidase–H_2O_2–halide system. *Clin. Res. 26*, 392A

DILWORTH, J. A. & MANDELL, G. L. (1977) Adults with chronic granulomatous disease of 'childhood'. *Am. J. Med. 63*, 233–242

HUBER, W. & SAIFER, M. G. P. (1977) Orgotein, the drug version of bovine Cu-Zn superoxide dismutase. I. A summary account of safety and pharmacology in laboratory animals, in *Superoxide and Superoxide Dismutases* (Michelson, A. M., McCord, J. M. & Fridovich, I., eds.), pp. 517–536, Academic Press, London & New York

JOHNSTON, R. B. JR., KEELE, B. B. JR., MISRA, H. P., WEBB, L. S., LEHMEYER, J. E. & RAJAGOPALAN, K. V. (1975) Superoxide anion generation and phagocytic bactericidal activity, in *The Phagocytic Cell in Host Resistance* (Bellanti, J. A. & Dayton, D. H., eds.), pp. 61–75, Raven Press, New York

LUND-OLESEN, K. & MENANDER, K. B. (1974) Orgotein: a new anti-inflammatory metalloprotein drug: preliminary evaluation of clinical efficacy and safety in degenerative joint disease. *Current Ther. Res. Clin. Exp. 16*, 706–717

MARBERGER, H., HUBER, W., BARTSCH, G., SCHULTE, T. & SWOBODA, P. (1974) Orgotein: a new anti-inflammatory metalloprotein drug; evaluation of clinical efficacy and safety in inflammatory conditions of the urinary tract. *Int. Urol. Nephrol. 6*, 61–74

MATSUMARA, G., HERP, A. & PIGMAN, W. (1966) Depolymerisation of hyaluronic acid by autoxidants and radiations. *Radiat. Res. 28*, 735–752

MCCORD, J. M. & SALIN, M. L. (1977) Self-directed cytotoxicity of phagocyte-generated superoxide free radical, in *Movement, Metabolism and Bactericidal Mechanisms of Phagocytes* (Rossi, F., Patriarca, P. L. & Romeo, D., eds.), pp. 257–264, Piccin, Padova

MENANDER-HUBER, K. B. & HUBER, W. (1977) Orgotein, the drug version of bovine Cu-Zn superoxide dismutase. II. A summary account of clinical trials in men and animals, in *Superoxide and Superoxide Dismutases* (Michelson, A. M., McCord, J. M. & Fridovich, I., eds.), pp. 537–549, Academic Press, London & New York

SALIN, M. L. & MCCORD, J. M. (1975) Free radicals and inflammation: protection of phagocytosing leukocytes by superoxide dismutase. *J. Clin. Invest. 56*, 1319–1323

VANE, J. R. (1976) Prostaglandins as mediators of inflammation, in *Advances in Prostaglandin and Thromboxane Research*, vol. 2 (Samuelsson, B. & Paoletti, R., eds.), pp. 791–801, Raven Press, New York

General discussion

CARBONYL COMPOUNDS AND CARCINOGENESIS

Goldstein: When cell membranes undergo lipid peroxidation there are three types of reactive agents which possibly could produce intracellular damage. The first consists of free radicals. I submit that free radicals formed in the cell membrane are unlikely to reach the nucleus because of efficient intracellular defence mechanisms. Let me further suggest that in model experiments indicating an effect of membrane-derived radicals on nuclear constituents the initial flux of free radicals is often so great that the resultant damage to the cell membrane causes immediate cell death without the need to implicate secondary intracellular radical effects. A second type of reactive compound is fatty acid hydroperoxides which I submit are probably too bulky and too hydrophobic to escape from the membrane and enter the intracellular milieu. A third type of membrane-derived product is represented by carbonyl end products of lipid peroxidation. The most frequently studied of these compounds, although not necessarily representative, is malonaldehyde. We have evaluated the potential effects of malonaldehyde on nuclear material in several different systems.

Using histidine-requiring auxotrophs of *S. typhimurium* we were able to show that malonaldehyde is mutagenic (Mukai & Goldstein 1976). A specific pattern was observed: mutagenesis was noted only in frameshift mutants with normal excision repair, and not in the ultraviolet-light-sensitive strains which tend to mutate in response to radicals and active forms of oxygen. This pattern is consistent with cross-linking of DNA by malonaldehyde leading to mutagenesis expressed through the error-prone repair system.

As I noted in my paper (pp. 295–314), others have demonstrated the cross-linking of DNA by malonaldehyde through Schiff base linkages between the carbonyl functions of malonaldehyde and amino groups of DNA (Reiss *et al.*

1972; Brooks & Klamerth 1968). We have shown, both in test tube and tissue culture systems, that malonaldehyde will even more readily cross-link DNA to histones, which are relatively rich in free amino groups (Hamburger & Goldstein 1978).

An even more indirect approach to this area has been to evaluate the effects of diets rich in polyunsaturated fatty acids in a two-stage mouse-skin carcinogenesis system in which tumours are initiated by either dimethylbenzanthracene or β-propiolactone and promoted by phorbol myristic acid. Previous reports suggest that lipid peroxidation causes cancer and that diets rich in polyunsaturated fatty acids lead to an increased incidence of cancer. (The relevance of these studies is highlighted by present medical advice to avoid heart attacks by eating such diets.) Malonaldehyde has been reported to be a tumour initiator in a mouse-skin carcinogenesis system (Shamberger *et al.* 1974). In our own studies mice were fed diets containing 4% or 15% fat, either predominantly saturated or unsaturated. The fat content of standard rodent laboratory chow is usually about 2%, and the average Western diet contains about 16% fat. At both 4% and 15% fat levels the tumour incidence was significantly higher in the groups fed unsaturated rather than saturated fat (Troll *et al.* 1978). As the unsaturated fat enhancement of tumour yield has been similar after initiation with both dimethylbenzanthracene, which requires metabolic activation to a proximal carcinogen, and β-propiolactone, which reacts directly with DNA, it is unlikely that the unsaturated fatty acid diet is simply altering microsomal metabolism. The connection between these two findings and lipid peroxidation, if any, is a matter for future study.

Dormandy: Many properties and actions long attributed to polyunsaturated lipids like arachidonic acid (e.g. bactericidal activity or action in clotting) may be due to traces of peroxidation products such as malonaldehyde. This probably accounts for the very unpredictable results obtained with different preparations of lipids; for example, the 'phospholipid' added to clotting mixtures is sheer cookery: some preparations work in one laboratory, some in another. The same applies to the bactericidal activity of polyunsaturated lipids. Evidence is accumulating which suggests that we have not always taken into account the fact that nearly all polyunsaturated lipid preparations contain traces of peroxidation products many of which are highly active biological agents (Barrowcliffe *et al.* 1975; Gutteridge *et al.* 1974, 1976).

Goldstein: We have gone to great lengths to try to make fresh preparations — we keep it under nitrogen and in the cold.

Fridovich: It has been suggested that many mutations are generated by errors in the repair process (Drake & Baltz 1976).

Goldstein: The chromosomal changes caused by ozone differ from radiation-induced changes in that they take a while to develop. This fits in with an error-repair system.

THE IDENTITY OF THE SUPEROXIDE RADICAL ANION SPECIES

Hill: The superoxide anion has been mentioned frequently in our discussions. However, there is not just one superoxide ion, there are many (see also p. 15). We may, occasionally, encounter the superoxide ion, O_2^-; it exists in the gas phase. Something like it might exist in aprotic media. Investigation of the reaction of superoxide and water in the gas phase has led to the proposal of the hydrogen-bonded structure (1). The reactivity of this hydrate is bound to

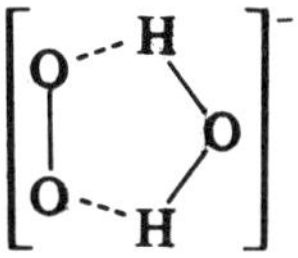

$$\left[\begin{matrix} O\cdots H \\ | \quad\quad \backslash \\ | \quad\quad\quad O \\ | \quad\quad / \\ O\cdots H \end{matrix}\right]^- \qquad (1)$$

differ from that of O_2^- in all sorts of ways. Other forms incorporate more water molecules (Arshadi & Kebarle 1970), e.g., $O_2(H_2O)_n^-$ whose properties will be different again. One important reaction of the superoxide ion is that with positively-charged species — these can vary from protons and metal ions to organic cations: $[M^{n+}O_2^-]^{(n-1)+}$, $R^+O_2^-$. Furthermore, each of those ion-pairs can be solvated, $[M^{n+}O_2^-]^{(n-1)+}(\text{solvent})_n$, or even solvent-separated. Already we are faced with a dozen possible species.

Another reaction that has been often alluded to is the Haber–Weiss reaction. Dr Cohen (1977) has given a spirited defence of this. Dr Bielski's results, of course, are incontrovertible. However, which superoxide species is one considering in the reaction with, e.g. hydrogen peroxide? The formation of ion-pairs such as $—NR_3^+\ O_2^-$ in a lipid membrane or $[CaO_2]^+$ with calcium ions is not improbable. To be absolutely sure that the Haber–Weiss reaction has no biological significance, we should need to know the reaction, if any, of each of the forementioned species, including $NR_3^+O_2^-$ and $[CaO_2]^+$, with hydrogen peroxide. Paraquat can be written as $(R_3N^+)_2$. What is the reaction of the paraquat superoxide ion-pair, if formed, with hydrogen peroxide generated in the same locality? We must qualify any statement we make about the superoxide ion with respect to its environment which may and, one can predict, will have a powerful effect on its reactions, including, perhaps, the so-called Haber–Weiss reaction. Although redox-active metal ions no doubt will catalyse this reaction, it may not be obligatory to have a redox-active centre; an organic or metal cationic centre could act just as well.

Segal: When one estimates superoxide production by a complex structure like a cell, by determining cytochrome *c* reduction, how much does the cytochrome *c* itself cause the throughput of superoxide, in other words by creating a gradient and pushing the reaction equilibrium to one side?

Hill: It depends very much on the conditions.

Obviously the superoxide ion, $O_2(H_2O)_n^-$, is the substrate for superoxide dismutase, but are all the superoxide species substrates?

Michelson: The metal complexes of superoxide are not destroyed by superoxide dismutase. The $[CuO_2]^+$ species Dr Bielski mentioned has tremendously different reactions.

Hill: That is very interesting: I chose calcium as a more 'innocent' cation.

Bielski: Pulse-radiolysis studies (Land & Swallow 1971; Nilsson 1972; Shafferman & Stein 1974) indicate that the interaction of cytochrome *c* with free radicals is complex. For example, both reducing and oxidizing radicals have been shown to cause an increase in absorbance at 550 nm. Although some reducing radicals such as the malate, lactate and ethanol species react with cytochrome relatively slowly but quantitatively, the hydrated electron reacts with it very rapidly showing intramolecular consecutive reactions and attack on the ferrocytochrome *c* protein. Also when hydroxyl radicals are added by pulse radiolysis in an amount equal to a $g(OH) = 5.7$, an increase in absorbance at 550 nm is observed corresponding to a *g* value of 1.9. In a related study with ribonuclease it was established that the sites of the intermediate radicals included bivalent sulphur and aromatic amino acids. It was implied that in time the electron on the oxidized terminus migrates, thereby resulting in heightened reduction of cytochrome *c*.

Michelson: Were any tunnelling effects of the electron seen?

Bielski: It was implied that the electron migrates through the protein.

Fridovich: But if that were so, one would get a product that is not identical with biologically-reduced cytochrome *c*; that is, something which, on reoxidation by cytochrome oxidase, will give you back the initial material. There might be residual oxidation somewhere. The O_2^--reduced product is the same as the biologically-reduced product: it can be oxidized back to the original substance by cytochrome oxidase.

Willson: But a small organic compound (such as formate or an alcohol) in the milieu may instead be oxidized by the hydroxyl radical and the resulting radical may then reduce the cytochrome *c*.

Fridovich: When O_2^- is the reductant, superoxide dismutase will prevent the reduction of cytochrome *c* but it will not if a reductant is being generated by an oxidative attack either on some residue of the protein or on formate or other small molecule. In that case superoxide dismutase could not intercede.

Willson: But is not Dr Segal suggesting that there is only a low concentration of O_2^- in the cell but that it is in equilibrium? By adding any reagent or scavenger one necessarily upsets the equilibrium and leads to more O_2^- becoming available. This is implied in the paper by Anclair *et al.* (1978) on NBT (see p. 51).

Fridovich: Some data that may illuminate this come from experiments in which the concentration of cytochrome *c* was varied and the rate of cytochrome *c* reduction, using xanthine oxidase as a source of O_2^-, was measured. We observed a saturation phenomenon; that is to say, the rate of cytochrome *c* reduction increased with cytochrome *c* concentration up to a limit; that limit was reached at about 1 μmol/l (depending on the pH). Further increases in concentration had no effect. If one were upsetting an equilibrium, I would have expected the rate to increase continually with concentration (McCord & Fridovich 1968).

Cohen: How do you explain the plateau in that case?

Fridovich: With insufficient cytochrome *c* to compete against spontaneous dismutation, each O_2^- is not reflected as a reduced cytochrome *c*. With enough cytochrome *c*, it competes overwhelmingly against the spontaneous dismutation and every O_2^- molecule results in one of cytochrome *c* reduced and no increase in cytochrome *c* level will further increase the rate of cytochrome *c* reduction.

Willson: It seems that in defining these reactions we should refer to O_2^- in water as $O_2^-{}_{aq}$. This might avoid some of the confusion in the literature. For example, Henry & Michelson (1977) discussed the possible reaction of O_2^- with CO_2, saying that it proceeds in a lipid environment. However, in water the reverse reaction ($CO_2^- + O_2$) proceeds very rapidly (p. 194). The medium should always be defined.

Hill: Agreed; but in biological systems the media are often heterogeneous. To interpret the reactions of biological systems one has to draw lessons from the chemistry in a wide variety of environments.

Michelson: Organic chemistry is replete with instances of solvent effects, so even different membranes will have different effects. We should almost have to define the dielectric constant of the environment.

Hill: Consider the Haber–Weiss reaction: in the environment in which it was studied by Dr Bielski it goes only at a very slow rate. In some of the environments I alluded to its rate may be much faster.

Bielski: That should be tested!

Michelson: We are touching a philosophical point; much of the argument has been that theoretically a reaction cannot go if in pure aqueous solution it has been shown to be kinetically impossible. But we never deal with a pure aqueous state.

Fridovich: We have always accepted that there are observations which we can explain in no other way except by such an iron-catalysed reaction. For a while we have been talking past each other rather than to each other; it would be much better to approach the problem by using systems and conditions in which the observations were made which called forth the Haber–Weiss proposal, in the first place.

Michelson: The attitude has recently changed markedly: various groups are now trying to find out the conditions in which the reaction can go, before considering the biological results, which are solid — in their own context.

Fridovich: Certainly in the sense that they have been repeated in many laboratories by different groups!

Cohen: McCord & Day (1978) recently demonstrated what appears to be a Haber–Weiss reaction with iron–EDTA as catalyst. The concentration of H_2O_2 needed to produce the phenomenon was low (viz. 10 μmol/l), but again this was a specific environment. Biologically a host of agents may catalyse the reaction: with regard to iron we should consider cytochrome P450, haem and other possibilities.

Hill: The same limitations that apply to superoxide are unfortunately applicable in the search for specific reagents for hydroxyl radical, superoxide anion and singlet oxygen. Most likely, they will behave specifically in particular environments.

Michelson: Doesn't the work on the iron–EDTA complex imply that the copper dityrosine complex destroys O_2^- because it produces hydroxyl radicals? O_2^- can reduce copper enzymes. In this step it is oxidized to O_2 giving Cu(I) which could then react with H_2O_2 to give $OH^\bullet$ radicals. Dr Hill, have you checked that it is not just O_2^- disappearing but hydroxyl radicals being produced?

Hill: No.

Cohen: I would like to generalize from $HO_2^\bullet$ to $RO_2^\bullet$. It has been asserted several times that superoxide dismutase is specific for O_2^-. However, I have heard from two sources (Peter Wardman and Martin Fielden, personal communications) that superoxide dismutase appears to catalyse the dismutation of $RO_2^\bullet$ as well.

McCord: I doubt that even $HO_2^\bullet$ is a substrate for superoxide dismutase.

Fridovich: I agree, because the rate of the enzymic reaction is unaffected by pH in the range 5–10. Since the pK_a is 4.8 the amounts of $HO_2^\bullet$ will drop by an order of magnitude for each unit rise in pH. Since the rate is constant, O_2^- must be the substrate for the enzyme.

Hill: The only substrate for the enzyme is likely to be $O_2(H_2O)_n^-$. We really need to know more about the interactions of superoxide with Brönsted and Lewis (e.g. CO_2) acids.

References

ANCLAIR, C., TORRES, M. & HAKIM, J. (1978) Superoxide anion involvement in NBT reduction catalysed by NADPH-cytochrome P450 reductase: a pitfall. *FEBS (Fed. Eur. Biochem. Soc.) Lett. 89*, 26–28

ARSHADI, M. & KEBARLE, P. (1970) *J. Phys. Chem. 74*, 1483–1485

BARROWCLIFFE, T. W., GUTTERIDGE, J. M. C. & DORMANDY, T. L. (1975) The effect of fatty acid autoxidation products in blood coagulation. *Thrombos. Diath. Haemorrh. 33*, 271–275

BROOKS, B. R. & KLAMERTH, O. L. (1968) Interaction of DNA with bifunctional aldehydes. *Eur. J. Biochem. 5*, 178–182

COHEN, G. (1977) In defense of Haber–Weiss, in *Superoxide and Superoxide Dismutases* (Michelson, A. M., McCord, J. M. & Fridovich, I., eds.), pp. 317–321, Academic Press, London & New York

DRAKE, J. W. & BALTZ, R. H. (1976) The biochemistry of mutagenesis. *Annu. Rev. Biochem. 45*, 11–37

GUTTERIDGE, J. M. C., LAMPORT, P. & DORMANDY, T. L. (1974) Autoxidation as a cause of antibacterial activity in unsaturated fatty acids. *J. Med. Microbiol. 7*, 387–389

GUTTERIDGE, J. M. C., LAMPORT, P. & DORMANDY, T. L. (1976) The antibacterial effects of water-soluble compounds from autoxidising linoleic acid. *J. Med. Microbiol. 9*, 105–110

HAMBURGER, S. J. & GOLDSTEIN, B. D. (1978) The cross-linking effects of malonaldehyde on DNA and chromatin protein, presented at the American Society of Biological Chemists' meeting, Atlanta, June, 1978 (abstr.)

HENRY, J. P. & MICHELSON, A. M. (1977) $O_2^{-\cdot}$ and chemiluminescence, in *Superoxide and Superoxide Dismutases* (Michelson, A. M., McCord, J. M. & Fridovich, I., eds.), pp. 283–290, Academic Press, London & New York

LAND, E. J. & SWALLOW, A. J. (1971) One-electron reactions in biochemical systems as studied by pulse radiolysis. V. Cytochrome *c*. *Arch. Biochem. Biophys. 145*, 365–372

MCCORD, J. M. & DAY, E. D. JR. (1978) Superoxide-dependent production of hydroxyl radical catalyzed by iron-EDTA complex. *FEBS (Fed. Eur. Biochem. Soc.) Lett. 86*, 139–142

MCCORD, J. M. & FRIDOVICH, I. (1968) The reduction of cytochrome *c* by milk xanthine oxidase. *J. Biol. Chem. 243*, 5753–5760

MUKAI, F. H. & GOLDSTEIN, B. D. (1976) Mutagenicity of malonaldehyde, a decomposition product of peroxidized polyunsaturated fatty acids. *Science (Wash. D.C.) 191*, 868–869

NILSSON, K. (1972) The reduction of ferricytochrome *c* studied by pulse radiolysis. *Isr. J. Chem. 10*, 1011–1019

REISS, U., TAPPEL, A. L. & CHIO, K. S. (1972) DNA-malonaldehyde reaction formation of fluorescent products. *Biochem. Biophys. Res. Commun. 48*, 921–926

SHAFFERMAN, A. & STEIN, G. (1974) Reduction of ferricytochrome *c* by some free radical agents. *Science (Wash. D.C.) 183*, 428–430

SHAMBERGER, R. J., ANDREONE, T. L. & WILLIS, C. E. (1974) Antioxidants and cancer IV. Initiating activity of malonaldehyde as a carcinogen. *J. Natl. Cancer Inst. 53*, 1771–1773

TROLL, W., BELMAN, S., GOLDSTEIN, B. D., MUKAI, F. & MACHLIN, L. (1978) Effect of feeding unsaturated or saturated fat on carcinogenesis on mouse skin, in *Proceedings of the American Association of Cancer Research, Cancer Res. 19*, 106 (abstr.)

Closing remarks

I. FRIDOVICH

Department of Biochemistry, Duke University Medical Center, North Carolina

It has been suggested that we who customarily work with aqueous solutions use the symbol $O_2{}^-{}_{aq}$ to indicate that we are dealing with a hydrated superoxide anion, whose reactivities will surely differ from the corresponding radical in a non-polar solvent. We must also bear in mind that a reaction between a radical and a stable scavenger will beget another radical. Consequently we must be concerned not only with the reactivities of the original radical, but with those of its ill-begotten descendants, as well. Since the radicals generated secondarily, from different scavengers, may differ greatly in reactivity, we can expect different effects from such scavengers. By way of example, Salin & McCord (1977) reported a marked protection by 1.0mM-mannitol of human neutrophils against free radical damage in a mixture containing 5% human serum and 55mM-glucose. The attacking radical was believed to be the hydroxyl radical, which reacts very rapidly with either glucose or mannitol; yet mannitol protected, whereas 55mM-glucose did not. In this case we may conclude that glucose, upon attack by $OH^{\bullet}$, generates a radical which was deleterious to the cells, whereas mannitol did not. It is clear that there are interpretative pitfalls in the use of radical scavengers, but they will be used because of their convenience and we must be aware of the possibilities for secondary reactions.

The photolytic production of $O_2{}^-$ described by Dr Bielski may well come into widespread use, since the equipment required is neither complex nor expensive, in contrast to that needed for pulse radiolysis. His suggestion that the copper-catalysed dismutation of $O_2{}^-$ proceeds through a $[CuO_2]^+$ intermediate is fascinating. We must now investigate the possibility that a similar mechanism is involved in the enzyme-catalysed dismutation of $O_2{}^-$.

Dr Flohé has presented data which remind us of the importance of glutathione peroxidase as a defence against the peroxide intermediates of oxygen reduction. Some of us who have a fixation upon $O_2{}^-$ and the superoxide dismutases must nevertheless admit that the threat of dioxygen is multifaceted and demands a comparably complex defence. Superoxide dismutases, catalases and peroxidases are all part of that defence.

Professor Hayaishi's report of an increase in lung indoleamine dioxygenase, after injection of endotoxin, may correlate with the reported increase in lung superoxide dismutase after injection of endotoxin and exposure to hyperoxia (Frank *et al.* 1978). We may confidently expect exciting new findings from the pursuit of these observations.

Dr Cohen's use of hydrocarbon exhalation by rats, as an index of *in vivo* lipid peroxidation, is exciting. Many gas chromatographs will surely be brought out of retirement and put to work for the quantitation of volatile hydrocarbons, as a direct result of his report.

The literature concerning the biology of oxygen radicals has been growing rapidly. The anti-inflammatory effects of superoxide dismutase derivatives, reported by Dr McCord, suggest that useful pharmacological applications are close at hand. I feel fortunate to have been involved in the inception of this field and I hope to remain active and alert long enough to see the fruition of the ideas we now find so heavy with promise.

References

Salin, M. L. & McCord, J. M. (1977) Free radicals in leukocyte metabolism and inflammation, in *Superoxide and Superoxide Dismutases* (Michelson, A. M., McCord, J. M. & Fridovich, I., eds.), pp. 257–270, Academic Press, London

Frank, L., Yam, J. & Roberts, R. J. (1978). The role of endotoxin in protection of adult rats from oxygen-induced lung toxicity. *J. Clin. Invest. 61*, 269–275

Index of contributors

Entries in **bold** *type indicate papers; other entries refer to discussion comments*

Indexes compiled by William Hill.

Subject index